Transforming Parks and Protected Areas

A rare collection of articles that fuses academic theory with a critique of practice and practical knowledge, *Transforming Parks and Protected Areas* looks in detail at the emerging issues in the design and operation of parks and protected areas. Addressing critical dynamics and current practices in parks and protected areas management, this extensive volume goes well beyond simple managerial solutions and descriptions of standard practice.

The protection of natural resources and biodiversity through protected areas is increasingly based on ecological principles. Simultaneously, the concept of ecosystem-based management has become broadly accepted and implemented over the last two decades. However, this period has also seen unprecedented rapid global social and ecological change, which has weakened many protection efforts. These changes have created an awareness of opportunities for innovative approaches to managing protected areas and of the need to integrate social and economic concerns with ecological elements in protected areas and parks management.

With contributions from leading academics and practitioners, *Transforming Parks and Protected Areas* will be of value to all those working within ecology, natural resources, conservation, and parks management as well as students and academics across the environmental sciences and land use management.

Kevin S. Hanna teaches environment and resource policy, environmental assessment, and land use planning at Wilfrid Laurier University. His research centres on integrated natural resource management, impact assessment, forestry communities, and regional planning approaches.

Douglas A. Clark studies and teaches governance for social-ecological systems, focusing on the circumpolar north. Doug spent 11 years with Parks Canada in a variety of postings, including as the first Chief Warden of Canada's Wapusk National Park, established in 1997. He has twice received Parks Canada's Award of Excellence from the Agency's CEO, and in 2004 he became a Canon National Parks Science Scholar.

D. Scott Slocombe has taught resource and environmental management at Wilfrid Laurier University since 1989. His research and consulting interests focus on protected areas, environmental planning, management, policy and education, and systems approaches.

Transforming Parks and Protected Areas

Policy and governance in a changing world

Edited by Kevin S. Hanna, Douglas A. Clark, and D. Scott Slocombe

NEW YORK AND LONDON

First published 2008
by Routledge
270 Madison Avenue, New York, NY 10016

Simultaneously published in the UK
by Routledge
2 Park Square, Milton Park, Abingdon, Oxon OX14 4RN

Routledge is an imprint of the Taylor & Francis Group, an informa business

© 2008 Kevin S. Hanna, Douglas A. Clark, and D. Scott Slocombe selection and editorial matter; individual chapters © the contributors

Typeset in Frutiger Light by
Florence Production Ltd, Stoodleigh, Devon
Printed and bound in Great Britain by
TJ International Ltd, Padstow, Cornwall

All rights reserved. No part of this book may be reprinted or reproduced or utilized in any form or by any electronic, mechanical, or other means, now known or hereafter invented, including photocopying and recording, or in any information storage or retrieval system, without permission in writing from the publishers.

British Library Cataloguing in Publication Data
A catalogue record for this book is available from the British Library

Library of Congress Cataloging in Publication Data
Transforming parks and protected areas: policy and governance in a changing world/[edited by] Kevin S. Hanna, Douglas A. Clark, and D. Scott Slocombe.
  p. cm.
  1. Protected areas – Management.   2. Parks – Mangement.
  3. Protected areas – Government policy.   4. Parks – Government policy.
  I. Hanna, Kevin S. (Kevin Stuart), 1961–.   II. Clark, Douglas A.
  III. Slocombe, D. Scott.
  S944.5.P78T73 2007
  333.78'3--dc22   2007005424

ISBN10: 0–415–37423–5
ISBN13: 978–0–415–37423–1

Contents

Notes on contributors vii
Acknowledgements xii

1 Introduction: protected areas in a changing world 1
Kevin S. Hanna, Douglas A. Clark, and D. Scott Slocombe

Part I
The challenges of governance 13

2 Evolution of contexts for protected areas governance 15
George Francis

3 Governance models for parks, recreation, and tourism 39
Paul F.J. Eagles

4 Information technology and the protection of biodiversity in protected areas 62
Michael S. Quinn and Shelley M. Alexander

5 Anthropological contributions to protected area management 85
Melissa J. Remis and Rebecca Hardin

6 Steering governance through regime formation at the landscape scale: evaluating experiences in Canadian biosphere reserves 110
Rebecca M. Pollock, Maureen G. Reed, and Graham S. Whitelaw

Part II
Critical perspectives 135

7 Conflict and protected areas establishment: British Columbia's political parks 137
Kevin S. Hanna, Roderick W. Negrave, Brian Kutas, and Dushan Jojkic

Contents

8 Deconstructing ecological integrity policy in Canadian national parks   154
Douglas A. Clark, Shaun Fluker, and Lee Risby

9 The science and management interface in national parks   169
R. Gerald Wright

10 Indigenous peoples and protected heritage areas: acknowledging cultural pluralism   181
David Neufeld

11 Political ecology perspectives on ecotourism to parks and protected areas   200
Lisa M. Campbell, Noella J. Gray, and Zoë A. Meletis

12 Summary and synthesis: observations and reflections on parks and protected areas in a changing world   222
Douglas A. Clark, Kevin S. Hanna, and D. Scott Slocombe

Index   227

Contributors

Shelley M. Alexander has conducted field-based and GIS research on large carnivores and their prey along the eastern front ranges of the Canadian Rockies since 1990. In 2001 she was hired as a faculty member in the Department of Geography, University of Calgary. She received her Ph.D. in 2002 in Spatial Ecology also at the University of Calgary. Shelley's research has included: GIS habitat suitability and movement models for wolves, spatial decision support systems, highway traffic barrier-effects on multi-species movement, multi-scalar predictive modelling, the use of GIS in linkage zone assessment for highway mitigating and spatio-temporal interactions in wildlife communities. She has published in a variety of peer-reviewed journals ranging from the *Canadian Geographer* to *Transportation Research,* the *Wildlife Society Bulletin* and *Biogeography.*

Lisa M. Campbell received her Ph.D. in geography from the University of Cambridge, UK, and is currently the Rachel Carson Assistant Professor of Marine Affairs and Policy at the Nicholas School of Environment and Earth Sciences, Duke University. Her work is broadly situated at the intersection of environment and development in rural areas of Latin America, the Caribbean, and Southern Africa, and is informed by political ecology. She focuses on conservation of endangered species both within and outside of protected areas, and on how conservation via ecotourism and other strategies conflicts with or enhances local community development.

Douglas A. Clark completed his Ph.D. in the Department of Geography & Environmental Studies at Wilfrid Laurier University, where he held a scholarship from The Canon National Parks Science Scholars Program. His research interests include policy processes and governance for social-ecological systems, and his dissertation research focused on local and regional-scale societal dynamics in grizzly bear conservation. He has 12 years of professional experience in six different national parks across Canada, which included serving as the first chief warden of newly established Wapusk National Park. He has twice been a recipient of Parks Canada's Award of Excellence, for training programme design and for search-and-rescue.

Paul F.J. Eagles has a B.Sc. in Biology from the University of Waterloo, an M.Sc. from the University of Guelph in Zoology and Resource Development, and a Ph.D. in Urban and Regional Planning from the University of Waterloo. Paul is an environmental planner with a broad interest in the application of applied ecology and planning theory to environmental conservation. Professionally, as a biologist and a planner, he has a long-standing interest in the planning of national parks and other forms of protected areas. Eagles' research interests lie within the wide range of subjects involved with environmental planning, which include specific interests in park planning and management, applied ecology, ecotourism, park tourism, outdoor recreation, environmental assessment and resource management, and environmental education.

Shaun Fluker is an Assistant Professor in resources law with the University of Calgary Faculty of Law. Shaun holds an LL.M. in environmental law from the University of Calgary. His LL.M. research explored issues in the legal implementation of ecological integrity as a decision-making principle both within and outside of protected areas.

George Francis has degrees in biology, zoology (ecology), economics and political science, and resource management, respectively from the Universities of Toronto, British Columbia, McGill, and Michigan. Prior to joining the University of Waterloo, George worked with the United Nations in New York; he conducted project and programme reviews for United Nations agencies throughout the world during the 1970s and 1980s. His areas of interest include issues of goverance in the context of bioregions and/or watersheds, and implications of complex systems theory for institutions and governance. George was involved with several inter-university cooperative studies of these issues for the Great Lakes and has been a member of advisory committees to the International Joint Commission and the Great Lakes Fishery Commission. The conservation of biodiversity is a key interest. In recent years, he has worked as an advisor on 'biosphere reserves' for the Canadian National Committee for the UNESCO 'Man and the Biosphere Programme', as a member of the Canadian Council on Ecological Areas, as a Trustee of the Nature Conservancy of Canada, and as a participant in various 'atlas' surveys and other ecological monitoring activities.

Noella J. Gray is a doctoral candidate at the Nicholas School of Environment and Earth Sciences, Duke University, and received her M.A. in geography from the University of Western Ontario. Her doctoral research focuses on the politics of co-management of marine protected areas in Belize, and looks specifically at the interplay of science and participation in protected areas design, designation, and management. She has also conducted research on volunteer ecotourism at a national wildlife refuge in Costa Rica, and its contributions to conservation and development.

Kevin S. Hanna's work focuses on issues in environmental policy, impact assessment, and land use planning. A strong thread in his research and writing is the analysis of public policy and the factors that affect implementation. He is an alumnus of the University of British Columbia and the University of Toronto, where he obtained his Ph.D. He is the editor of *Environmental Impact Assessment, Process and Practice* and *Fostering Integration: Concepts and Practice in Resource and Environmental Management* (with D. Scott Slocombe), both from the Oxford University Press. Dr Hanna has served as a policy adviser and analyst for Environment Canada. He has been a faculty member at the University of Toronto and now at Wilfrid Laurier University in Waterloo, Ontario, where he teaches environment and resource policy, impact assessment, and sustainability theory.

Rebecca Hardin is an assistant professor jointly appointed in the University of Michigan's Department of Anthropology and School of Natural Resources and Environment. She has worked in the equatorial forests of Central Africa, as a Peace Corps volunteer and as an anthropologist. Her research focuses on social relations of forest use in the Sangha River region, where Cameroon, Central African Republic, and Congo meet. Her Ph.D. is from Yale University. Hardin's postdoctoral research projects focus on health issues as they relate to environmental management practices in mining and logging concessions in Central African Republic and the Republic of South Africa. She has been a lecturer in

Anthropology at Yale, a visiting professor in Political Science at the Sorbonne, and an assistant professor of Anthropology and Environment at McGill University. She has also been an Academy Scholar at the Harvard Academy for International and Area Studies.

Dushan Jojkic is a graduate of York University's Master of Environmental Studies programme with a focus on urban ecosystem management; he currently works for the Toronto and Region Conservation Authority as project manager for restorative projects. Beyond his current work in local and community-based watershed management and his past research in the Carmanah region of British Columbia, Dushan has an interest in protected areas inspired by travels through national parks in Venezuela, Malaysia, Nepal, and Cambodia.

Zoë A. Meletis is a doctoral candidate at the Nicholas School of Environment and Earth Sciences, Duke University, and received her M.Sc. in planning at the University of Toronto. Her doctoral research looks at the difference between ecotourism in theory and ecotourism in practice, using the case study of Tortuguero, Costa Rica, with special attention to the local solid waste management crisis and its relationship to tourism. Her thesis incorporates theory from geography and other social sciences, such as environmental justice, theories of resistance, and political ecology. Her Master's work on the lives of informal recyclers in Vietnam informed a micro-credit project directed at women workers in the informal recycling industry.

Roderick W. Negrave was born and raised in the southern interior of British Columbia, Canada. His formal education includes a B.Sc. (Agr) and Ph.D. from the University of British Columbia and an M.Sc. from the University of Alberta. Negrave's formal studies have focused on applied plant ecology, including range science management, forest ecology, and silviculture. Currently employed as a research silviculturist with the British Columbia Forest Service, Negrave has previously worked as a parks biologist for the Alberta government, college professor, consulting ecologist, and agronomist.

David Neufeld is the Yukon & Western Arctic Historian for Parks Canada and an adjunct faculty member at Yukon College. He is a former director of the National Council on Public History. His research focus is indigenous, especially Athabascan and Inuvialuit, cultural identity and the forms and character of the connections between Indigenous communities and the Settler state. This includes both oral history and fieldwork in indigenous communities and investigations into the history and sociology of Western scientific knowledge. Much of this work is expressed through the analysis of cultural landscapes and the forms in which they are understood and managed. His teaching at Yukon College addresses both the practical application of this research in the Renewable Resource Managers diploma programme and in undergraduate northern Canada history courses.

Candace Newman is a Ph.D. candidate and adjunct professor at the University of Waterloo, where she is the research manager for the Waterloo Laboratory for Earth Observations. Over the past five years she has been researching communication of remotely sensed information to Marine Protected Area managers in Indonesia. She spent over a full year living in Manado and on Bunaken Island, living amongst the local people and learning about the challenges they face balancing conservation and development. Using the words and experiences of the local communities, she has developed communication guidelines to appropriately integrate remotely sensed biophysical information into local projects. Most importantly, the remotely sensed information has been used to answer critical questions posed by local managers.

Rebecca M. Pollock is a doctoral candidate in the joint Trent University-Carleton University Canadian Studies Ph.D. programme in which she holds a Pierre Elliot Trudeau Foundation scholarship. The focus of her research is on the experience of UNESCO biosphere reserves

Contributors

in governing sustainability at the landscape scale. Rebecca lives in the Georgian Bay Littoral biosphere reserve and serves as Vice-President to the Canadian Biosphere Reserves Association.

Michael S. Quinn is an associate professor in the Faculty of Environmental Design at the University of Calgary and the Director of Research and Liaison for the Miistakis Institute – a research support organization specializing in spatial data and analysis. He holds a B.Sc. in Forest Science from the University of Alberta, an M.Sc. in Forest Wildlife from the University of Alberta and a Ph.D. in Environmental Studies from York University. Michael's teaching and research interests are in the areas of ecosystem management, watershed management, landscape ecology, land-use planning, protected areas management, community-based natural resource management, and urban ecology. He co-manages the Transboundary Environmental Policy, Planning, and Management initiative between the University of Calgary and the University of Montana.

Maureen G. Reed is a professor in the Department of Geography at the University of Saskatchewan, with interests in the social sustainability of resource economies; community capacity; and the role of property regimes in habitat protection. Maureen's current research assesses the place of Canadian biosphere reserves in community-based ecosystem management. She is also a member of the Redberry Lake Biosphere Reserve committee.

Melissa J. Remis earned her doctorate in biological anthropology at Yale University in 1994. For her dissertation research, she initiated one of the first long-term studies on the feeding ecology and locomotion of western lowland gorillas. Remis has continued her field research on gorillas in the Central African Republic (CAR) and has conducted research on the evolution of feeding strategies in African apes and their conservation. Since 1997, Remis has been engaged in a long-term collaborative project with Rebecca Hardin (UMICH) on human impacts on African apes and other key mammals in the Dzanga-Sangha Reserve, CAR. Their work represents an effort to span divides of social science and ecological approaches to conservation problems. Efforts have included field workshops and training for Central African and US students, population surveys and ethnographic research, most recently published in *American Anthropologist*. Remis has received funding from the National Science Foundation, National Geographic, Fulbright IIE, Wenner-Gren, World Wildlife Fund, and other institutions. She was recently named a Purdue University Faculty Scholar 2006-2011.

Lee Risby is an Evaluation Officer with the Global Environment Facility (GEF) Evaluation Office. He holds a Ph.D. and a Masters in Geography from Cambridge University and a BA in Social Anthropology from Keele University. He has conducted research and programme evaluations on donor-funded projects in the areas of biodiversity conservation, natural resource management, and energy conservation in Africa, Asia, Eastern Europe, Latin America, and the Middle East.

D. Scott Slocombe is Professor of Geography & Environmental Studies at Wilfrid Laurier University, where he has been since 1989. Prior to that he studied ecology, environmental studies, and planning at the Universities of Waterloo and British Columbia. His research focuses on environmental planning and management, systems approaches, integrated and ecosystem approaches, and environmental policy and assessment, particularly in northern and western Canada, Australia, and southern Europe.

Graham S. Whitelaw is Assistant Professor of Environmental Studies, Queen's University, and an adjunct lecturer in the Department of Environment and Resource Studies at the University of Waterloo. He holds a post-doctoral fellowship for research on 'Citizen Engagement in Governance for Socio-Ecological Sustainability'. Graham is also an honorary director of the Canadian Biosphere Reserve Association.

R. Gerald Wright was a Wildlife Professor at the University of Idaho. His landscape analysis projects on land use and planning in the West have been used to plan national parks and other protected areas. He was one of the first researchers to conduct studies of visitor interactions in park wildlife viewing areas. His book, *Wildlife Research and Management in the National Parks*, is recognized as the definitive historical book on wildlife management in parks and is used in National Park Service (US NPS) resource management training courses and in college classrooms across the USA and abroad. Wright served as a research ecologist for the US NPS in Anchorage, as an NPS ecologist in Denver and as an assistant director for the International Biological Program, Grassland Biome, also in Colorado. He has more than 100 professional publications and technical papers and is currently conducting large mammal research projects in several western parks and writing a book on how humans have evolved with wild animals. Professor Wright was recently recognized for meritorious service by the US Department of the Interior for his contributions to landuse planning and wildlife management in national parks.

Acknowledgements

The editors gratefully acknowledge support from the Social Sciences and Humanities Research Council of Canada and The Canon National Parks Science Scholars Program, which enabled us to undertake this project. The editors would like to thank the four peer reviewers who provided valuable comments to the authors; their insight and time are greatly appreciated.

Photo credits: Douglas A. Clark, Kevin S. Hanna (cover), Rick Harris, Clint Kendrick, Melissa J. Remis, D. Scott Slocombe (cover), Redberry Lake Biosphere Reserve; Douglas Clark (Figure 8.2); Clint Kendrick (Figure 8.3). Permission to reproduce the graphic in Figure 4.1 was granted by Nature GIS.

Cover images: Temperate rainforest, western Vancouver Island; St Elias Range, Yukon.

Chapter 1
Introduction:
protected areas in a changing world

Kevin S. Hanna, Douglas A. Clark, and D. Scott Slocombe

Worldwide, ideas about what parks and protected areas[1] are and how best to manage them have been completely transformed within the span of most contemporary park managers' careers. The protection of natural and cultural resources, human heritage, and even entire ecosystems through protected areas is increasingly based on the application of ecological principles. The concept of ecosystem-based management has become broadly accepted and widely implemented over the last 20 years, though with sometimes differing interpretations (Agee and Johnson, 1988; Slocombe, 1993, 1998; Grumbine, 1994, 1997; Wright, 1996). This time period has also seen unprecedented, rapid social and ecological change at a range of scales, from local to global. Results of these changes include human domination of Earth's terrestrial and coastal ecosystems (Vitousek *et al.*, 1997; Jackson *et al.*, 2001; Millennium Assessment, 2005), anthropogenic climate warming (IPCC, 2001), and, arguably, a single globalized capitalist economy (Friedman, 2005). Considered together, these changes have created a novel state of global vulnerability for the planet's social and ecological systems (Homer-Dixon, 2001, 2006; Diamond, 2005; Millennium Assessment, 2005).

The consequences of these changes for protected areas are both potentially serious and incompletely understood. A workshop at the 2003 World Parks Congress generated a set of scenarios for protected areas in 2023 that illustrated a highly divergent array of plausible potential futures, given current trends and conditions (McNeely and Schutyser, 2003). Recent changes have also created awareness and opportunities for newly innovative and traditional approaches to protected areas management. Attempts at integrating social and economic concerns with ecological elements in protected areas and parks management have grown steadily more numerous in recent years (e.g. Hulme and Murphree, 1999; Western, 2000; Borrini-Feyerabend *et al.*, 2005), and this approach has, broadly, been labelled the 'new paradigm' for protected areas (Beresford and Philips, 2000). Businesses and civil society institutions are assuming a significantly larger role in the governance of protected areas. Funding for protected areas now comes from a wider variety of sources, and the proportion of that funding provided by governments is decreasing (Dearden *et al.*, 2005).

This book explores the transformation of parks in this context by bringing together a variety of perspectives on current theories, practices, philosophies, and emerging issues in the design and operation of parks and protected areas. We believe that this is an opportune time for such an assessment, for two main reasons. First, there are many urgent problems of environmental stress and failing human livelihoods in and around protected areas worldwide. Such complex problems are qualitatively different from those that protected area regimes traditionally dealt with (e.g. Sellars, 1997). Second, there is now a deep schism in the conservation community about the relationship between people and nature, especially regarding their relative importance within protected areas. A brief examination of both of these propositions follows.

Managing protected areas is becoming more complex

In the late 1990s, a photocopied list of park wardens' duties from the early part of the twentieth century made the round of warden offices in Canadian national parks. It was one page long, spoke mostly about putting out fires and arresting poachers, and ended with the exhortation 'to fulfill all these duties in a manner which will leave you both feared and respected'. The nostalgic appeal of this old document in those offices was obvious when compared against the tasks that present-day Canadian wardens – and park staff worldwide – must now perform. A comparable job description today runs many pages long, and describes technical qualifications and educational standards in immense detail and breadth. The fear and respect that old-time wardens were instructed to instil has been replaced by lists of 'personal suitability factors' that say much more about working collaboratively with others, relationship-building, and seeking consensus among diverse stakeholders. Wardens these days are far more likely to spend their time monitoring and conducting studies on changing park ecosystems and human uses than chasing poachers, and, ironically, may even find themselves setting fires to restore previous ecosystem states.

Such comparisons between the romanticized past and the seemingly mundane, complicated present, bear witness to the struggles involved in adapting to change. Protected area management has hardly remained a static profession since the days of that early Canadian job description, and many recent trends in protected area governance are without precedent (Dearden et al., 2005). Learning and implementing new approaches is not easy for conservation organizations (even assuming that problems have been accurately identified), nor are new adaptations always perfect solutions. As Adams and Hulme (2001) note, the task for protected areas is not supplanting one governance paradigm with another but, rather, finding the best combination of them by identifying the most complementary roles and contributions of existing and new approaches.

The challenges, complexities, and pitfalls of early twenty-first-century reality in protected areas are illustrated by two case studies (Boxes 1.1 and 1.2). The planning process in Bunaken National Park, Indonesia (Box 1.1) appears to have tried to engage multiple stakeholders and accommodate diverse interests, but did so within the constraints of a pre-existing park management plan developed through a traditionally top-down, rational-comprehensive planning approach (Stone, 2002). There was unresolved conflict between bureaucratic norms (urgency to protect marine biodiversity and to complete and implement their plans) and community norms (about the appropriate nature of involvement and problem-solving processes). This conflict led to goal substitution in the zonation plan: demarcating zones to manage stakeholder disputes rather than to manage biodiversity sustainability needs became the objective. Rather than clarifying matters, scientific input into the process was characterized by strong and seemingly valuable information, yet complicated by uncertainty and contestation. This, most likely, was unsurprising to the scientists involved, who were aware of the limitations and uncertainties of their data. For stakeholders though, this meant confusion and further uncertainty, given that scientific information often originates with input and direction from bureaucratic institutions. Overall the intentions on behalf of stakeholders, decision-makers, and scientists were good. 'Best practices' in park planning were adopted from the Bunaken case and transferred to other regions within Indonesia faced with similar issues, and yet the outcomes appear to have been universally frustrating and have fallen far short of their objectives. Bunaken is, unfortunately, far from an isolated case in terms of any of these specific problems.

Box 1.1 The case of Bunaken National Park

Bunaken Island is situated within the World's Coral Triangle, a global centre of marine biodiversity, and, consequently, one of the world's top priorities for marine conservation. In 1991, Bunaken Island, together with four surrounding islands and two mainland sections, became Indonesia's first National Marine Park, encompassing 89,000 hectares, as well as 30,000 local people who reside in 22 villages. Following recognition of the Park, a US-based conservation organization – USAID Natural Resources Management Project (NRM) – partnered with Indonesia's National Development Planning Agency and the Ministry of Forestry to develop a 25-year National Park Management Plan. This plan set in motion a series of legally mandated projects, including the preparation of a zonation plan for each island and mainland area.

Establishing a zonation plan immediately raised questions about the validity of restricting activities that maintain traditional lifestyles, beliefs, and practices. In addition, management authority for the Park remained in the hands of the provincial government who had resisted attempts toward collaborative management. These concerns echoed throughout the Park when in 1993, NRM began their efforts to design a zonation plan that both conserved the existing coral reef biodiversity and minimized conflict between villagers and dive operators, the two primary stakeholder groups. After a three-year process, whereby scientific input from the local university (UNSRAT) and local knowledge from Park leaders was drawn out and amalgamated, the first zonation plan, designating eight different zone types, was created. Unfortunately, each zone recognized separate portions of the coral reef that could not be easily demarcated. In addition, most zones were given generic names such as 'water support zone' and 'limited use zone' that revealed little about their purpose. Shortly afterwards, the government offered a second zonation plan that conflicted with the first, further compounding confusion. The conflicting nature of these two zonation plans and Indonesia's transition to democracy and decentralized natural resource management, however, provided a significant opportunity to draft a third zonation plan.

Efforts to generate the third zonation plan focused on involving multiple stakeholders, who had not been involved in the past. Reasons for this included improving compliance and agreement, accommodating existing enforcement capabilities, and satisfying minority groups who did not agree with the concept of any restriction on fishing activities. Over the years, this disagreement had manifested itself in damaging ways: mooring buoys marking the division between zone boundaries had been cut, park notice boards had been removed, and no-catch regulations in the tidal flat had been disobeyed. Satisfying a greater number of stakeholders while creating the third zonation plan has involved a trade-off. With greater weight placed on stakeholder interests, strict biodiversity conservation has been de-emphasized. As a result, the design of the final zonation plan represents a compromise between the interests of the local people and coral reef biodiversity conservation. Another outcome is that the information needs and wishes of stakeholders are unmet: biodiversity indicators and measures were never established, and therefore cannot be used to determine whether the zonation plan has enhanced biodiversity.

The story of Bunaken National Park illustrates the realities of implementing conservation strategies in a culturally varied environment, and the process modifications that are required to generate compliance among Park stakeholders. The traditional park planning model implies that success comes with compliance to scientific, institutional, and formal governance processes; yet deviation from these processes

> was necessary to secure a workable solution. Equally importantly, this story illustrates how our measures of success must be re-evaluated to accommodate the realities of implementing conservation strategies in dynamic and constantly changing cultural environments.
>
> Candace Newman, University of Waterloo

Churn Creek (Box 1.2) symbolizes a shift in conceptualizing and defining protected areas. Far from creating a designation that shuts out most uses and seeks to preserve a static landscape, Churn Creek provides for the integration of people, land, and natural resources, while protecting specific places and important ecological values. The Churn Creek approach also supports broader land use planning objectives; it flows from a larger process of land use planning, one that integrates parks and protected areas into a comprehensive view of land and resource use. As such, it is an example of the new paradigm (Beresford and Philips, 2000) in application – whether or not such was the original intent of the participants.

The 'people versus nature' controversy

In the last few years a deep tension over how parks and protected areas should be managed has come into the open. Essentially (and somewhat simplistically), proponents of what has become generally known as community-based conservation advocate for the inclusion of local people into both decision-making and as recipients of benefits from protected areas, especially in developing nations. Countering this movement are those who fear that such inclusion will lead to unsustainable development by those people and a drift away from protecting ecological values. Such a simplified and binary description portrays the essence of the argument, but at the risk of overlooking much of the context, detail, and subtleties around each camp's propositions and critiques of the other. We briefly summarize some of the key aspects of the development of this debate below.

By the 1980s, century-old 'fortress conservation' approaches were being increasingly overshadowed in the developing world (though not replaced) by a wave of what have become known as Integrated Conservation and Development Projects (ICDPs) (Adams and Hulme, 2001). The concurrent rise of diverse social trends such as economic neoliberalism and the growing power of indigenous peoples' movements worldwide were likely influences of this development. The late 1990s saw a resurgence of the protectionist paradigm in response to perceived failures of ICDPs specifically, and collaborative approaches to conservation more generally, with a focus on conserving tropical biodiversity in protected areas through more traditional enforcement-focused techniques (e.g. Brandon et al., 1998). Pointed critiques of that protectionist movement followed (Wilshusen et al., 2002; Brechin et al., 2003), and the back-and-forth continues, with criticism going in both directions (e.g. Chapin, 2004 and responses; Locke and Dearden, 2005). Increasingly, though, commentators are pointing out the dangers of such an oversimplified debate (Adams and Hulme, 2001; Wilshusen et al., 2002; Redford et al., 2006).

Nevertheless, the current academic discourse on protected areas is polarized. As others have noted, although this characterization is simplistic (e.g. Brandon et al. 1998; Brockington et al. 2006), that does not necessarily mean it is inaccurate. Recent exchanges in the academic literature (e.g. Chapin, 2004 and responses) have created considerable acrimony within the conservation community (Brockington et al., 2006). Tensions are also vividly apparent in the linkage of some of these ideas with specific institutional agendas, such as concerns about the World Conservation Union (IUCN) advancing the 'new paradigm' for protected

Box 1.2 Churn Creek Protected Area

The Churn Creek Protected Area is located in the south central (Cariboo) region of British Columbia (BC), Canada's westernmost province. The area is remote, rugged, and sparsely populated. In many respects, Churn Creek's history and designation represent a transition in thinking about defining and managing protected areas.

In 1995 Churn Creek was designated a protected area as a result of a larger land use planning process – the Cariboo-Chilcotin Land Use Plan process (CCLUP), one of several similar processes completed and ongoing across BC. The approach was highly participatory; those who contributed, including the affected communities, have a strong sense of ownership over process outcomes. The overall plan emphasizes a multiple-use approach with a key role for protected areas – the CCLUP created 17 new parks and protected areas, Churn Creek among them.

Churn Creek is notable for two reasons: the landscape it protects, and the management approach and 'thinking shift' it represents. The size of the protected area is about 37,750 hectares of mid-, low- and high-elevation grassland and mountains, making it the most significant grassland protected area west of the Rocky Mountains (BC Parks, 2000). Land forms include terraces, hoodoos, gullies, lakes, and wetlands, and the vast Fraser River canyon (BC Parks, 2000). It is a spectacular landscape.

While Aboriginal peoples have long lived in the area, European settlement began only in the late 1800s. Ranching and timber have formed the economic base of the region and have had the most notable impacts on the landscape. Grazing impacts have been relatively minimal, while logging has had a more discernable influence on the area's ecological qualities. Other activities include hunting, trapping, fishing, and some mining (placer).

As planning for Churn Creek unfolded, the integration of grassland and range values emerged as an important objective. In 1998 two entities were created to advise agencies and develop a Management Plan: a Steering Committee composed of First Nations and representatives of BC Parks and other agencies, and a Local Advisory Group (LAG) – which was more inclusive and was made up of about 100 local people representing a broad range of interests. The LAG considered a variety of management options for varied landscape and use values, preferences and recommendations were sent to the Steering Committee. While both the Steering Committee and the LAG worked to develop the Management Plan, the Steering Committee had primary responsibility – especially for addressing 'specifics' (e.g. use zoning, operations, and Plan implementation). Outlining the details would become the primary objective in developing the Management Plan, because the larger CCLUP process had already determined macro policy objectives.

The CCLUP already provided clear direction for key aspects of any management regime for Churn Creek. While the CCLUP sought to protect regional ecological values, it also explicitly allowed long-established uses such as recreation, grazing, hunting, trapping, backcountry tourism, and some mining (though tenures within the protected areas were discontinued). Private land could only be included in a protected area if the province purchased it, and existing grazing levels were to be maintained.

In 1998 the BC government purchased the Empire Valley Ranch and incorporated it into the Churn Creek Protected Area. The ranch had previously been owned by a logging company that coveted it for its timber value; their logging practices were controversial. The province bought the Empire Valley to insure that remaining ecological and ranching values would be preserved, and to guarantee that ranching and grazing would remain an important activity in the Churn Creek area – this was an objective under the provisions of the CCLUP. Today the Empire Valley is a working

> cattle operation, managed by a local rancher under a lease agreement with the province of BC; it functions within the management framework of the protected area.
>
> With the addition of the Empire Valley and the gradual development of the Management Plan, Churn Creek emerged as a unique example of a protected area that has maintained a range of existing uses while accounting for the protection of ecological values and unique landscapes. About 82 per cent of the area is managed as 'natural environment' with no vehicle access, six per cent is zoned as 'natural environment with vehicle access' for areas with established high recreation potential, four per cent is 'special feature' with no grazing, and two per cent is open to 'intensive use and recreation' which allows for specific ranching activities (BC Parks, 2000).
>
> Kevin S. Hanna, Merritt BC

areas at the expense of stricter protection for biodiversity and wilderness (Locke and Dearden, 2005). Returning briefly to the Bunaken case, it would likely be characterized by one camp as having failed because it tried to mix incompatible social and ecological goals, and raised unrealistic expectations. The other camp would probably emphasize the very real questions of social justice involved in supplanting a collaborative regime with coercive state enforcement practices, arguing that such concerns cannot be swept away by simply declaring a conservation 'emergency' (Peluso, 1993). Good arguments could no doubt be made from either perspective, and such open debate is healthy; as long as it is genuinely aimed at resolving social-ecological problems in the field and not simply advancing narrow interests within the conservation community.

Both parks and people would be poorly served by pretending that this fundamental conflict of values doesn't exist, that it can be definitively won, or that finding common ground is always possible. Indeed, the social construction of protected areas embeds that conflict into their very existence as institutions (Hermer, 2002; Jones and Wills, 2005), and protected area managers worldwide likely recognize this as an enduring dynamic tension that they must constantly navigate. Any 'final resolution' to this conundrum that exclusively favours either the short-term interests of society or purist biodiversity protection is unlikely to persist without considerable ecological and/or social costs. The latter are often borne locally and by those who can least afford them (GEF, 2005). Further, empirical examples from protected areas (Pimbert and Pretty, 1997), large-scale state development schemes (Scott, 1998), and resource management systems (Holling and Meffe, 1996; Folke et al., 2002) all suggest that in a world of accelerating change such inflexible institutions are likely to hasten their own demise.

Redford et al. (2006) make an impassioned plea for depolarizing this debate by recognizing the complexity of protected area management in practice, and they observe: 'The discourse on parks is being driven towards brittleness (i.e. lack of resilience) – bad news for both protected areas and people living in and near them.' Their invocation of resilience provides an interesting way to consider how the park discourse could be enriched. Resilience can be defined as the capacity of a system to undergo disturbance while maintaining both its existing functions and controls and its capacity for future change (Holling, 1973; Gunderson, 2000; Carpenter et al., 2001). Resilience is determined not only by a system's ability to buffer or absorb shocks, but also by its capacity for learning and self-organization to adapt to change (Gunderson and Holling, 2002). Applying this metaphor to the current discourse on parks leads rapidly to consideration of the participants' collective ability to learn and adapt to change (Hulme and Murphree, 2001). How can such adaptive capacity be enhanced? As protected areas' social and ecological circumstances change, the dialogue between theory and practice must remain tightly coupled, responsive, dynamic, and productive to remain useful. As such, lessons from the field and from academe must be communicated, integrated, adapted, and constantly challenged.

This book provides ideas aimed toward enhancing the resilience of the discourse on parks, and so helping to prevent its degeneration into a sclerotic argument that serves neither human nor ecological needs. Park managers and policy makers, researchers, and civil society movements would all benefit immensely from a critical examination of the rapidly evolving field of protected area management, and an assessment of recent experiences worldwide. Comprehensive analysis and critique is largely missing from the protected areas literature. This volume is an attempt to fill that gap, and meet the hunger that colleagues have expressed for a more judicious, less descriptive treatment of the protected areas design, management, and operations milieu. Chapter authors were encouraged to critically assess current practices and challenge conventional ideas in protected area management. These authors bring a wide array of disciplinary and interdisciplinary approaches – and a substantial wealth of experience – to current theories, practices, philosophies, and emerging issues in the design and management of parks and protected areas, with a global focus.

This book is organized into two parts, each of which represents a different aspect of the complexity facing parks and protected areas. Part I provides an overview of emerging practices and governance approaches for protected areas. Chapter 2 (Francis) examines broad social and ecological trends that have influenced the management of protected areas over the last three decades. Changes in protected area governance are examined from a complex systems perspective in light of issues such as climate change, political-economic cycles and world history, and cultural change. Governance for protected areas is situated within wider issues of governance more generally, and prospects are assessed for applying emerging systems approaches to current and future thinking about protected areas.

Chapter 3 (Eagles) considers management models for protected areas. Globally, there is a wide range of administrative arrangements for protected area management. Up to six different approaches are visible: government agency, parastatal, private for-profit company, public–private mix, private non-profit corporation, public contract to private companies, and mixed groups of institutions. This chapter outlines the various management models and compares the benefits and drawbacks of each approach from a number of points of view, including conservation, financial effectiveness, and tourism management.

Chapter 4 (Quinn and Alexander) examines the growing role of information technology in protected area management, particularly geographic information systems (GIS). The key roles of information technology (data acquisition, storage and display, retrieval and analysis), as well as accessibility and implementation of the products, are reviewed, and issues associated with each of those roles is discussed in specific international and Canadian contexts. Most conservation efforts involve habitat identification and/or the delineation of areas needing protection – all of which are spatial in nature. Nevertheless, future information technology needs are systemic and cultural as much as they are technical.

Chapter 5 (Hardin and Remis) investigates social/ecological interactions across park boundaries. This chapter integrates approaches from physical and cultural anthropology to describe changing relationships between humans and animals in the Central African Republic. Their integrated social-ecological research approach implies alternatives to the simple divisions between core and buffer zones for conservation outcomes, and illustrates recent challenges to the heavily gendered and ethnically differentiated monitoring and management practices to date in this setting. As such, their research points to interesting gaps in the literature on integrated conservation and development projects, and on social inequity and ecological and economic complexity as they shape emerging conservation strategies.

Chapter 6 (Pollock et al.) considers the role of local communities and community-scale institutions in governance processes for protected areas. Comparing three case studies of UNESCO Man and the Biosphere Reserves in Canada, they consider under what conditions communities can help to shape the governing regimes for protected areas at a landscape scale and some of the factors that allow environmental organizations, in particular, to steer governance directly. They also assess local community capacity to address two particular governance challenges: open systems and institutional fragmentation.

Part II provides critical perspectives on the dynamics that impact policy and management approaches for protected areas. Chapter 7 (Hanna et al.) considers the role of conflict in protected area establishment. The designation of protected areas often reflects complex social and political processes where conservation science, economic agendas, and political imperatives frequently interact in conflict settings. This chapter explores the important role of place-centred conflict in protected area designation and larger policy processes. In cases examined from western Canada, conflict events have not only led to the creation of several wilderness parks, but also influenced public policy at a greater scale.

Chapter 8 (Clark et al.) critically assesses the success of a major international protected area goal: ecological integrity. In practice, ecological integrity often resembles the old-style preservationist model dressed in more contemporary ecological terminology. This chapter describes the transformation of ecological integrity theory into policy and law in the Canadian national park system, and examines specific cases of park management problems that might be more effectively addressed through alternative framings of the concept of ecological integrity.

Chapter 9 (Wright) addresses the question of whether or not scientific knowledge and the informed opinions of scientists play an important role in the management decisions made in protected areas; both in a historical and contemporary context. This topic is examined in terms of decisions made with respect to natural resource management, as well as the influence of management decisions on natural resources that are made in the context of accommodating visitor use. Expectations are important: those that protected area managers have of scientists (both positive and negative) and how scientists deal with such expectations. Finally, this chapter assesses the effectiveness of the ways that scientific knowledge is typically communicated to managers.

Chapter 10 (Neufeld) investigates the nature of the modern State's understandings of its citizens' identities; particularly those of indigenous people, who are playing an increasingly central role in protected area management worldwide. The evolving character of this recognition is examined through case studies that illuminate different aspects of the challenge of reconciling governmental mandates for heritage protection with demands for cultural pluralism. The author draws upon his 20 years of experience as a Parks Canada historian working with First Nation communities in the Yukon Territory to trace the detailed and often bumpy path to renovating the national narrative.

Chapter 11 (Campbell et al.) applies political ecology to develop an improved theoretical understanding of ecotourism to parks and protected areas. As such, it addresses a major gap in the related literature that treats ecotourism primarily on a case study basis. The latter approach tends to emphasize 'getting ecotourism right' through the use of evaluative frameworks, for example, rather than situating ecotourism in a broader understanding of human–environment relations or political economy. With its emphasis on both discursive and material practice, political ecology can enhance our understanding of ecotourism as a preferred conservation alternative. Political ecology helps to situate the impacts of ecotourism on parks and protected areas and the human communities living with them, within broader social and political processes.

The concluding chapter (Clark et al.) summarizes the major themes from the two preceding sections and discusses the 'lessons learned' for protected area management. Here the authors also outline areas for further research and application.

There are several things that readers should keep in mind. First, while this volume provides examples from around the globe that help illustrate themes and trends, it is not the intent to provide a comprehensive survey of park and protected area management practices and experiences worldwide. Nor does this book try to include every trend or driver of change affecting protected areas; for example, climate change is one such topic that this volume does not address in detail. The world is a complex and varied place, and our sampling and synthesis tries to recognize that complexity. We have tried to steer a pragmatic course between sweeping (and most likely illusory) simplification, and drowning in a mass of place-specific and contingent details. Finally, although the chapters that follow were largely (though not entirely) contributed by academic authors, a resilient and productive discourse on parks urgently requires input from the diverse peoples who live in and around protected areas, and the staff of park agencies worldwide. Indeed, engaging those people should be a central characteristic of whatever form the emergent 'new paradigm' for protected areas eventually takes, and the as-yet unseen paradigm that will inevitably challenge that one in the years to come.

Note

1   In this volume we use 'park' to refer to areas so designated by constituted governments at various levels, and 'protected area' more broadly and in the sense of the IUCN's current definition: 'An area of land and/or sea especially dedicated to the protection and maintenance of biological diversity, and of natural and associated cultural resources, and managed through legal or other effective means' (www.unep-wcmc.org/protected_areas/categories/index.html, accessed 3 July 2006).

Literature cited

Adams, W. and Hulme, D. (2001). Changing narratives, policies and practices in African conservation. In D. Hulme and M. Murphree (eds), *African Wildlife and Livelihoods: the promise and performance of community conservation*, pp. 24–37. Oxford: James Currey.

Agee, J.K. and Johnson, D.R. (eds) (1988). *Ecosystem Management for Parks and Wilderness*. Seattle, WA: University of Washington Press.

BC Parks (2000). *Management Plan for Churn Creek Protected Area*. Williams Lake: BC Parks, Cariboo District.

Beresford, M. and Philips, A. (2000). Protected landscapes: a conservation model for the 21st century. *George Wright Forum* 17, 15–26.

Borrini-Feyerabend, G., Pimbert, M., Farvar, T., and Kothari, A. (2005) *Sharing Power: learning by doing in co-management of natural resources throughout the world*. Gland: CEESP and IUCN.

Brandon, K., Redford, K.H., and Sanderson, S.E. (eds) 1998. *Parks in Peril: people, politics, and protected areas*. Washington, DC: Island Press.

Brechin, S.R., Wilshusen, P.R., Fortwangler, C.L., and West, P.C. (eds) (2003). *Contested Nature: promoting international biodiversity with social justice in the twenty-first century*. Albany, NY: State University of New York.

Brockington, D., Igoe, J., and Schmidt-Soltau, K. (2006). Conservation, human rights, and poverty reduction. *Conservation Biology* 20 (1), 250–252.

Carpenter, S., Walker, B., Anderies, J.M., and Abel, N. (2001). From metaphor to measurement: resilience of what to what? *Ecosystems* 4, 765–781.

Chapin, M. (2004). A challenge to conservationists. *WorldWatch* November/December 2004, 17–31.

Dearden, P., Bennett, M., and Johnston, J. (2005). Trends in global protected area governance, 1992–2002. *Environmental Management* 36 (1), 89–100.

Diamond, J. (2005). *Collapse: how societies choose to fail or succeed*. New York: Viking.

Folke, C., Carpenter, S., Elmqvist, T., Gunderson, L., Holling, C.S., *et al.* (2002). Resilience and Sustainable Development: building adaptive capacity in a world of transformations. Scientific Background Paper on Resilience for the Process of the World Summit on Sustainable Development on behalf of the

Environmental Advisory Council to the Swedish Government. www.resalliance.org/reports/resilience_and_sustainable_development.pdf (accessed 1 November 2002).

Friedman, T.L. (2005). *The World is Flat: a brief history of the twenty-first century.* New York: Farrar, Strauss, & Giroux.

Global Environmental Facility (2005). *The Role of Local Benefits in Global Environmental Programs.* Washington, DC: GEF Office of Monitoring and Evaluation.

Grumbine, R.E. (1994). What is ecosystem management? *Conservation Biology* 8, 27–38.

—— (1997). Reflections on 'What is ecosystem management?' *Conservation Biology* 11, 41–47.

Gunderson, L.H. (2000). Ecological resilience – in theory and application. *Annual Review of Ecology and Systematics* 31, 425–439.

—— and Holling, C.S. (eds) (2002). *Panarchy: understanding transformations in human and natural systems.* Washington, DC: Island Press.

Hermer, J. (2002). *Regulating Eden: the nature of order in North American parks.* Toronto: University of Toronto Press.

Holling, C.S. (1973). Resilience and stability of ecological systems. *Annual Review of Ecology and Systematics* 4, 1–23.

—— and Meffe, G.K. (1996). Command and control and the pathology of natural resource management. *Conservation Biology* 10, 328–337.

Homer-Dixon, T.A.D. (2001). *The Ingenuity Gap: can we solve the problems of the future?* Toronto: Random House.

—— (2006). *The Upside of Down: catastrophe, creativity, and the renewal of civilization.* Toronto: Knopf.

Hulme, D. and Murphree, M. (1999). Communities, wildlife, and the 'new conservation' in Africa. *Journal of International Development* 11, 277–285.

—— and —— (eds) (2001). *African Wildlife and Livelihoods: the promise and performance of community conservation.* Oxford: James Currey.

IPCC (2001). *The Third Assessment Report of the Intergovernmental Panel of Climate Change.* Cambridge: Cambridge University Press.

Jackson, J.B.C., Kirby, M.X., Berger, W.H., Bjorndal, K.A. and Botsford, L.W. (2001). Historical overfishing and the recent collapse of coastal ecosystems. *Science* 293, 629–638.

Jones, K.R. and Wills, J. (2005). *The Invention of the Park: from the Garden of Eden to Disney's magic kingdom.* Cambridge: Polity.

Locke, H. and Dearden, P. (2005). Rethinking protected area categories and the new paradigm. *Environmental Conservation* 32 (1), 1–10.

McNeely, J.A. and Schutyser, F. (eds) (2003). *Protected Areas in 2023: scenarios for an uncertain future.* Gland: IUCN.

Millennium Ecosystem Assessment – MEA (2005). *Living Beyond Our Means: natural assets and human well-being.* Washington, DC: Island Press.

Peluso, N.L. (1993). Coercing conservation? The politics of state resource control. *Global Environmental Change* 3 (2), 199–217.

Pimbert, M.P. and Pretty, J.N. (1997). Parks, people and professionals: putting 'participation' into protected area management. In K.B. Ghimire and M.P. Pimbert (eds), *Social Change and Conservation*, pp. 297–330. London: Earthscan.

Redford, K.H., Robinson, J.G., and Adams, W.M. (2006). Parks as shibboleths. *Conservation Biology* 2, 1–2.

Scott, J. (1998). *Seeing Like a State: how certain schemes to improve the human condition have failed.* New Haven, CT: Yale University Press.

Sellars, R.W. 1997. *Preserving Nature in the National Parks: a history.* New Haven, CT: Yale.

Slocombe, D.S. (1993). Implementing ecosystem-based management. *BioScience* 43, 612–622.

—— (1998). Lessons from experience with ecosystem-based management. *Landscape & Urban Planning* 40, 31–39.

Stone, D. (2002). *Policy Paradox: the art of political decision making* (Revised Edition). New York: W.W. Norton.

Vitousek, P., Mooney, H.A., Lubchenco, J., and Melillo, J.M. (1997). Human domination of earth's ecosystems. *Science* 277, 494–499.

Western, D. (2000). Conservation in a human-dominated world. *Issues in Science and Technology*, Spring 2000, 53–60.

Wilshusen, P.R., Brechin, S.R., Fortwangler, C.L., and West, P.C. (2002). Reinventing a square wheel: critique of a resurgent 'protection paradigm' in international biodiversity conservation. *Society & Natural Resources* 15, 17–40.

Wright, R.G. (ed.) (1996). *National Parks and Protected Areas: their role in environmental protection*. New York: Blackwell Scientific.

Part I
The challenges of governance

Chapter 2
Evolution of contexts for protected areas governance

George Francis

> ... how stubborn our ignorance has proven as it relates to the major events that will determine our future.
> Jeffrey A. McNeely, IUCN Chief Scientist, commenting on experiments with scenarios for exploring the larger contexts of changes and threats to protected areas, *World Conservation* 34 (2), 21 (2003)

> We cannot, however, fulfill our duties as stewards of the Earth's last natural ecosystems if we plan and manage for a world that no longer exists.
> Barber *et al.* (2004: xv)

Introduction

Evolution of the contexts for governance over protected areas is a matter about governance generally. While local situations will remain critically important for dealing with such issues, everything that happens locally isn't caused there. The importance of 'global change' that sets the larger contexts for local governance has also to be considered. But how? The two lead-in epigraphs capture the larger quandary underlying this situation. Similar concerns are being addressed in most countries either in the context of 'governability' generally, or in other contexts, such as managing resource systems for sustainability that include a strong conservation component.

This chapter sets out the intellectual challenge that reflective practitioners may wish to consider from perspectives that have been developed in the literature about complex systems. These perspectives can guide ways to make sense of, and obtain insights into, global change dilemmas. The purpose is not to debate particular issues raised by proponents of protected areas (of whom I am one) but to enrich thinking about issues of governance, using protected areas as examples.

A first section sketches the broad outlines of the subject of 'governance' and sets it in the context of a narrative about 'globalization' over the past 30 years or so. It is followed by a summary of issues of governance for protected areas that have been addressed by the World Conservation Union (formerly, the International Union for the Conservation of Nature and Natural Resources – IUCN) and other international bodies. As the number, kinds, and spatial extent of protected areas increased, and reflective thinking about them evolved, issues of management for such areas became of increasing concern and importance. As more organizations and groups engaged with protected areas management, arrangements for local governance to foster the necessary consultation and cooperation emerged as central issues.

With these considerations as examples, the concept of 'complex systems', and approaches taken towards understanding them, are briefly summarized. 'Social-ecological systems' (SESs) are understood to be complex systems that function across a wide range of spatial and temporal scales. They illustrate the kinds of systems in which all of us reside, and their dynamics raise issues about vulnerability and resiliency. Note is then made of 'global change' associated with SESs, using climate change as an example.

Interpretations of global change have mainly concerned the biophysical manifestations of the evolving future. For a historical overview at some of the same large scales, a 'world-systems' perspective is broadly sketched. Historical interpretations are always subject to controversy including assertive counter-interpretations, and world-systems analysis is no exception. It is, however, based on impressive scholarship informed by systems thinking. As a major contributor to this approach has noted, we are now at 'the end of the world as we know it', the title of a book by Wallerstein (1999), in two different senses of the phrase. Transformations of global capitalism in the present world of nation-states are well under way towards some unknown future. Complexity perspectives also reveal limitations of relying upon isolated areas of disciplinary knowledge for understanding this new era of world history and issues raised by it.

The chapter concludes with an invitation for reflective practitioners to contemplate a 'thought experiment'. It draws on complex systems thinking in order to stimulate 'sense-making' insights into the changing contexts for governing protected areas and whatever new beliefs and values might have to become rationales for continuing to protect them.

An overview perspective on 'governance'

The predicament

An enormous amount of interest in 'governance' has developed over the past 20 years or so, stimulated by recurring questions about 'governability' in large modern and complex societies. As more demands or expectations are placed on governments, the limitations or failures of what can be done effectively by the 'command and control' procedures of government become widely evident. Different limitations and failures come from placing too much reliance on market managerialism in a global economy dominated by the private sector. It can be adept at providing commodified goods and services for those who can afford them, but otherwise ignores non-market needs and impacts from externalized costs. At the same time, with a growing complexity of institutional systems, social-political processes continually identify new objects and categories of people in need of government or governance to manage or control them (Swyngedouw, 2005). Familiar examples for protected areas are more species to protect, ecosystems to be managed, and human activities to be restricted or controlled to achieve this.

Heavy dependence is placed on the reliability of large-scale, sophisticated technologies in all sectors of society to the point of creating a pervasive public sense of vulnerability and risk in the whole system. This sense is not easily alleviated by more management or control in the name of 'security'. Environmental issues often reflect this predicament. A sense of vulnerability also arises from extensive functional interdependencies among the dense populations of organizations found in modern societies. While a growing abundance of organizations might enhance societal capacity generally, it also creates mutual interferences in whatever any one of them is striving to do. This organizational complexity generates the phenomena of tangled hierarchies in administrative systems, 'externalities' (or negative interdependencies) in socio-economic systems, and environment impacts in ecosystems.

Yet more challenging, many questions to be dealt with cannot be handled from the ideal of science-based rational planning and management. Instead, complicated science-based issues often arise in situations where controllability is low (rational planning and central management are out), facts are uncertain, and usable science is scarce (knowledge is not readily at hand). Decisions remain urgent, but the stakes are high in terms of their consequences (sound judgement is needed). These situations, called 'post-normal science' by Ravetz (1999, 2004), are increasingly the rule rather than the exception. Many environmental issues exemplify this. The decision process, then, is best served by expanded peer groups that extend beyond technical experts in the employ of particular organizations. This is needed in order to pose questions about the (in)completeness of evidence and arguments, their underlying assumptions, and the ethical implications of choices to be made.

One main response to this complexity has been to engage 'civil society' much more in the process. In some countries, this 'third sector' is perceived to be distinct from governments and corporate businesses, but otherwise linked to, or dependent strongly on, both. Civil society is represented by a variety of non-governmental organizations (NGOs). Not all countries view the NGO world this way. However, most recognize the importance of securing the consent of the governed as citizens or customers. To the extent this relies upon elements of democracy, it is likely to be more effective. Civil society is then encouraged to become active in different partnership arrangements that constitute 'hybrid organizations' composed of government agencies, private businesses, NGOs, and sometimes local citizen groups. They collectively strive to achieve what no one of them can do on its own. This should be quite familiar to people associated with protected areas who might, themselves, have tried to organize such arrangements. There is considerable discourse about citizen engagement, popular participation, stakeholder processes, accountability, trust-building, social capital, and other similar or overlapping ideas that are now recognized as necessary to put protection for protected areas on a sounder basis.

An interpretation of 'governance'

The term 'governance' is used in a number of different ways. Torfing *et al.* (2003) reported nine competing definitions and interpretations. Kooiman (2003) elaborated a conceptual framework for governance and governing to clarify a number of issues; Kjær (2004) reviewed different interpretations of governance in major sub-disciplines of political science and public administration; and Paquet (2005) summarized a substantial body of literature about the subject published over the years. Generally, 'governance' refers to networked hybrid organizations. Jessop (2003: 1) for example, defined it as:

> the reflexive self-organization of independent actors involved in complex relations of reciprocal interdependence, with such self-organization being based on continuing dialogue and resource-sharing to develop mutually beneficial joint projects and to manage the contradictions and dilemmas inevitably involved in such situations. Governance organized on this basis need not entail a complete symmetry in power relations or complete equality in the distribution of benefits: indeed, it is highly unlikely to do so almost regardless of the object of governance or the 'stakeholders' who actually participate in the governance process.

These hybrid organizations may be government-driven arrangements or NGO-led. Small interpersonal networks are largely based on mutual trust among individuals, continual discussion about means and ends, accommodation of interests, and a large degree of voluntary participation. Larger inter-organizational arrangements, such as coalitions or alliances, may cooperate through formal partnerships with mutual obligations set out in legal terms. In some countries, intersectoral (or intersystemic) coordination is sought through a 'corporatism' arrangement whereby centralized negotiation and mutual adjustments are made among elite representatives of governments, corporate leaders, trade unions, and professional associations.

The functions of governance networks are seldom all-encompassing. Instead, they focus on one or a few advisory, advocacy, educational, or task-related roles directed primarily to processes of policy formation, planning, or programme implementation. The networks span different kinds of boundaries, such as jurisdictional, administrative, and proprietary boundaries that limit the scope of most organizations, as well as boundaries of knowledge ('silos') maintained by disciplines and professions. Their span may be 'horizontal' over a large geographic area, and/or 'vertical' across scales to connect local groups to national and international arenas. The 'steering' of networks in some collaborative manner is a challenge, as is finding how best to organize their 'self-organization' as they go along. Success in this is one source of their effectiveness. Governance arrangements associated with protected areas exhibit these phenomena, with IUCN the most prominent among the global to local sharing of information, experience, and expertise.

Networks for governance are both guided and constrained by basic rules within nation-states established by constitutional divisions of authority among jurisdictions, distribution of legal authority among administrative agencies, and property rights. Property may be state-owned, privately owned, or community-owned as 'common property' where the community allocates rights to use among its members. At the international level, rules are set out in treaties and conventions that create management regimes for special geographic regions (e.g. Antarctica) or subjects for international collaboration (e.g. Convention on Biological Diversity).

The framework of rules for the development and functioning of networks for governance is sometimes referred to as 'meta-governance'. Kooiman (2003) distinguishes three levels of this 'governing how to govern'. The first addresses the effectiveness of networks for problem-solving and creating new opportunities; the second reviews the effectiveness of institutions in facilitating and supporting network capabilities; while the third examines the inherent rationality and ethical responsibility of what is done. Obviously, there are many things that can, and do, go wrong. 'Governance' is a response to complexity in society, but not always the answer. It, too, can show limitations and failures. Jessop (2003) interprets meta-governance as organizing the conditions for self-organization of the networks, including balancing ('collibrating') the conditions under which networks have to operate, and maintaining sufficient flexibility to switch to other methods of coordination if networking relationships are failing. Some underlying basic issues have also been discussed, for example, questions about the democratic accountability and legitimacy of networked governance, the inclusion or exclusion of certain groups, poorly defined responsibilities or obscure objectives, and co-optation of networks by more powerful interests (e.g. Skelcher, 2005; Swyngedouw, 2005).

Good governance at the global level has formidable difficulties (Koehn and Rosenau, 2002; Rosenau, 2003). The world has become a highly integrated global capitalist economy set within the organizations of nation-states. There is a deeply entrenched functional and spatial differentiation of roles and privileges in this world society. A small set of core countries have the most advanced technologies and organizational know-how for different economic sectors, while a much larger number of peripheral countries supply little more than resources and cheap labour. The semi-peripherals in between usually have an urban sector more closely linked to core economies, and rural areas that remain peripheral. Significant transitions have taken place over the past 30 years or so under the general rubric of 'neoliberalism', and have had major influences over the contexts for protected area governance.

Emergent neoliberalism

A general narrative about the most recent neoliberal era (which can be drawn from many sources, e.g. Jessop, 2002a, b; Loughlin, 2004) usually begins with the ideal of independent nation-states, each having a largely self-contained and balanced economic structure supported by strong social policies. They engaged in trade with other national economies through state-mediated rules about exchange rates, tariffs, limits on foreign ownership, and other measures all intended to protect the integrity of national economies and polities. This ideal has been called a 'Keynesian welfare national state with a Fordist mode of economic organization' (Jessop, 2002a). The reference is to mass production and consumption of goods and services (based on assembly-line technologies pioneered by Ford for automobiles in the early twentieth century) and the adoption of Keynes' economic theories whereby the State tried to maintain full employment and consumption levels (hence also production) through public spending during lags in business cycles.

These arrangements were associated with the post-Second World War economic expansion in the core economies, characterized by relatively widespread material prosperity and state enhancement of wealth distribution within countries. It was also believed that every nation-state in the world could aspire to this ideal provided they followed the right policies. Intergovernmental assistance for development was spurred by 'cold war' competition between the US and USSR to win converts to policies for either market or state capitalism. The 'developmentalism' espoused by the United Nations and other intergovernmental bodies reflects this belief (exemplified most recently by the UN Millennium Development Goals) along with a conviction that core–periphery distinctions could gradually disappear.

This ideal for each nation-state became increasingly unworkable by the 1970s, especially with a concurrent downswing in the global economy. Most industrialized countries then entered a period of high unemployment, low growth with high inflation ('stagflation'), and increased public debt. This led to the corporate capitalist economic (and 'neoconservative' political) backlash to remove or undermine the policies and regulations that nation-states had in place. In some countries, two discrete phases have been discerned, the first a 'rollback' of the welfare state ideal starting in the early 1980s, and the second a 'roll-out neoliberalism' in the 1990s to obtain dominance for corporate models over other institutional forms, and compliance with corporate economic interests (Peck and Tickell, 2002). The easing of state controls of the private sector facilitated the rapid expansion of global capitalism, with innovative and specialized production for increasingly diversified markets (vs mass sameness), and the strengthened influence of transnational corporations. Financial markets became integrated globally and increasingly devoted to a continuous and massive speculation in currencies, stocks and bonds, a plethora of investment funds, and other financial products. Some national government powers were redistributed 'upward' to international institutions that fostered global capitalism, 'downward' to more local levels of government under the rubric of subsidiarity or debt reduction, or just abandoned as no longer necessary ('de-regulation').

In the roll-out phase, States promoted the interests of global capitalism by reducing taxes, cutting welfare entitlements, promoting 'flexible workforces' ('workfare'), restricting union powers to protect wages and working conditions, privatizing public utilities and services, and promoting 'competitiveness' and 'productivity' as defined by the private sector. Governments also competed with one another to attract knowledge-intensive, innovation-oriented 'new economy' investments in urban regions. These oases of growth helped exacerbate patterns of uneven development and widened inequalities between rich and poor. Debt crises for a number of countries led to the imposition of 'structural adjustments' on behalf of neoliberal agendas by the International Monetary Fund and World Bank before new borrowing could occur. As the social and environmental costs and contradictions of this neoliberal alternative became increasingly apparent, so did attempts to find ways to rebalance capitalism with broader societal goals and curb its more destructive tendencies. There are extensive debates about interpreting these events (see Broad, 2004).

Initiatives to reorganize or strengthen governance responded in many ways. Some sought to facilitate these changes and capture benefits from them. Others sought to soften the effects of regional economic declines, ameliorate increasing poverty, and deal with social justice questions. Network arrangements were viewed as a means to contain socio-economic contradictions of the transitions as best they could. New institutional arrangements sometimes extended across old borders within or between nation-states that were no longer seen as the primary 'containers' for economic and other relationships. The 'governance of complexity and the complexity of governance' (Jessop, 1997) are mutually interdependent and continuing to evolve. Overall, this constitutes the larger contexts for thinking about long-term governance of protected areas.

On protected areas and the governance challenge

Expanded coverage

Over the past half century, the IUCN has become the widely recognized world authority for biodiversity conservation. Its mission 'is to influence, encourage and assist societies throughout the world to conserve the integrity and diversity of nature and to ensure that any use of natural resources is equitable and ecologically sustainable' (www.iucn.org/en/about/). It does this by preparing extensive databases, assessments, guidelines, and case studies based on conservation science and 'convening dialogues between government, civil society and the private sector' through major conferences and workshops. As an organization, IUCN exemplifies impressive governance capacity to mobilized networks of collaboration globally and across geographic scales from field projects to international arenas.

There have been many accomplishments in establishing systems of protected areas throughout the world over the past several decades. This success has raised issues about on-site management and arrangements for governance. The IUCN has devoted increasing attention to governance, especially before and during the Fifth IUCN World Parks Congress (2003), and through their series of publications: the 'Ecosystem, Protected Areas and People (EPP)' project (see Barber et al., 2004); the 'Best Practice Protected Area Guideline' series; and *Parks*, the international journal for protected areas managers. It has also worked closely with the Convention on Biological Diversity (CBD) to gain recognition of the importance of protected areas. This was achieved in 2004 with the CBD/Conference of the Parties Decision VII/28 on matters pertaining to 'Governance, Participation, Equity and Benefit Sharing' associated with biodiversity conservation. The CBD will rely upon IUCN expertise, data, and information to advise it on these issues. Close links have also been formed with the UNESCO Man and the Biosphere (MAB) Program on 'biosphere reserves' (UNESCO, 2002), the Millennium Ecosystem Assessment Program (MEA, 2003, 2005), and indirectly with the United Nations Millennium Development Goals (www.un.org/millenniumgoals).

In 1994, the IUCN adopted an international classification for protected areas based on six categories of management objectives. This provided guidance for the 'World Database on Protected Areas Consortium' to elaborate global inventories of such areas. It also allowed for international comparisons to be made among some 1,388 different administrative terms used around the world to designate protected sites for different purposes (Sheppard, 2000). The most recent and comprehensive inventory based on this system lists 102,102 protected sites covering 17.1 million square kilometres, or 11.5 per cent of the global land surface, and about 1.64 million square kilometres of marine protected areas, or about 0.5 per cent of the world's oceans (Chape et al., 2003). For terrestrial areas it seems to approximate a goal set by the Fourth IUCN World Parks Congress (1992) to have at least 10 per cent of each major biome represented in protected areas. It seems also to meet the earlier suggestion by the Brundtland Commission to strive for a threefold increase in the total expanse of protected areas in order to constitute a representative sample of the Earth's ecosystems (WCED, 1987: 166), a target subsequently interpreted widely to mean 12 per cent of the territory of individual nation-states.

There has also been a considerable expansion in thinking about protected areas in the past 30 years, described as a 'paradigm shift' by Phillips (2003, 2004). This shift is reflected in the IUCN management categories (Ravenel and Redford, 2005). The first three categories represent a strict protectionist philosophy that strives to safeguard designated landscapes from direct human disturbances. The other three categories, in effect, recognize that this approach is not feasible, or not sufficient in many heavily populated parts of the world. Instead, it is a matter of recognizing situations where long-established patterns of resource use have co-existed with considerable biodiversity, some of it dependent on the resource use practices. Protection in these situations is intended to foster conservation and sustainable development through good resource stewardship. In principle, the newer IUCN categories apply to an extraordinarily wide range of human-dominated landscapes throughout the world (IUCN, 2004). Clarifications about their application can become particularly important when they affect policy applications, and are used as frameworks for the monitoring and assessment of management actions (Dudley et al., 2004).

The need for an expanded concept for protected areas was anticipated in the early 1970s by the 'biosphere reserve' ideals introduced by UNESCO. This concept has also evolved to the point where UNESCO views it to be central to issues of sustainable development:

> Biosphere reserves can be platforms for building place-specific, mutually reinforcing policies and practices that facilitate (i) conservation and sustainable use of biodiversity (ii) economic growth and other needs and aspirations of local communities and (iii) the emergence of knowledge-based governance and management arrangements at local, provincial and national levels. In this regard, biosphere reserves could serve as learning laboratories for local, national and international sustainable development agendas.
>
> (UNESCO MAB, 2006a)

As of 2006, 507 biosphere reserves had been recognized in 102 countries (UNESCO MAB, 2006b). They experience the same kinds of management and governance challenges as do other configurations of protected areas in regional landscapes. Pollock, Reed, and Whitelaw (this volume) describe some experiences with biosphere reserves in Canada.

Considerable debate is associated with these developments. The 'coverage' provided by all of these protected areas does not mesh very well with different assessments of ecological needs. Sites that are strictly protected according to IUCN categories constitute about 5.1 per cent of the world's terrestrial surface (Brooks et al., 2004). In addition to marine ecosystems, where protection strategies are still much debated (see Agardi et al., 2003), there are shortfalls for major biomes such as grasslands and undisturbed tropical forests. Global gap analyses can help set priorities for expanding protected area networks for major groups of biota (Rodrigues et al., 2004). Some detailed studies have suggested that up to 50 per cent of regions (within countries) would need protection measures to maintain representative populations of biodiversity, including wide-ranging animal species, and the 'paltry tithe to nature' of 10–12 per cent serves to create complacency rather than urgency about what still needs to be done (Soulé and Sanjayan, 1998).

The 'how much is enough?' question returns as a contrast between 'policy-based approaches' based on administrative targets, and 'evidence-based approaches' based on ecological and biodiversity studies (Tear et al., 2005). It is recognized that more must be done to assess the biological effectiveness of protected areas (Chape et al., 2005). Similarly, much has yet to be done to lay the basis for meeting the CBD target of achieving a 'significant reduction' in the current loss of biodiversity by 2010 (Balmford et al., 2004). Assessments of this target open up issues of taxonomy for biota that so far lack it, and the development of composite indices drawn from suites of taxa in order to detect trends in biodiversity (see Horlyck and de Heer, 2004; Sutherland et al., 2004). Continual concerns are expressed about whether human considerations are being given too much attention at the expense of ecological and biodiversity conservation (Gartlan, 1998; Synge, 2004; Terborgh, 2004; Locke and Dearden, 2005; Wright, this volume). However, either–or arguments seem to miss the mark. In many situations, recognition of human needs should be a prerequisite for conserving biodiversity. Local communities following customary practices often conserve biodiversity better than remote bureaucracies would do, especially if the former have become well adapted to the ecosystems in which they live (see Stevens, 1997; Borrini-Feyerabend et al., 2004).

Evolving concepts

A major feature of the 'modern paradigm' (Phillips, 2003), informed by conservation biology and landscape ecology, is to design protected areas as network configurations of nodes, buffers, and corridors of relatively undisturbed habitats set in different landscape matrices. Ecological networks are sometimes conceived as nested sets across a range of scales from the local to the continental and international. Boundaries of all kinds are crossed in this vision, and governance issues multiply accordingly. One especially creative use of this concept is in transboundary protected areas ('peace parks') which have been seen as a means to build mutual trust and cooperation in areas that have a history of conflict, as well as adding to the extent of protected areas for conservation purposes. As of 2004, there were 169 of these kinds of protected areas in 113 countries (WCPA, 2004). Some of them involved portions of three countries, and all of them involved local communities on each side of a border.

Bennett and Wit (2001) identified over 150 proposed ecological networks at regional scale or higher, and compiled information for 38 of them. Two main kinds of networks were evident. One consisted of key sites at distant locations (such as the Western Hemisphere Shorebird Reserve Network) used by highly migratory species. The other consisted of 'greenway' corridors forming mosaics of protected sites and utilized lands. While all of the examples sought protection for habitats and species, most corridor-linked configurations recognized cultural landscapes and opportunities to promote stewardship for sustainable use of ecosystems for resources. These initiatives were generally in the early collaborative planning stages.

In part to fulfil CBD commitments, a Pan-European Biological and Landscape Strategy (2004) is guiding implementation of a Pan-European Ecological Network, defined as:

> a system of representative core areas, corridors, stepping stones and buffer zones designed and managed in such a way as to preserve biodiversity, maintain or restore ecosystem services and allow a suitable and sustainable use of natural resources through interconnectivity of its physical elements with the landscape and existing social/institutional structures.

This requires a nested set of land use and protection configurations at the sub-national to local community scales. Some experiences from striving to implement these kinds of initiatives in the European context have been noted, for example in the Carpathian Ecoregion Initiative (Nelson, 2004), the design of 'territorial systems of ecological stability' in the Czech Republic (Kubes, 1996), and in managing conflicts between human activities and biodiversity conservation in Europe generally (Young et al., 2005). Similar issues arise in North America, especially for initiatives such as the proposed 'Rewilding North America Megalinkages' along three north–south continental mountain ranges and a west–east connection across the northern boreal forest (Forman, 2004).

IUCN has addressed local management issues associated with protected areas. Where management capabilities are weak, a protected area may be little more than a 'paper park' which undermines the reliability of the actual 'coverage' and conservation effectiveness of protected areas listed in inventories. To help foster improvements, Thomas and Middleton (2004) have summarized best practices in management planning for protected areas, while Hockings et al. (2006) have summarized different approaches used to assess the management effectiveness of protected areas. In addition, IUCN recognized four broad types of governance structures, and some mixed models based on them, to reflect different principles and values along with different collaborative arrangements among government agencies, NGOs, and community groups for managing protected areas (see Eagles, this volume). Dearden et al. (2005) conducted a survey of changes in governance for protected areas in 41 countries during the period 1992–2002 and reported substantial changes towards increased participation by stakeholder groups, the use of a wider range of participatory techniques, and greater use of formal accountability mechanisms.

But there are larger-scale and longer-term factors at work as well. It has been noted that the actions of intergovernmental agencies and international NGOs in the context of 'globalization' have had an effect on the extent, type, and geographic distribution of protected areas as well as the organizational arrangements for them (Zimmerer et al., 2004). Much less positive are direct or indirect threats to protected areas associated with economic and political transitions in recent decades. Discussion of these and other factors such as degradation of ecosystem services and rapid loss of biodiversity, anticipated as a future prospect for protected areas, can be found in Alcorn et al., (2003), IUCN (2004), and McNeely (2005). The Living Planet reports strive to summarize impacts in terms of a 'living planet index' and operational definitions of 'humanity's ecological footprint' (WWF, 2004).

How might one strive to grasp all this? Enter 'complexity'.

Perspectives from complex open systems

An introductory sketch

'Systems' are generally defined as a set of parts and their interrelationships functioning as a whole with the properties of the whole being more than, or quite different from, the sum of the parts. System boundaries, which are sometimes fuzzy and subject to interpretations, distinguish a system from its environment, with 'environment' being defined as anything external to the system that constrains or otherwise impacts on it. 'Complex systems' are ones in which the different kinds of parts can, and do, change in relation to one another over time, sometimes to the point they reconfigure the system and its properties. Living systems, which are open to, and completely dependent on, external energy sources, are the most germane examples of complex open systems.

A number of approaches have been taken over the last 30 years or so towards understanding complex systems. For the most part, the work has been done independently by groups of natural scientists and mathematicians interested in different biophysical systems, or by social scientists and historians interested in humans and societal phenomena. The divide between the two cultures remains deep but there are some recent attempts to bridge understanding from both sides (Scoones, 1999; Moore, 2003; Luckett, 2004; Warren, 2005).

Three overlapping approaches have been taken by both groups. One is to search for effective processes through collaboration among people with quite different knowledge and experience who then strive to make sense of the complexities of some particular system of shared interest. A second is to conduct detailed case studies of similar kinds of systems as they change over several decades or more in order to discern underlying patterns that may have causal importance. The third is to elaborate and refine conceptual frameworks as a prerequisite for trying to apply them. Each entry point soon has to deal with issues addressed by the other two.

Complex open systems can be viewed as variants of the general open systems theory originally associated with L. von Bertalanffy and colleagues in the 1950s and 1960s.

Perspectives on them constitute 'complex systems thinking', with the term 'thinking' used in place of 'theory'. There is considerable debate about whether strong theory, applicable across the different kinds of complex systems that have been studied, exists at this point (Chu *et al.*, 2003), at least at the scales, or for the phenomena of greatest interest to various scholars. It is acknowledged that laws of thermodynamics ultimately underlie all complex systems phenomena (see Schneider and Sagan, 2005), and it is generally recognized that strong collective principles of organization are major emergent phenomena (see Laughlin, 2005). Consistent with these underlying premises, an overall 'gestalt' of these systems can be discerned from writings by various authors from either natural science or social science backgrounds (see Schneider and Kay, 1994; Chase-Dunn and Hall, 1997; Kay *et al.*, 1999; Carpenter and Gunderson, 2001; Gunderson and Holling, 2002; Walby, 2004; Wallerstein, 2005). The main attributes and processes exhibited by complex open systems are:

'Exergy' driven, with multiple domains of stability. This acknowledges the underlying thermodynamics whereby systems are driven by high-quality energy ('exergy'), some of which is used to maintain and build structures while the rest is dissipated as heat ('dissipative structures'). Systems can maintain dynamic equilibria far from a 'normal' equilibrium of maximum 'entropy' as called for by the second law of thermodynamics in a closed system (a finite system running down). There is no fixed equilibrium point. With sufficient pressures (steep gradients) systems can 'flip' from one configuration to another, then hold in a new dynamic equilibrium position. Human systems are also driven by energy, with solar energy augmented by fossil fuels (embodied solar energy) and nuclear power, and they, too, can be interpreted as structures maintained in various dynamic equilibria configurations.

Self-organization with emergent properties. These phenomena arise from changing internal relationships among the parts of a system, and sometimes as a response to external perturbations which, in turn, can lead to changed functions or behaviours. 'Self-organized' is contrasted with 'other-organized' by external managers. 'Emergent' properties are a characteristic of all complex systems, and can appear rather suddenly as phase transitions occurring at critical points along gradients and/or as threshold effects manifested at particular scales. In human systems, emergence can be thought of as the cumulative effects from many human decisions or actions that become combined synergistically, or from continual dialectic interactions between structures (rule systems) and agents. Phenomena such as urban sprawl and stockmarket fluctuations can be interpreted this way.

Holarchical organization as systems within systems, within . . . etc. Complex systems tend to display organization at different scales to form 'layers' with tighter interconnections within any given scale combined with fewer or looser links across scales. Systems constituting each layer can function independently to some extent, but cross-scale entanglements from the linkages can transmit changes that cascade across scales. Entities that can be interpreted both as a system and a part, depending on how they are viewed, are called 'holons', hence 'holarchy'. They are sometimes called 'hierarchies' but this term connotes top-down direction and control, which is an exception, not the rule in complex systems.

Co-evolutionary development sensitive to conditions of place, and exhibiting phase cycles. Systems (co-)evolve through mutual adjustments among their parts and with their environment over time, and hence have a 'history'. But they are also sensitive to the physical constraints of location in terms of resources needed for development. Human systems have overcome many such constraints through technological innovations, trade, and domination of some societies over others but with serious questions about longer-term sustainability at larger scales. Systems also exhibit phase cycles (not periodicities) where slower transformations from one phase to another lead to qualitative change in structure and functions, and can include episodic collapses and starting over.

Many unknowns and some unknowables. These come from the inherent indeterminacies in the systems themselves as well as many uncertainties associated with limited understanding of them. This assures modest predictability. Epistemological and ontological issues continually arise, especially among groups of people with diverse backgrounds striving to understand the 'same' system of mutual interest.

Social-ecological systems (SESs)

Interpretations of SESs

The rather awkward term 'social-ecological systems' (SESs) captures the sense of interdependencies among human systems and ecosystems, and their 'openness' to inputs of energy. They give rise to behaviours or functions of greatest interest for understanding system dynamics affecting protected areas. The many issues about the design, implementation, and management of protected areas, especially in larger regional landscapes, are embedded in these kinds of systems.

Ecological approaches that use complex systems thinking have been explored concurrently by several groups of scholars, using the three overlapping approaches noted previously. Collaboration in cross-cultural situations drawing upon different kinds of knowledge and methodologies were reported by Waltner-Toews *et al.* (2003); Waltner-Toews (2004), and Waltner-Toews and Kay (2005). Elaborations of concepts and conceptual frameworks have been discussed by Allen and Hoekstra (1992); Ahl and Allen (1996); and Kay and Regier (2000). Insights from detailed case studies of regional-scale ecosystems that have evolved under different management regimes have been summarized by Holling (1996, 2001); Elmqvist *et al.* (2003); Walker *et al.* (2004); and Folke *et al.* (2004).

The extensive studies by C.S. Holling and colleagues have spanned over some 30 years. They are particularly relevant for issues concerning protected areas. Publications summarizing this work in considerable detail include Holling (1978); Clark and Munn (1986); Gunderson *et al.* (1995); and Gunderson and Holling (2002). Their studies viewed regional ecosystems such as forests, agricultural watersheds, semi-arid savannas, and large wetland complexes to have evolved over a number of decades. During this time they formed holarchies (or 'panarchies'), exhibited four-phase adaptive cycles with some cross-scale interactions, and demonstrated varying degrees of 'resilience' to survive collapse and reorganization phases. As summarized by Walker *et al.* (2004: 6–7):

> The dynamics of SESs can be usefully described and analyzed in terms of a cycle, known as an adaptive cycle, that passes through four phases. Two of them – a growth and exploitation phase merging into a conservation phase – comprise a slow, cumulative forward loop of the cycle, during which the dynamics of the system are reasonably predictable. As the [conservation] phase continues, resources become increasingly locked up and the system becomes progressively less flexible and responsive to external shocks. It is eventually, inevitably, followed by a chaotic collapse and release phase that rapidly gives way to a phase of reorganization, which may be rapid or slow, and during which, innovation and new opportunities are possible. The [release and reorganization] phases together comprise an unpredictable backloop. The reorganization phase leads to a subsequent growth and exploitation phase, which may resemble the previous [growth and exploitation] phase or be significantly different. This metaphor of the adaptive cycle is based on observed system changes, and does not imply fixed, regular cycling. Systems

can move back [from one phase to the earlier phase]. Finally, (and importantly), the cycles occur at a number of scales and SESs exist as 'panarchies' – adaptive cycles interacting across multiple scales. These cross-scale effects are of great significance in the dynamics of SESs.

This interpretation of ecosystem dynamics forms the basis for extended critiques of resource management practices which focus on resource extraction rather than on maintaining the resilience of the ecosystems growing the resources. The paradox (or 'pathology') of resource management arises from contradictions between management actions which 'produce' the resources and encourage local economic dependency on their continued provision, and the fact that the management actions also stress ecosystems to a point where, in extreme circumstances, some unanticipated event triggers their sudden collapse, along with the dependent local economy. An 'adaptive management' strategy is advocated for these largely unknown situations. This strategy treats management as a kind of ongoing experiment that should be monitored widely, especially in terms of changing ecosystem conditions, in order to give the signals for changing management approaches before they drive systems to collapse. Protected areas may help long-term recovery phases.

It has been assumed that this general interpretation of SESs can be applied directly to human systems. Redman and Kinzig (2003) used it to interpret archaeological studies of ancient Mesopotamia and Aboriginal settlement patterns in the south-western United States. Allen et al. (2003) emphasize the thermodynamics associated with ecosystems and the crucial importance of energy for the sustainability of human systems. Examples are cited from historical and archaeological studies of collapsed societies that were unable to maintain their energy and other resource bases (Tainter, 1988, 2000). Allen et al. (2003) also advocate 'supply-side' ecosystem management which focuses on the restoration and maintenance of entire ecosystems in place of devoting attention only to the extraction of resources from them, a prevailing practice in resource management.

A general protocol for applying the panarchy and resilience perspectives is given by Walker et al. (2002) and is being explored for a number of areas in the world by The Resilience Alliance (http:// www.resalliance.org). Management, and the social learning needed to apply it, have to address scale mismatches between ecosystems and management units (e.g. Cumming et al., 2006), determine key attributes of resilience in any given SES (e.g. Bennett et al., 2005; Walker et al., 2006), and promote different approaches for governance that differ substantially from conventional administrative organizations commonly associated with protected areas and other resource management sectors (e.g. Lebel et al., 2006).

Resilience, protected areas, and global change

Special attention has been given to phenomena of resilience and adaptability in SESs which focus on the 'backloop' from a collapse phase to a new beginning in the four-phase cycle. 'Resilience' generally refers to the 'capacity of a system to absorb disturbance and reorganize while undergoing change so as to still retain essentially the same function, structure, identity and feedbacks' (Walker et al., 2004: 7). For the human components of SESs, processes of social learning, social capital formation, trust building, access to different knowledge systems, and cross-scale institutional support arrangements are key factors in maintaining adaptive capacity in the face of change or sudden disruptions (see Berkes et al., 2003). This overall approach to understanding SESs was submitted to the 2002 World Summit on Sustainable Development (Folke et al., 2002), and it is conceptually related to vulnerability analysis that is a feature of recent proposals for 'sustainability science' (e.g. Turner et al., 2003).

Biodiversity has a critical role for ecological resilience and the maintenance of 'ecosystem services' (Folke et al., 1996). Resilience entails a 'response diversity' among species that contribute to the same ecological functions in ecosystems. Different functional groups of species must be available for renewal processes to occur. The emphasis is not on species richness per se, but on the 'insurance metaphor' of how ecosystems cope with, and adapt to, change (see Elmqvist et al., 2003; Folke et al., 2004). This, in turn, raises issues about the roles of protected areas. For ecosystems to respond to long-term and large-scale changes, protected areas must provide the 'ecological memory' (in the form of biodiversity as biological legacy), but not only in the familiar systems of 'static reserves'. These could be supplemented by 'dynamic reserves' in the larger landscape that can serve as 'ecological fallows' for low-intensity managed successions; 'ephemeral reserves' to protect species in the early successional or exploitation phases; and 'mid-succession reserves' that can be left alone and used in sustainable ways with only occasional management interventions. All of these kinds of reserves would be within the mosaics of the larger landscapes, but the locations of the dynamic sites would shift over time (Bengtsson et al., 2003). The governance issues associated with this would be quite demanding. Francis (2003) suggested ways to analyse these kinds of situations using the concepts of institutions, actors, domains, and regimes as these could apply to conservation issues.

Climate change will pose new stresses on the dynamics and resiliency of SESs and protected areas within them. An impressive volume of documentation on climate change, and on probable or possible effects of it, has been prepared by extensive international collaboration leading up to, and following, the 1992 United Nations Framework Convention on Climate Change (UNFCCC) which entered into force in 1994. The stated goal of the Convention was to stabilize greenhouse gas (GHG) concentrations in the atmosphere at levels that would prevent serious disruption to agricultural production or to major ecosystems so as to allow sustainable development to occur. The work of the Convention is supported by the Intergovernmental Panel on Climate Change (IPCC), established in 1988 to provide advice through authoritative assessments and technical reports. In addition, research on a range of related issues has been coordinated through the International Council of Science (formerly the International Council of Scientific Unions – ICSU).

Note should also be made of the creation of an 'Earth System Science Partnership' (ESSP) in 2001 as a formal alliance of four international global change research programmes associated with ICSU (e.g. Brasseur et al., 2005). The ESSP's purpose is to foster collaboration on scientific questions at the scale of the entire planet, viewed as 'a coupled interactive human-environment system' (Schellnhuber, 1999; Schellnhuber et al., 2004). Related work has been initiated to identify 'syndromes of global change' over large areas of the world (Schellnhuber et al., 1997; Lüdeke et al., 2004). The resilience of SESs will be tested in a number of locations, especially those most vulnerable to climate change or degrading conditions in 'syndrome' regions. This will include loss of biodiversity. ESSP and related collaborative work holds promise of informing broad regional strategies for maintaining protected areas under these changing conditions.

Gitay et al. (2002) give authoritative overviews of interrelationships between climate change and biodiversity, along with trends globally, and for major regions of the world. Sala et al. (2000) developed global scenarios for biodiversity change in ten terrestrial biomes and in freshwater ecosystems for the year 2100. Root et al. (2003: 57) undertook meta-analyses of 143 studies on species and global warming and concluded that:

> the balance of evidence from these studies strongly suggests that a significant impact of global warming is already discernable in animal and plant populations. The synergism of rapid temperature rise and other stresses, in particular habitat destruction, could easily disrupt the connectedness among species and lead to a reformulation of species communities, reflecting differential changes in species, and to numerous extirpations and possibly extinctions.

There have been a number of attempts to model vegetation response to climate change. They entail construction of linkages across several orders of magnitude in order to 'scale-up' data from field sites to connect with the 'scale-down' scenarios derived from global climate models. These necessarily require simplifications and assumptions about changes in vegetation associations, dispersal capabilities, future climates, and new equilibria conditions. There are extensive critiques of these endeavours on both technical and scientific grounds. Earlier models that implicitly assume ecosystems can and will move virtually intact to new locations having climate regimes similar to where they are now, are generally deemed to be unrealistic. Instead, it is recognized that species, and even distinct populations of them, have their own tolerance levels, adaptive capacity, and dispersal mechanisms. Old biotic associations will unravel, and new, perhaps strikingly different, associations will form. Implicit assumptions that climate change will level off to new dynamic equilibrium conditions soon after a doubling of atmospheric CO_2 concentrations (from pre-industrial levels) is reached, are also questionable. Mitigation measures would have to be much more forceful than they have been so far. When and where new equilibrium conditions will form is simply unknown.

A number of implications for protected areas arise from expectations that climate and biodiversity will change (Welch, 2005). One is the question of what, specifically, should be protected. Within decades, a number of protected areas may no longer have the suites of biota that are there now, and some of the new components on site may be 'alien' or 'pest' species which have better adaptive capacities than some valued native species. 'Ecological representativeness' (Scott et al., 2002) and 'ecological integrity' (Clark, Fluker, and Risby, this volume) will need to be re-examined, unless these terms come to mean only that whatever is in a protected area has these attributes by definition. Conservation emphasis might place more importance on protection for the enduring topographic features in order to support whatever biotic assemblages eventually evolve there in response to changed local climate regimes. This would also support initiatives such as the UNESCO Global Geoparks Network launched in 2004 (www.unesco.org/science/earth/gparks/geoparks.html). If new favourable habitats for valued native species are likely to become available at higher latitudes or altitudes, the possibility of easing dispersal might be explored in terms of corridor connections, removing some human barriers to dispersal, provision of 'stepping stone' habitats in the larger landscape mosaic, and, possibly, establishing new refugia in the newer locations of anticipated future favourable habitats. But these are large tasks indeed.

Conservation priorities set in terms of species richness need to be reconsidered in light of new biological debates about species and evolutionary entities. Taxonomies based on a phylogenetic concept of species rather than a more traditional biological species concept tend to distinguish more species, each with smaller (and potentially more 'endangered') populations (Agapow et al., 2004; Isaac et al., 2004). Taxonomies that rely on genetic markers as definitive diagnostic tools, such as nucleotide sequences coded by particular mitochondrial genes, are developing very rapidly (see www.barcodinglife.org), and have become matters for growing scientific debate (see Savolainen et al., 2005; Ebach and Holdrege, 2005). Other questions arise about selecting 'evolutionary significant units' (Moritz, 1994), protecting processes that generate 'taxonomic biodiversity' (Ennos et al., 2005), and giving priority to phylogenies (Mace et al., 2003).

Given all the unknowns inherent in these situations, rationales for protected areas in terms other than biodiversity conservation with recreational opportunities for humans might be in order.

Complex systems thinking and 'globalization'

Evolving human systems and global change

Climate change and its implications for SESs are the major biophysical components of global change. At a global scale, over periods of centuries and even millennia, the human dimensions of systems change have been investigated by a number of scholars under the rubric of 'world-systems analysis'. There are debates about the temporal scales to adopt, some favouring millennia over centuries (e.g. Frank and Gill, 1996; Chase-Dunn and Hall, 1997) and others the past five centuries to concentrate more on the rise and possible demise of global capitalism and the nation-state system (e.g. Wallerstein, 1998; Arrighi and Silver, 2001).

'Globalization' is not a new phenomenon once attention is drawn to overlapping and interconnected trade and other social interaction networks that have existed in many forms for centuries (Chase-Dunn and Grimes, 1995). As Chase-Dunn and Hall (1997: 108) summarize:

> Economic processes operate in a context of demographic forces and ecological constraints to produce technological change and hierarchy formation. This feeds back into demographic processes and forward into social organizational potentials through four general problems: new forms and levels of competition, new scarcities, new risks, and new demands for savings and investments. Population pressure, emigration, and the search for new resources all contribute to the periodic expansion and contraction, or pulsations, that occur in all world-systems.

Over time, these systems became larger and more complex, their modes of accumulation (or system logic) changed, and they exhibited rise-and-fall patterns. Beginning some five centuries or so ago, the capitalist mode of production for exchange and profit emerged to become the dominant one in Europe. It has subsequently become global in its extent and now consists of a highly integrated global capitalist economy set within a world of almost 200 nation-states.

Continuity in this world-system entails a relentless drive for the 'unceasing accumulation of wealth' which is the fundamental logic of capitalism, and the purpose of existence for corporations, the most common form of business organization. There is no inherent 'enoughness' in this logic. This drive exists concurrently (and in combination) with related struggles for domination over the governmental structures for control over territory, resources, and people within nation-states. Rivalries and conflicts abound. But the wealthy need the powerful to protect their wealth and the conditions under which they can continue to accumulate it, and the powerful need the wealthy in order to maintain their governmental control. The never-ending political and economic dynamics of all this generates most of the daily 'news' from around the world.

The 'trajectories' of development over this long historical period have given rise to extensive population growth; material well-being for many people along with growing inequalities in its distribution; a large increase in the number of nation-states; a huge increase in the number of transnational corporations and international 'civil society' organizations; a massive increase in economic production combined with increasing capital intensity of production (labour displacement); intrusion of commodified goods and services into all spheres of modern life; and world-scale environmental degradation. It has created the relative inequalities among those countries with the core, semi-peripheral, and peripheral economies (as previously noted). Although membership in these categories shifts over time, and with episodic hegemonic reorganization, this overall pattern remains much the same. It is currently kept in place by threats of military intervention from powerful states, by structural agreements for trade relations; financial arrangements for borrowing, foreign investment and debt creation; internal pricing practices of transnational corporations; and by less than transparent accountability for much of what transpires.

Longer-term system dynamics

Phase cycles or transitions of particular interest in the much longer world-systems perspective are all multi-decadal. Economic phases of expansion, diffusion, and contraction over periods in the order of 50–60 years are associated with access by core countries to major new resources and/or technological innovations which can undermine and replace economies based on older technologies. Over much longer periods of a century or more, the 'systemic cycles of wealth accumulation' in the core economies shift their geographic location when overseas investments and/or growing financial speculation provided by institutions in the old core region generate better rates of return from elsewhere. The new regional centres begin to outcompete the old, and interregional disputes become increasingly politicized. The unfolding example is the shift of the centre of the global economy away from North America (and Europe) to Asia (Arrighi, 1994; Arrighi and Silver, 2001; Wallerstein, 2003). It is laying the economic basis for a new 'hegemonic sequence' and 'revolutionizing' the historical geography of capitalism (Arrighi, 2004).

Political cycles are loosely associated with these longer phased economic shifts and can lead to the rise and demise of hegemonic powers that preside over the world order. Drawn-out international warfare has been associated with the rise of a new hegemonic power which then can set or enforce the international groundrules to direct some new world order for security and trade (e.g. Modelski, 1996; Bornschier and Chase-Dunn, 1999).

There is some question about whether or not the major phase cycles of old can still play themselves out under conditions of approaching global ecological or other limits. If not, then some entirely new 'system flip' may be in the offing. There is much debate over this (e.g. Wallerstein, 1998; Hopkins and Wallerstein, 1998; Arrighi and Silver, 1999; Boswell and Chase-Dunn, 2000; Chase-Dunn, 2005). These world-system processes do not unfold without considerable protest and resistance. 'Anti-systemic' movements organized around different cultural ideals have always tried to modify or transform dominant political-economic trends (e.g. Arrighi *et al.*, 1989; Wallerstein, 2005).

Sustainable alternatives and protected areas?

Arguably, the 'sustainable development' narratives reflect reformist strategies in protest to the present form of global capitalism and supporting market-states. 'Sustainability' has received rhetorical support on many occasions from the UN system, most recently at the World Summit on Sustainable Development in 2002. Market-oriented versions of sustainability, reflected by 'ecological modernization' (improvements in the efficiencies of resource use), 'industrial ecology' (wastes from one enterprise used as raw materials by others), and markets for pollution rights or credits (the most acceptable approach for corporate capitalism), could be viewed as adaptive management strategies that otherwise help maintain market dominance.

These historical and contemporary phenomena have been interpreted in terms of a historical dialectic between globalization, resistance, and democratization (Chase-Dunn and Gills, 2003). They have also been viewed as recurring phenomena of the embedding and disembedding of capitalism in other dimensions of society, ones that provide essential support for capitalism while also being exploited by capitalism unless it can be reigned in. The re-embedding gives rise to new regulatory frameworks that can be quite location-specific. They appear to be emerging for large urban regions rather than being recreated at the level of nation-states.

By operating at the scale they do, these ongoing changes in political economy and the governance to implement them alter the contexts within which protected areas issues have to be addressed. These forces are not always remote and they have environmental consequences:

> These include enclosures accompanying genetic engineering and bioprospecting; the creation of private property rights to pollute; the growth of user fees for 'public' nature reserves; and the privatization of all manner of natural resources, from fisheries to forests to water.
>
> (McCarthy and Prudham, 2004: 277)

With governments increasingly subordinated to private economic interests, expenditures for public infrastructures and services, including many protected areas, have been reduced and often devolved to local levels in the name of subsidiarity, but with insufficient financial resources to maintain them. Without some adaptive response, many protected areas agencies would become increasingly reliant on benign neglect as their management strategy, with relevant sciences marginalized or abandoned, and the prospect of 'paper parks' as an undesirable future legacy. This is a concern in 'developed' countries as well as others (Shultis, 2005).

Two major adaptive responses are apparent. One is the promotion of tourism in close cooperation with the private sector, using major protected areas as destination sites (Campbell *et al.*, this volume). The challenge has been to find ways to direct some of the expenditures towards maintaining protected areas and appropriate tourism facilities to receive them (see Eagles, 2002). This response remains oriented towards clientele in or from the core economies, but it strives to draw visitors to sites throughout the world. Continuing success depends upon maintaining high-quality facilities and security for visitors. This may not always be possible in the future, especially under conditions of 'failed states'.

A second response has given more attention to ways for making protected areas more accepted, and of benefit to people living in or near them, if for no other reason than as a kind of insurance policy to protect the protected areas. Over the last 20 years or so, many attempts have been made to reconcile conservation objectives with local community needs through 'integrated conservation and development projects – ICDPs'. This approach was used especially in the more economically peripheral regions of the world, and is consistent with the sustainability theme recognized by the IUCN in their classification of protected areas. The difficulties that ICDPs have experienced are not with the principle of linking conservation and sustainability needs for local people, but in practical problems of designing, funding, and managing projects in collaborative ways, and at appropriate scales, to meet expectations and overcome implementation problems (see Newmark and Hough, 2000; Wells and McShane, 2004; Hardin and Remis, this volume). Shortcomings of ICDPs have also led to resurgent demands for authoritarian protectionist policies for biodiversity that ignore the 'deeply politicized nature of nature protection', and the likely generation of resistance and conflict that would result (Wilshusen *et al.*, 2002). There are similar needs to protect protected areas in industrialized countries, but the means to do so often entail easing economic transitions out of dependencies on declining agricultural or other resource extraction industries and towards some other alternative, often sought by tourism development. Michaelidou *et al.* (2002) discuss these kinds of issues.

So where does this leave us? – facilitating the transition of the world as we used to know it into an unknown and uncertain 'what next?'. It is neither predetermined nor necessarily catastrophic. Shifting beliefs and values associated with this might well be one of the main effects on protected areas. One should recall the different cultural interpretations of human–nature relationships that have been associated with protected areas in the past. They include sacred sites, specific species taboos, royal hunting grounds, and general respect for ancestral lands. Over the past two centuries or so, the history of protected areas has been closely related to European imperialism with imposed values, including the imposition of parks on people (e.g. Grove, 1995; Jepson and Whittaker, 2002). The Euro-American influence on thinking about protected areas for the past several decades is a carryover from these earlier times. It is quite likely that within the next generation or two, current debates over biodiversity, ecological integrity, and ecosystem services as reasons for protecting protected areas will seem as quaint or archaic as nineteenth-century cultural constructs of the picturesque and the sublime as reasons to protect scenery, and aristocratic safari hunts as reasons to protect wildlife.

Thinking beyond a home box

In conclusion, I invite reflective practitioners to contemplate 'thought experiments' that draw upon complex systems thinking for investigating protected area situations of special interest. An exercise might be attempted alone, or with a small group of friends before trying it out with a local stakeholder group. Three main questions need to be addressed, and it will become an iterative process.

First, given the situation of interest, how can some overall system that encompasses it best be characterized? This might depend on issues of special concern, such as ones associated with maintaining a protected area or policies to promote sustainable livelihoods for people in the region. Should the protected area and its surroundings be depicted as an SES nested in distinctive landforms or watersheds; or as a component part of an economically peripheral resource-based region linked to a dominant primary industry such as agri-business or mining; or as an area enmeshed in a rapidly urbanizing region that is putting pressures on fast-diminishing 'green space'; or as a situation caught up in a political system where issues of property rights, social justice, and factional struggles for control are being played out? Several of these or other depictions might apply, but one, more so than the others, might best characterize the system that is driving the changes impacting upon the protected area. The organizational arrangements through which these issues are to be dealt with need to be identified as this will constitute the governance overlays.

A second, related set of questions invites one to inspect the SES of primary interest from a number of different perspectives in order to identify self-organizational phenomena operating at different scales and the main factors that they seem to 'drive' or that seem to be 'driving' them. There is no set classification for this but the perspectives collectively should cover human population and demographic characteristics (which might reflect both driving factors and responses to other changes) and a mix of human systems phenomena including the rule systems and politics of governance; changes in the economic base and social consequences; urbanization and regional development; information and communications networks; technological and other innovations for problem-solving; and changing entrenched cultural beliefs and values. Biophysical phenomena to review include weather patterns and climate; geomorphological processes, biogeochemical recycling of nutrients and contaminants; evolving landscape mosaics; phase cycles of ecosystems; and population dynamics of biota. Not all of these will matter for a given situation, and only a few may be drivers for what is happening. The classification can be expanded, reduced, or recombined in whatever ways seem appropriate for the situation.

The last set of questions poses the challenge of trying to describe the current status of the system of primary interest, possibly by sketching narratives (future histories) of the ways in which the system could develop. What desirable traits might be strengthened, what undesirable ones reduced or de-activated, and who ultimately should make these kinds of decisions? What kind of 'interventions' would be needed? Can they be set out as actions directed to particular organizations of governance, or do institutional gaps become apparent at this stage? There might be a poor match between need and institutional capacity, and much may depend on whether national and international bodies take important initiatives that are not directed to protected areas as such. From then on, it would be a matter of monitoring, adaptive management, or pursuing some other strategy deemed necessary and appropriate for the situation.

None of the above is easy. There will be disagreements within a group that attempts it, and the kinds of knowledge required will probably range beyond what is readily available for participants. While answers might be elusive, insights are guaranteed. In a coming world of change this might be the most practical thing one can do. And, perhaps the best part, it could probably be done in the contemplative surroundings of a protected area.

Literature cited

Abel, N. (2002). *Sustainable Use of Rangelands in the 21st Century: on the road to a better future for the Western Division of NSW*. Canberra: CSIRO.

Agapow, P.-M., Bininda-Emonds, O.R.P., Crandall, J.L., Gittleman, J.L., Mace, G.M., Marshall, J.C., and Purvis, A. (2004). The impact of species concept on biodiversity studies. *The Quarterly Review of Biology* 79 (2), 161–179.

Agardi, T., Bridgewater, P., Crosby, M.P., Day, J., Dayton, P.K. *et al.* (2003). Viewpoint: Dangerous targets? Unresolved issues and ideological clashes around marine protected areas. *Aquatic Conservation: Marine and Freshwater Ecosystems* 13 (4), 353–367.

Ahl, V. and Allen, T.F.H. (1996). *Hierarchy Theory: a vision, vocabulary, and epistemology*. New York: Columbia University Press.

Alcorn, J.B., Luque, A., and Valenzuela, S. (2003). *Global Governance and Institutional Trends Affecting Protected Areas Management: challenges and opportunities arising from democratization and globalization*. Washington, DC: World Resources Institute.

Allen, T.F.H. and Hoekstra, T.W. (1992). *Toward a Unified Ecology*. New York: Columbia University Press.

——, Tainter, J.A., and Hoekstra, T.W. (2003). *Supply-Side Sustainability*. New York: Columbia University Press.

Arrighi, G. (1994). *The Long Twentieth Century: money, power, and the origins of our times*. London: Verso.

—— (2004). Spatial and other 'fixes' of historical capitalism. *Journal of World-Systems Research* X (2), 527–539.

—— and Silver, B.J. (1999). *Chaos and Governance in the Modern World System*. Minneapolis, MN: University of Minnesota Press.

—— and —— (2001). Capitalism and world (dis)order. *Review of International Studies* 27, 257–279.

——, Hopkins, T.K., and Wallerstein, I. (1989). *Antisystematic Movements*. London: Verso.

Balmford, A., Crane, P., Dobson, A., Green, R.E., and Mace, G.M. (2004). The 2010 challenge: data availability, information needs, and extraterrestrial insights. *Philosophical Transactions of The Royal Society B* 360, 221–228.

Barber, C.V., Miller, K.R., and Boness, M. (eds) (2004). *Securing Protected Areas in the Face of Change, Issues and Strategies*. A report by the Ecosystem, Protected Areas and People project. Cambridge, UK: IUCN Publication Services.

Bengtsson, J., Angelstam, P., Elmqvist, T., Emanuelsson, U., Folke, C., Ihse, M., Moberg, F., and Nyström, M. (2003). Reserves, resilience and dynamic landscapes. *Ambio* 32 (6), 389–396.

Bennett, E.M., Cumming, G.S., and Peterson, G.D. (2005). A systems model approach to determining resilience surrogates for case studies. *Ecosystems* 8, 945–957.

Bennett, G. and Wit, P. (2001). *The Development and Application of Ecological Networks: a review of proposals, plans and programmes*. Amsterdam: AIDEnvironment and IUCN.

Berkes, F., Colding, J., and Folke, C. (eds) (2003). *Navigating Social-Ecological Systems: building resilience for complexity and change*. Cambridge, UK: Cambridge University Press.

Bornschier, W. and Chase-Dunn, C. (eds) (1999). *The Future of Global Conflict*. London: Sage Publishers.

Borrini-Feyerabend, G., Kothari, A., and Oriedo, G. (eds) (2004). *Indigenous and Local Communities and Protected Areas: toward equity and enhanced conservation*. WCPA Best Practice Protected Area Guideline Series No. 11. Gland, Switzerland and Cambridge, UK: IUCN Publication Services.

Boswell, T. and Chase-Dunn, C. (2000). *The Spiral of Capitalism and Socialism: toward global democracy*. London: Lynne Rienner.

Brasseur, G., Steffan, W., and Noone, K. (2005). Earth systems focus for Geosphere-Biosphere Program. *EOS Transactions, American Geophysical Union* 86 (22), 209–213.

Broad, R. (2004). The Washington Consensus meets the global backlash: shifting debates and policies. *Globalizations* 1 (2), 129–154.

Brooks, T.M., Bakarr, M.I., Boucher, T., Da Fonseca, G.A.B., Hilton-Taylor, C. *et al.* (2004). Coverage provided by the global protected area-system: is it enough? *BioScience* 54 (12), 1081–1091.

Carpenter, S.R. and Gunderson, L.H. (2001). Coping with collapse: ecological and social dynamics in ecosystem management. *BioScience* 51 (6), 451–457.

Chape, S., Blyth, S., Fish, L., Fox, P., and Spalding, M. (compilers) (2003). *2003 United Nations List of Protected Areas*. Gland, Switzerland and Cambridge, UK: UNEP-World Conservation Monitoring Centre.

——, Harrison, J., Spalding, M., and Lysenko, I. (2005). Measuring the extent and effectiveness of protected areas as an indicator for meeting global biodiversity targets. *Philosophical Transactions: Biological Sciences* 360 (1454), 443–455.

Chase-Dunn, C. (2005). Social evolution and the future of world society. *Journal of World-Systems Research* XI (2), 171–192.

—— and Gills, B. (2003) Understanding waves of globalization and resistance in the capitalist world-system: social movements and critical global(ization) studies. Paper available at: http://irows.ucr.edu/papers/irows12/irows12.htm.

—— and Grimes, P. (1995). World-systems analysis. *Annual Review of Sociology* 21, 387–417.

—— and Hall, T.D. (1997). *Rise and Demise: comparing world systems*. Boulder, CO: Westview.

Chu, D., Strand, R., and Fjelland, R. (2003). Theories of complexity: common denominators of complex systems. *Complexity* 8 (3), 19–30.

Clark, W.C. and Munn, R.E. (eds) (1986). *Sustainable Development of the Biosphere*. London: Cambridge University Press.

Crandell, K.A., Bininda-Emonds, O.R.P., Mace, G.M., and Wayne, R.K. (2000). Considering evolutionary processes in conservation biology. *Trends in Ecology & Evolution* 15 (7), 290–295.

Cumming, G.S., Cumming, D.H.M., and Redman, C.L. (2006). Scale mismatches in social-ecosystems: causes, consequences, and solutions. *Ecology and Society* 11 (1), art 14. Available at http://www.ecologyandsociety.org/vol1/art14/.

Dearden, P., Bennett, M., and Johnston, J. (2005). Trends in global protected area governance, 1992–2002. *Environmental Management* 36 (1), 89–100.

Dudley, N., Harrison, J., and Rosabel, P. (2004). The future development of the categories system. *Parks* 14 (3), 72–81.

Eagles, P.F.J. (2002). Trends in park tourism: economics, finance and management. *Journal of Sustainable Tourism* 10 (2), 132–153.

Ebach, M.C. and Holdrege, C. (2005). More taxonomy, not DNA barcoding. *BioScience* 55 (10), 822–823.

Elmqvist, T., Folke, C., Nyström, M., Peterson, G., Bengtsson, G., Walker, B., and Norberg, J. (2003). Response diversity, ecosystem change, and resilience. *Frontiers in Ecology and Environment* 1 (9), 488–494.

Ennos, R.A., French, G.C., and Hollingsworth, P.M. (2005). Conserving taxonomic complexity. *Trends in Ecology & Evolution* 20 (4), 164–168.

Folke, C., Holling, C.S., and Perrings, C. (1996). Biological diversity, ecosystems, and the human scale. *Ecological Applications* 6 (4), 1018–1024.

——, Carpenter, S., Elmqvist, T., Gunderson, L., Holling, C.S., and Walker, B. (2002). Resilience and sustainable development: building adaptive capacity in a world of transformations. *Ambio* 31 (5), 437–440.

——, ——, Walker, B., Scheffer, M., Elmqvist, T., Gunderson, L., and Holling, C.S. (2004). Regime shifts, resilience, and biodiversity in ecosystem management. *Annual Review of Ecology, Evolution, and Systematics* 35, 557–581.

Forman, D. (2004). *Rewilding North America: a Vision for Conservation in the 21st Century*. Washington, DC: Island Press.

Francis, G. (2003). Governance for conservation. In F.R. Westley and P.S. Miller (eds), *Experiments in Consilience: integrating social and scientific responses to save endangered species*, pp. 223–243. Washington, DC: Island Press.

Frank, A.G. and Gill, B.K. (eds) (1996). *The World System: five hundred years or five thousand?* London: Routledge.

Gartlan, S. (1998). Every man for himself and god against all: history, social science and the conservation of nature. In: Resource Use in the Tri-National Sangha River Region of Equitorial Africa, *Yale F & ES Bulletin* 102, 216–226.

Gitay, H., Suarez, A., Dekken, D.J., and Watson, R.T. (eds) (2002). *Climate Change and Biodiversity*. WMO and UNEP: IPCC Technical Paper V.

Grove, R.H. (1995). *Green Imperialism: colonial expansion, tropical island Edens, and origins of environmentalism 1600–1860*. Cambridge, UK: Cambridge University Press.

Gunderson, L.H. and Holling, C.S. (eds) (2002). *Panarchy: understanding transformations in human and natural systems*. Washington, DC: Island Press.

——, ——, and Light, S.S. (eds) (1995). *Barriers and Bridges to the Renewal of Ecosystems and Institutions*. New York: Columbia University Press.

Holling, C.S. (ed.) (1978). *Adaptive Environmental Assessment and Management*. Laxenburg, Austria: International Institute for Applied Systems Analysis, and New York: John Wiley & Sons.

—— (1996). Surprise for science, resilience for ecosystems, and incentives for people. *Ecological Applications* 6 (3), 733–735.

—— (2001). Understanding the complexity of economic, ecological and social systems. *Ecosystems* 4, 390–405.

Hopkins, T.K. and Wallerstein, I. (1998). *The Age of Transition: trajectory of the world system 1945–2025*. London: Zed Books.

Horlyck, V. and de Heer, M. (2004). Development of biodiversity indicators for Europe: lessons learned. Presentation document for a *Joint Meeting on Development of Plan and Guidelines for Indicators and Monitoring to Achieve the 2010 Target for Biodiversity in Europe*. Copenhagen. 21–23 April 2004.

International Geosphere-Biosphere Program – IGBP (2004). *Toward an IGBP Project on Analysis, Integration and Modeling of the Earth System*. Draft concept paper (on-line).

International Union for the Conservation of Nature and Natural Resources (IUCN-World Conservation Union) (2004). *Speaking a Common Language: summary*. IUCN World Conservation Congress, Bangkok, 2004.

Isaac, N.J.B., Mallet, J., and Mace, G.M. (2004). Taxonomic inflation: its influence on macroecology and conservation. *Trends in Ecology & Evolution* 19 (3), 464–469.

Jepson, P. and Whittaker, R.J. (2002). Histories of protected areas: internationalization of conservation values and their adoption in the Netherlands Indies (Indonesia). *Environment and History* 8, 129–172.

Jessop, B. (1997). The governance of complexity and the complexity of governance. In A. Amin and J. Hausner, *Beyond Market and Hierarchy: interactive governance and social complexity*, pp. 95–129. Cheltenham, UK: Edward Elgar Publishing.

—— (2002a). *The Future of the Capitalist State*. Cambridge, UK: Blackwell Publishing Polity Press.

—— (2002b). Liberalism, neoliberalism and urban governance: a state-thematic perspective. *Antipode* 34 (3), 452–472.

—— (2003). *Governance and metagovernance: on reflexivity, requisite variety, and requisite irony*. Department of Sociology, Lancaster University. Paper available at: www.comp.lancs.ac.uk/sociology/papers/Jessop-Governance-and-Metagovernance.pdf.

Kay, J.J. and Regier, H.A. (2000). Uncertainty, complexity, and ecological integrity: insights from an ecosystem approach. In P. Crabbe, A. Holland, L. Ryszkowski, and L. Westra (eds), *Implementing Ecological Integrity: restoring regional and global environmental and human health*, Chapter 8. Dordrecht, NL: NATO/Kluwer Academic Publishers.

——, ——, Boyle, M., and Francis, G. (1999). An ecosystem approach for sustainability: addressing the challenge of complexity. *Futures* 31 (7), 721–742.

Kjær, A.M. (2004). *Governance*. Cambridge, UK: Polity Press.

Koehn, P.H. and Rosenau, J.N. (2002). Transnational competence in an emergent epoch. *International Studies Perspectives* 3, 105–127.

Kooiman, J. (2003). *Governing as Governance*. London: SAGE Publications.

Kubes, J. (1996). Biocentres and corridors in a cultural landscape: a critical assessment of the 'territorial system of ecological stability'. *Landscape and Urban Planning* 35, 231–240.

Laughlin, R.B. (2005). *A Different Universe: reinventing physics from the bottom down*. Cambridge, MA: Perseus Books.

Lebel, L., Anderies, J.M., Campbell, B., Folke, C., Hatfield-Dodds, S., Hughes, T.P., and Wilson, J. (2006). Governance and the capacity to manage resilience in regional social-ecological systems. *Ecology and Society* 11 (1), 19. Available at www.ecologyandsociety.org/vol11/iss1/art19/.

Locke, H. and Dearden, P. (2005). Rethinking protected areas categories and the new paradigm. *Environmental Conservation* 32 (1), 1–10.
Loughlin, J. (2004). The 'transformation' of governance: new directions in policy and politics. *Australian Journal of Politics and History* 50 (1), 8–22.
Luckett, S. (2004). Environmental paradigms, biodiversity conservation, and critical systems thinking. *Systemic Practice and Action Research* 17 (5), 511–534.
Lüdeke, M.K.B., Petschel-Held, G., and Schellnhuber, H.-J. (2004). Syndromes of global change: the first panoramic view. *Gaia* 13 (1), 42–49.
Mace, G.M., Gittleman, J.L., and Purvis, A. (2003). Preserving the tree of life. *Science* 300, 1707–1709.
McCarthy, J. and Prudham, S. (2004). Neoliberal nature and the nature of neoliberalism. *Geoforum* 25, 275–283.
McNeely, J.A. (2005). Protected areas in 2023: scenarios for an uncertain future. *George Wright Forum* 22 (1), 61–74.
Michaelidou, M., Decker, D.J., and Lassoie, J.P. (2002). The interdependence of ecosystem and community viability: a theoretical framework to guide research and application. *Society and Natural Resources* 15, 599–616.
Millennium Ecosystem Assessment – MEA (2003). *Ecosystems and Human Well-being: a framework for assessment*. Washington, DC: Island Press.
——— (2005). *Living Beyond Our Means: natural assets and human well-being*. Washington, DC: Island Press.
Modelski, G. (1996). Evolutionary paradigm for global politics. *International Studies Quarterly* 40, 321–342.
Moore, J. (2003). The modern world-system as environmental history? Ecology and the rise of capitalism. *Theory and Society* 32, 307–377.
Moritz, C. (1994). Defining 'evolutionary significant units' for conservation. *Trends in Ecology & Evolution* 9 (10), 373–375.
Nelson, J.G. (2004). *The Carpathians: assessing an EcoRegional Planning Initiative*. Report prepared at the request of the Carpathian EcoRegion Initiative, Vienna. Waterloo, ON: Faculty of Environmental Studies, University of Waterloo.
Newmark, W.D. and Hough, J.L. (2000). Conserving wildlife in Africa: integrated conservation and development projects and beyond. *BioScience* 50 (7), 584–592.
Pan-European Biological and Landscape Diversity Strategy (2004). *Protected Areas and Ecological Networks*. Geneva and Strasbourg: Eighth meeting of the Council (for the Strategy), 19–21 January 2004.
Paquet, G. (2005). *The New Geo-Governance: a baroque approach*. Ottawa: University of Ottawa Press.
Peck, J. and Tickell, A. (2002). Neoliberalizing space. *Antipode* 34 (3), 380–404.
Phillips, A. (2003). A modern paradigm. *World Conservation* 2, 6–7.
——— (2004). The history of the international system of protected area management categories. *Parks* 14 (3), 4–14.
Ravetz, J. (1999). What is post-normal science? *Futures* 31, 647–654.
——— (2004). The post-normal science of precaution. *Futures* 36, 347–357.
Ravenel, R.M. and Redford, K.H. (2005). Understanding IUCN protected areas categories. *Natural Areas Journal* 25 (4), 381–389.
Redman, C.L. and Kinzig, A.P. (2003). Resilience of past landscapes: resilience theory, society and the *longue durée*. *Ecology and Society* 7 (1), 14. Available at www.ecologyandsociety.org/vol7/iss1/.
Rodrigues, A.S.L., Akcakaya, H.R., Andelman, S.J., Bakarr, M.I., Boitani, L., et al. (2004) Global gap analysis: priority regions for expanding the global protected-area network. *BioScience* 54 (12), 1092–1100.
Root, T.L., Price, J.T., Hall, K.R., Schneider, S.H., Rosenzweig, C., and Pounds, A. (2003). 'Fingerprints' of global warming on wild animals and plants. *Nature* 421 (6918), 57–59.
Rosenau, J.N. (2003). Globalization and governance: bleak prospects for sustainability. *Internationale Politik und Gesellschaft* 3, 11–29.
Sala, O.E., Chapin, F.S., Armesto, J.J., Berlow, E., Bloomfield, J., et al. (2000). Global biodiversity scenarios for the year 2100. *Science* 287, 1770–1774.
Savolainen, V., Cowan, R.S., Vogler, A.P., Roderick, G.K., and Lane, R. (2005). Towards writing the encyclopaedia of life: an introduction to DNA barcoding. *Philosophical Transactions of The Royal Society B* 360, 1805–1811.

Schellnhuber, H.J. (1999). 'Earth System' analysis and the second Copernican revolution. *Nature* 402 (Supp.) c19–c23, 2 December 1999.

——, Crutzen, P.J., Clark, W.C., Claussen, M., and Held, H. (eds) (2004). *Earth System Analysis for Sustainability*. Cambridge, MA: The MIT Press in cooperation with the Dahlem University Press.

——, Block, A., Cassel-Gintz, M., Kropp, J., Lammel, G., *et al.* (1997). Syndromes of global change. *Gaia* 6 (1), 19–34.

Schneider, E.D. and Kay, J.J. (1994). Life as a manifestation of the second law of thermodynamics. *Mathematics and Computer Modelling* 19 (6–8), 25–48.

—— and Sagan, D. (2005). *Into the Cool: energy flow, thermodynamics, and life*. Chicago, IL: University of Chicago Press.

Scoones, I. (1999). New ecology and social sciences: what prospects for a fruitful engagement? *Annual Review of Anthropology* 28, 479–507.

Scott, D., Malcolm, J.R., and Lemieux, C. (2002). Climate change and modelled biome representation in Canada's national park system: implications for system planning and park mandates. *Global Ecology & Biogeography* 11, 475–484.

Sheppard, D. (2000). Conservation without frontiers: the global view. *The George Wright Forum* 17 (2), 70–80.

Shultis, J. (2005). The effects of neo-conservatism on park science, management, and administration: examples and a discussion. *The George Wright Forum* 22 (2), 51–58.

Skelcher, C. (2005). Jurisdictional integrity, polycentrism, and the design of democratic governance. *Governance: an international journal of policy, administration, and institutions* 18 (1), 89–110.

Soulé, M.E., and Sanjayan, M.A. (1998). Conservation targets: Do they help? *Science* 279 (5359), 2060–2061.

Stevens, S. (ed) (1997). *Conservation through Cultural Survival: Indigenous Peoples and Protected Areas*. Washington, DC: Island Press.

Sutherland, W.J., Pullin, A.S., Dolman, P.M., and Knight, T.M. (2004). The need for evidence-based conservation. *Trends in Ecology & Evolution* 19 (6), 305–308.

Swyngedouw, E. (2005). Governance innovation and the citizen: the Janus face of governance-beyond-the-state. *Urban Studies* 42 (11), 1991–2006.

Synge, H. (2004). Summary of a working group session, World Conservation Congress, Bangkok. *WCPA News* 93.

Tainter, J.A. (1988). *The Collapse of Complex Societies*. Cambridge, UK: Cambridge University Press.

—— (2000). Problem solving: complexity, history, sustainability. *Population and Environment* 22 (1), 3–41.

Tear, T.H., Kareiva, P., Angermeier, P.L., Comer, P., Czech, B., *et al.* (2005). How much is enough? The recurrent problem of setting measurable objectives in conservation. *BioScience* 55 (10), 835–849.

Terborgh, J. (2004). Reflections of a scientist on the Vth IUCN World Parks Congress. *Parks* 14 (2), 55–57.

Thomas, L. and Middleton, J. (2004). *Guidelines for Management Planning of Protected Areas*. Best Practice Protected Area Guidelines Series No. 10. Gland, Switzerland and Cambridge, UK: IUCN Publications Services.

Torfing, J., Sørensen, E., and Christensen, L.P. (2003). *Nine Competing Definitions of Governance, Governance Networks, and Meta-governance*. Working Paper 2003: 1, Centre for Democratic Network Governance, Roskilde University, Denmark.

Turner, B.L., Kasperson, R.E., Matson, P.A., McCarthy, J.J., Corell, R.W., *et al.* (2003). A framework for vulnerability analysis in sustainability science. *Proceedings of the National Academy of Sciences* 1000 (14), 8074–8079.

UNESCO (2002). *Biosphere Reserves: special places for people and nature*. Paris: UNESCO.

UNESCO MAB (2006a). *Panels on Biosphere Reserves*. International Co-ordinating Council on the Man and the Biosphere (MAB) Programme, 19th session. SC-06/CONF.202/INFO 4: para 2.

—— (2006b). Meeting of the International Co-ordinating Council for the Man and the Biosphere (MAB) Programme, 19th session. 23–29 October 2006. Paris.

Walby, S. (2004). *Complexity Theory, Globalization and Diversity*. Paper presented to the British Sociological Association. Published by the Department of Sociology, Lancaster University. Available at: www.comp.lancs.ac.uk/sociology/papers.

Walker, B., Holling, C.S., Carpenter, S.R., and Kinzig, A. (2004). Resilience, adaptability and transformability in social-ecological systems. *Ecology and Society* 9 (2), 5. Available at www.ecologyandsociety.org/vol9/iss2/art5/.

——, Gunderson, L., Kinzig, A., Folke, C., Carpenter, S., and Schultz, L. (2006). A handful of heuristics and some propositions for understanding resilience in social-ecological systems. *Ecology and Society* 11 (1), 13. Available at www.ecologyandsociety.org/vol11/iss1/art13/.

——, Carpenter, S., Anderies, J., Abel, N., Cumming, G.S., *et al.* (2002). Resilience management in socio-ecological systems: a working hypothesis for a participatory approach. *Conservation Ecology* 6 (1), 14. Available at www.consecol.org/vol6/iss1/art14.

Wallerstein, I. (1998). *Utopistics, or Historical Choices for the Twenty-First Century*. New York: W.W. Norton & Co.

—— (1999). *The End of the World as We Know It: social science for the twenty-first century*. Minneapolis, MN: University of Minnesota Press.

—— (2003). *The Decline of American Power*. New York: W.W. Norton & Co.

—— (2005). *World-System Analysis: an introduction*. Durham, NC: Duke University Press. 2nd printing.

Waltner-Toews, D. (2004). *Ecosystem Sustainability and Health: a practical approach*. Cambridge, UK: Cambridge University Press.

—— and Kay, J. (2005). The evolution of an ecosystem approach: the diamond schematic and an adaptive methodology for ecosystem sustainability and health. *Ecology and Society* 10 (1), 38. Available at www.ecologyandsociety.org/vol10/iss1/art38/.

——, Kay, J.J., Neudoerffer, C., and Gitau, T. (2003). Perspective changes everything: managing ecosystems from the inside out. *Frontiers in Ecology and the Environment* 1 (1), 23–30.

Warren, W.A. (2005). Hierarchy theory in sociology, ecology, and resource management: a conceptual model for natural resource or environmental sociology and socioecological systems. *Society and Natural Resources* 18 (5), 447–466.

Welch, D. (2005). What should protected areas managers do in the face of climate change? *The George Wright Forum* 22 (1), 75–93.

Wells, M.P. and McShane, T.O. (2004). Integrating protected area management with local needs and aspirations. *Ambio* 33 (8), 513–519.

Wilshusen, P.R., Brechin, S.R., Fortwrangler, C.L., and West, P.C. (2002). Reinventing a square wheel: critique of a resurgent 'protection paradigm' in international biodiversity conservation. *Society and Natural Resources* 15, 17–40.

World Commission on Environment and Development (WCED) (1987). *Our Common Future*. Oxford: Oxford University Press.

World Commission on Protected Areas (WPCA) (2004). *Workshop Report*. Transboundary Protected Areas Task Force Meeting on La Maddalena Island. Appendix 2: Background.

World Wildlife Fund for Nature (WWF) (2004). *Living Planet Report 2004*. With UNEP World Conservation Monitoring Centre and the Global Footprint Network. Cambridge, UK: Banson.

Young, J., Watt, A., Nowicki, P., Alard, D., Clitherow, J., *et al.* (2005). Towards sustainable land use: identifying and managing the conflicts between human activities and biodiversity conservation in Europe. *Biodiversity and Conservation* 14, 1641–1661.

Zimmerer, K.S., Galt, R.E., and Buck, M.V. (2004). Globalization and multi-spatial trends in the coverage of protected-area conservation (1990–2000). *Ambio* 33 (8), 520–529.

Chapter 3
Governance models for parks, recreation, and tourism

Paul F.J. Eagles

Introduction

Governance, the means for achieving direction, control, and coordination, determines the effectiveness of management. There is much diversity and scope in the governance models employed to deliver parks, recreation, and tourism (PRT) services. In examining these assorted approaches, Glover and Burton (1998: 143) proposed a typology of institutional arrangements for the provision of PRT services: (1) *governmental arrangements* whereby public sector agencies alone provide a public service; (2) *cross-sector alliances* consisting of a contractual relationship between a public sector agency and a profit-making or not-for-profit organization (e.g. partnerships and contracts); (3) *regulated monopolies* whereby a non-public organization is granted a monopoly to directly provide public services (e.g. franchise); and, (4) *divestiture* whereby public services, lands, or facilities are sold or leased to profit-making or not-for-profit agencies. All these typologies of service delivery can be found among the outsourcing practices used in parks and protected areas. Some agencies use the profit-making commercial sector or the non-profit private sector to deliver some services. In the latter practice, the park agencies act as supervisory bodies. The Glover and Burton approach assumes that the starting point of discussion is the ownership of PRT services by the public sector, without fully exploring the obvious alternative, the ownership and operation by the private sector, either profit-making or non-profit (Harper, 2000).

More (2005) proposed five models, which he called: (1) fully public model; (2) public utility model; (3) outsourcing; (4) private, non-profit ownership; and, (5) private, for-profit ownership. The *fully public model* has a government agency operate all services. The *public utility model* functions with a government agency functioning much like a private corporation. *Outsourcing* involves the contracting out of some or all services to private companies. *Private, non-profit ownership* describes parks being owned and operated by a non-government organization while *private, for-profit ownership* involves a park being owned and operated by a private company. In addition to Glover and Burton (1998) and More (2005), Graham *et al.* (2003) developed a classification based upon the literature on governance. They suggested that there are four governance types for protected areas. These are: (1) *government management*; (2) *multi-stakeholder management*; (3) *private management*; and (4) *traditional community management*. They suggested that *government management* can occur with two approaches: (1) a national, provincial, state, or municipal government agency, or (2) delegated management from government to some other body. *Multi-stakeholder management* can occur as: (1) collaborative management; or (2) joint management. *Private management* can occur as: (1) individuals; (2) not-for-profit organizations; or (3) for-profit corporations. *Traditional community management* can occur with: (1) indigenous peoples; or (2) local communities. The major addition by Graham *et al.* (2003) to the discussion is the concept of traditional community management.

Glover and Burton (1998), More (2005), and Graham *et al.* (2003) treat the land ownership and park operation as unitary actions with one actor undertaking all activities. In this chapter I propose that it is useful to investigate governance by looking separately at three independent approaches: (1) the identity and role of the owner of the land and resources; (2) the source of the income for management; and (3) the type of management body. Evaluation of governance models can be enhanced by looking at each approach independently. In this chapter, I first explain each approach, and then discuss commonly used governance models within the context of these three approaches. Finally, I propose a structure for the further investigation of protected area governance.

Ownership of the resources

There are three alternatives for resource ownership for parks and protected areas: (1) a government agency; (2) a non-profit institution; or (3) a for-profit corporation (Figure 3.1). It is possible to conceive of a situation involving joint ownership, but I am not aware of any examples.

Government agencies can function at any level of administration, such as the national government, a provincial government, a regional government, or a municipal government. Canada, for example, has major park management agencies at each of these levels of administration.

Non-profit institutions are public organizations which, by law, must operate in a non-profit manner. They are typically independent of governments. Typically, they are social, cultural, legal, and environmental *advocacy groups* with goals that are primarily non-commercial. *Non-profit organizations* typically gain most of their funding from private sources and are found in the arts, charities, education, politics, religion, research, and environmental protection.

For-profit corporations are legally defined companies that can be owned: (1) widely by many individuals; (2) by other corporations; or (3) by private individuals. For-profit corporations are often heavily involved in the provision of tourism services in parks and protected areas that are owned by government. A few own and manage conservation lands outright, both for conservation and for a combination of conservation and income through the institution known as an ecolodge.

Graham *et al.* (2003) suggest that individuals, Aboriginal governments, and traditional communities are also ownership alternatives. For the purposes of this chapter, I consider individuals as being in the private sector, in either profit or non-profit roles, but typically in the profit-making capacity. The word community can refer to a group of people in a geographical area, or a group of people linked by a common interest. This group can function

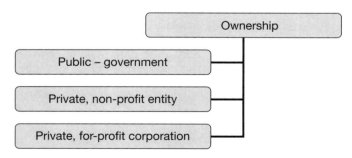

Figure 3.1 Alternatives for ownership of resources

Governance models for parks, recreation, and tourism

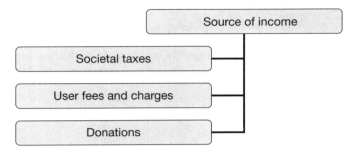

Figure 3.2 Alternatives for sources of income

like a private profit-making corporation, as a non-profit organization, or as a government body. Traditional Aboriginal communities typically function like a private-sector agent from the point of view of those outside such communities. Within the communities they may appear more like a local government.

Sources of income

There are three broad categories of sources of income for parks and protected areas: (1) societal taxes; (2) user fees; and (3) donations (Figure 3.2). There are many forms of taxes and many types of user fees available for consideration. Van Sickle and Eagles (1998) documented the many income sources utilized by parks in Canada in the mid-1990s and showed an ongoing shift from taxes to user fees. Landrum (2005) showed a similar trend for US State Parks, with a change from 38 per cent of operating income obtained from user fees in 1984 to 46 per cent in 2004.

Management body

For this analysis, it is important to consider the management institution that controls activities separately from the resource owner and the source of the income. There are four alternatives for this institution: (1) a government agency; (2) a parastatal, which is a corporation owned or wholly controlled by government; (3) a non-profit corporation; or (4) a for-profit corporation, either public or private (Figure 3.3). The parastatal can be either a for-profit or a non-profit entity, depending upon its legal constitution.

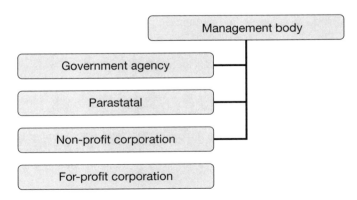

Figure 3.3 Alternatives for management body

41

Graham et al. (2003) suggested that there are four types of government options for government management. One is a park management function fully integrated into another agency, such as a wildlife agency or a forestry agency. A second is a parks agency as a distinct unit within a larger government agency. A third is a separate agency reporting directly to a cabinet member. A fourth is a parastatal, a semi-autonomous body reporting to a board of directors. The board members are typically appointed by the government. This chapter merges the first three into one approach, and keeps the parastatal separate.

An emerging approach, often called co-management, occurs when two or more bodies share decision-making powers. The most frequently used model involves a combination of a government agency and an Aboriginal group, a local community, or a collection of local communities. There are a large number of variations on this approach under development.

Purpose of the enterprise

It is necessary to understand the purpose of the enterprise of protected areas. There are two, overarching, intertwined, and well-recognized goals (Eagles and McCool, 2002). One is the conservation of natural and cultural resources. The other is the provision of education and recreation services (Eagles and McCool, 2002; Eagles et al., 2002). Wilkinson (2003) discusses the Canadian experience of attempting to achieve both sustainable tourism development and high levels of ecological integrity in the national parks. In this chapter, I assume that all aspects of governance and management are undertaken with these two goals in mind.

Typical models

The various approaches proposed by Glover and Burton (1998), Graham et al. (2003), and More (2005) are useful, but do not fully explore all the implications of the many combinations of land ownership, management, and income source. There are 36 combinations possible and each can be seen to be useful for some unique set of circumstances. However, from my experience only a few of these 36 are commonly used. Presumably, these have advantages that lead to their more widespread adoption. The seven most common combinations will be described in detail in subsequent sections of this chapter. Each model is numbered sequentially and given a name that describes the approach. A few combinations are seldom, if ever, used, and are not discussed, except to suggest reasons for lack of use.

Model 1: Golden Era National Park Model

Model 1, the Golden Era National Park Model, has government ownership of the resources, the vast majority of funding coming from societal taxes, and a government agency as manager. This model was personified by the National Park Service of the USA before the early 1950s. In this early formative period, the charging of fees by the government agency was not allowed by US law which made management entirely dependent upon annual allocations from the central government (Figure 3.4). The term Golden Era is appropriate because it is highly lauded by some scholars, such as Wade (2005). Model 1 is very similar to More's (2005) fully public model.

The Swedish Environmental Protection Agency manages the national parks and the nature reserves of Sweden. The current Swedish approach fits Model 1 well as the land ownership is with the government, the funds come almost entirely from the central government, and the managers are government staff members. Significantly, the Swedish central government and the relevant local government authority in the area of the national park share many of the management responsibilities.

In the USA and elsewhere, the ongoing use of this model for tourism management revealed structural impediments. The restricted ability of the government agency to charge

Governance models for parks, recreation, and tourism

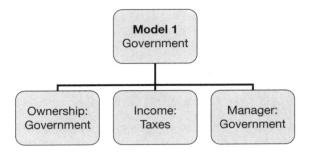

Figure 3.4 Golden Era National Park Model

and utilize use fees, combined with increasing levels of tourism, resulted in a search for an alternative institution for the delivery of tourism services. Often, this led to the use of for-profit corporations funded by use fees. Occasionally, non-profit institutions were used. This situation sometimes moved much of the tourism service delivery into the private sector.

This reliance on government taxes for most of the income is becoming less viable and therefore less common due to the inherent limitations in finance. As the world's protected area estate continues to grow at a high rate, it appears that this growth is financially unsustainable as current levels of tax-based income do not increase accordingly. Seldom do government budget allocations keep pace with increases in land area. This forces the further use of various user fees and charges. A major decision is required in the identification of the institution that charges and retains these charges. The next models briefly discuss the various institutional structures used.

Model 2: Parastatal Model

Model 2, the Parastatal Model, has government ownership of resource, the majority of funding from user fees, and a government-owned corporation as the manager. In Canada this is known as the Crown Corporation approach. Within the parks world, this model is personified by the parastatal park agencies of eastern and southern Africa. Examples include TANAPA in Tanzania, SANParks in South Africa, and the Kenya Wildlife Service in Kenya. Frequently, but erratically, donations from foreign aid bodies provide money for capital needs, such as roads, visitor centres, and communications networks (Figure 3.5). Resource management is typically the full responsibility of the parastatal park agency. This model sometimes includes high levels of service provision by for-profit corporations. Therefore, tourism service delivery is often a combination of parastatal service delivery and for-profit corporation service delivery. This model is similar to More's (2005) public utility model.

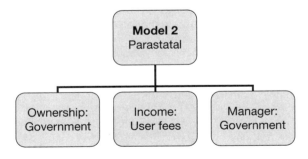

Figure 3.5 Parastatal Model

This Parastatal Model is common and increasing globally. The success of management is highly dependent upon sufficient levels of income from tourism and occasionally from donations. It is possible to discern two different approaches to implementation with this model. One approach sees the park agency as being relatively unsuccessful in the direct provision of tourism services by its own staff and programmes. Due to this lack of success, for-profit companies are used by the parastatal for this purpose. The other approach sees the parastatal as successful in providing tourism by its own staff and institutions and therefore has little need for outsourcing.

In Tanzanian national parks, some tourism services are provided by both the park agency and by private companies. For example, there are campgrounds provided by TANAPA, and camping facilities provided by private companies. Eagles and Wade (2006) found dissatisfaction by tourists with the government-operated facilities and programmes. The same survey showed that the for-profit sector was much more effective in delivering successful tourism services. One outcome of this situation was an ongoing shift towards increasing the role of the private sector and decreasing the role of the parastatal in direct tourism service delivery. When all management activities are dependent upon user fees, and if the parastatal is weak in earning income directly by its own service provision, the agency often relies on high levels of gate fees. For example, TANAPA in Tanzania increased the daily use fee of national parks on 1 January 2006 to US$50 per day for adults and US$10 per day for children aged between five and 16 years (TANAPA, 2005). This fee structure makes the Tanzanian national parks some of the most expensive park destinations in the world for foreign tourists. In order to encourage Tanzanian citizen use of the parks, national citizens pay a much lower entrance fee.

Eagles and Wade (2006) showed that in Tanzania the management of tourism directly by parastatal staff and facilities was ineffective. However, the management of tourism by the private, profit-making organizations was effective. It is worthy of note that from 1998 to 2000 TANAPA (2001) took in more money than it spent each year. Even with the ineffective tourism operation of its own services, its overall financial operations were successful, largely through reliance on gate fees.

However, in contrast, there are examples of parastatals with effective tourism management. The classic and long-standing example of this occurs on the Canadian side of the Niagara River in the Province of Ontario. The Niagara Parks Commission was created by Ontario statute in 1885 with the explicit responsibility of owning and managing the land adjacent to the Niagara Falls and the Niagara Gorge. This major effort was undertaken to replace chaotic and exploitive private tourism operations by government institutions. In the late 1880s, government money was used to purchase large amounts of private land to create parkland beside Niagara Falls, and to create a parkway along the Niagara River and Gorge from Lake Erie to Lake Ontario. From its inception, the park agency was required by provincial government policy to cover all operating costs by user fees. This was done successfully, and since 1945, this park agency has earned a profit every year but one. The exception was 2003, the year when tourism was devastated by the SARS disease outbreak in nearby Toronto. This agency, in contrast to the TANAPA example, is successful in providing tourism services by its own staff, on its own land, and with its own facilities. Within the land owned by The Niagara Parks Commission, all public services are provided by the park's agency, including viewing areas, roads, restaurants, stores, banks, transit, and policing. The agency does not charge entrance fees for the parkland or the parkway, a major difference to the TANAPA example. Rather, it charges for parking. It also earns income from all the food services, all the merchandise stores, and most of the specialized recreation services on its land. All of these services, with the exception of the sight-seeing boat tour at the base of the falls, are provided by agency employees within agency facilities, without the use of a concessionaire. The agency also earns royalty income from a major hydroelectric utility located on its property.

The Niagara Parks Commission has a board of directors composed of appointees. One half of the board is appointed by the Government of Ontario. The other half is appointed by the local municipal councils. This shared supervision ensures that both provincial issues and municipal issues are represented on the board.

The Niagara Parks Commission operates the largest single tourism destination in Canada, with up to 16 million visitors per year. Financial success is ongoing and is dependent upon income from large flows of tourism and many different sources of income. There are other examples of parastatals that are successful in income generation through tourism fees and charges, such as Ontario's Conservation Authorities and South African National Parks.

Each of Ontario's 36 Conservation Authorities, with both watershed management and outdoor recreation responsibilities, operate with a board structure similar to that of The Niagara Parks Commission. The operations of Conservation Authority tourism programmes are financed entirely by a combination of gate fees for day-users, and camping fees for overnight stays. These agencies operate 487 Conservation Areas, encompassing 74,799 hectares, and providing a total visitation of 5,798,566 people (Baldin, 2003). The Ontario Conservation Authority example shows that financial sustainability is possible using the parastatal approach with lower levels of recreational use than occurs in Niagara.

South African National Parks, SANParks, also operates under a parastatal form of management, with an independent financial operation and a board of directors. During the 1990s, the government of South Africa directed SANParks to earn substantially larger amounts of income from fees and charges as government monies were to be diverted from parks to other government priorities. SANParks was largely successful in this venture, and by 2005 had 75 per cent cost recovery of the agency budget through fees and charges (SANParks, 2006).

These Canadian and South African examples of effective and financially successful park and tourism management by parastatal agencies show that there is nothing inherent in this approach that limits effective tourism management, as one might have assumed from the Tanzanian example. In fact, these examples suggest that in terms of the percentage of total tourism expenditures that flow through to the government agency for management, the parastatals are probably more effective than the Golden Era Model. Nevertheless, the use of this model is surprisingly rare given its financial and political successes in several countries.

Model 3: Non-profit Organization (NPO) Model

Model 3, the Non-profit Organization (NPO) Model, has resource ownership by a non-profit corporation, the majority of funding coming from donations and the manager being a non-profit organization. Model 3 is close to More's (2005) private, non-profit model.

This model typically includes large amounts of volunteer time and volunteer donations. Income is also earned from commodity sales, specialized service programmes, and fund-raising efforts (Figure 3.6). This NPO model is common and increasing in use globally.

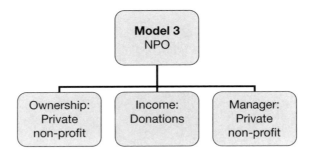

Figure 3.6 Non-profit Organization (NPO) Model

The model is personified by many environmental membership organizations in Europe, with very strong representation in the UK. The second largest land owner in Britain is the Royal Society for the Protection of Birds (RSPB), a non-profit organization. The RSPB owns, manages, and provides public recreation services for 150 nature reserves in the UK, covering approximately 68,000 hectares (RSPB, 2006). The RSPB reserves provide critical habitat for many species of birds in the UK.

This NPO approach is typically used for small sites that are geographically close to a support community that provides the volunteers. One major exception to this rule of small sites is the large Monteverde Cloud Forest Reserve in Costa Rica. This 5,000-hectare site is owned and managed by two NPOs, the Tropical Science Center and the associated Conservationist League of Monteverde. It has high levels of conservation effectiveness and now services a major ecotourism industry with over 50,000 visitors per year (TSC, 2005).

These examples illustrate that the NPO approach has been shown to be viable in certain locales and circumstances. The heavy reliance on donations, at least until a major ecotourism activity is developed, limits its application to situations that have sufficient human capital and donation potential.

Model 4: Ecolodge Model

Model 4, the Ecolodge Model, has resource ownership by a for-profit corporation, funding from user fees, and management by a for-profit corporation (Figure 3.7). Alderman (1990) provided an early glimpse into the operation of ecolodges in Africa and Latin America. This was followed by Langholz (1996) who evaluated the economics and management success of these organizations in sub-Saharan Africa and Central America. Saunders and Halpenny (2001) provided a review of the economics and finance of ecolodges in 52 countries. Model 4 is very similar to More's (2005) private, for-profit model.

This model is used by thousands of ecolodges in many parts of the world, with much sophistication in approaches in South Africa. Ecolodges are geographically concentrated in a few countries, with Australia, Belize, and Costa Rica having strong and increasing numbers of such institutions. Ecolodges are frequently located in the vicinity of a government-owned protected area. An example of this type of concentration occurs along the western border of Kruger National Park in South Africa. In this locale, the ecolodges' land is privately owned but all the wildlife is owned by the government of South Africa. The animals move back and forth across the land of the national park and the ecolodges, making for a large, contiguous ecosystem. An example is the Sabi Sabi Private Game Reserve with 5,000 hectares of land bordering on Kruger National Park (Loon et al., 2007). This is a financially successful example of a private operation that conserves the natural environment while providing tourism services for the upscale tourism market. Kwan (2005) found that the ecolodges in Belize provide high levels of service quality to their visitors.

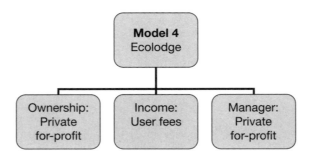

Figure 3.7 Ecolodge Model

Overall, success is spotty and is dependent upon successful tourism marketing and management. Most ecolodge properties are small, but may provide valuable and compatible ecological buffer zones around publicly owned, protected areas. There is evidence of increasing financial success (Osland and Mackay, 2004), but insufficient documentation outlining the significance of the ecological protection of these areas and their buffer roles.

Model 5: Public and For-profit, Private Combination Model

There are many approaches that are best described as combination models. One is the Public and For-profit, Private Combination Model, Model 5. For this discussion, I chose a model that has government ownership of all resources, with management and finance undertaken by a combination of public and private organizations.

The first combination approach explored involves public ownership of all land, a split of management between public and for-profit private organizations, and a split of finance between taxes and user fees. This is probably the most common, combined approach used today (Figure 3.8) and deserves the largest discussion. This is similar to More's (2005) outsourcing model.

The split in management between public and for-profit private entities typically involves public management of resources and private management of tourism services. In stark terms, the management of resources is paid for by taxes, while the private management of tourism is paid for by user fees. However, the financial lines are often blurred with some income from taxes going into tourism management and some tourism income going to resource management.

Canada illustrates a fascinating example of the use of this combination model in the contrasting approaches of the provincial parks of Ontario and British Columbia. In both provinces all resources are owned by government. In both cases large, mature park agencies supervise management. In both cases most of the resource management is done by government agencies. But the similarities end there. There are stark differences in the approaches to finance and tourism management.

Ontario has 329 provincial parks covering 7.8 million hectares of land. These parks serviced around 10.5 million visitor-days of use in 2005 (Ontario Parks, 2005). British Columbia has 830 protected areas covering 11,424,389 hectares of land. These parks serviced 18.3 million visits in 2004 (BC Parks, 2006).

In Ontario, slightly under 80 per cent of all provincial park budgets comes from tourism fees and charges, with the remainder coming from the provincial government. Government policy directs that the agency must earn income from as many sources as possible. This leads to entrepreneurial behaviour by government staff in tourism. The tourism income comes from many sources including: entrance fees, camping charges, concessionaire payments, souvenir

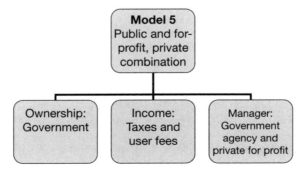

Figure 3.8 Public and For-profit, Private Combination Model

sales, and food sales. Therefore, the vast majority of both resource management and tourism management is financed by tourism-based income. The tourism services are operated by a complicated combination of public and private operations. Some campgrounds are park-agency operated. Others are operated by contractors. Some stores are operated by contractors, others by park management. For example, in 2001 Algonquin Provincial Park in Ontario had 103 concession agreements (Bowman, pers. comm.), but this number subsequently dropped. As the government agency gained more experience in tourism management and saw the potential for increasing income, it took back some services from private operators when the contracts ended.

In the late 1980s the Province of British Columbia in Canada started to transfer all tourism services in provincial parks to the private, profit-making sector (Segal, 2005). Currently in British Columbia, about 20 private companies provide all the tourism services in the provincial parks. In this province, the private sector operators earn income from user fees and also get a subsidy from the government agency. The total income of the private operators is hidden from public view, so it is not possible to give a definitive statement on the percentage of the total budget coming from tax-based grants versus user fees.

How did this unique Canadian situation come about, where one provincial park agency uses the private sector exclusively and another uses a combined model that is moving towards government agency operation? Initial investigation of the situation suggests that the answer appears to be in the realm of dogma concerning the suitability of the relative role of the public and private sector in park tourism. I can find no evidence that the adoption of these two different approaches was based upon an objective review of their pros and cons according to criteria for good governance.

Successive British Columbia governments adopted the premise that the private, for-profit sector is the most appropriate approach for the delivery of tourism services. There is evidence that this policy adoption was undertaken due to aggressive private-sector lobbying. These governments accepted the premise that these private companies should be subsidized by government in the delivery of these services. Importantly, these premises held under right-wing (Social Credit), left-wing (NDP), and mid-range (Liberal) governments.

In Ontario prior to 1995 all budgets for provincial parks came from government. All income was given to central government coffers. However, in Ontario in the early 1990s the Parks Director, Norm Richards, decided to adopt a governance model that was more financially effective and efficient than the current government grant model. He utilized consulting reports, key contact analysis, and political lobbying to re-evaluate the current management structure and to propose an alternative. This culminated in 1996 when the Conservative Government of Premier Mike Harris adopted a new structure for provincial park operations (Ontario Parks, 1996). The Ontario Provincial Park Agency was allowed, for the first time, to retain all of its income in a special-purpose account. It was also allowed to carry over this income from one fiscal year to the next. The Ontario Cabinet approved seven business objectives for the new Ontario Parks Agency:

- to provide strong leadership in natural and cultural protection;
- to move towards financial self-reliance in the parks system, with a first-year target of reducing government allocation by $9 million;
- to operate like a business;
- to build management and workforce excellence in business operations;
- to be a leader in building a new business model in government;
- to involve the private sector in programme delivery, from service contracts to park contracting;
- to improve customer service and market products and services effectively.

A board of directors was appointed to provide overall policy management advice to the Minister responsible for the agency. The agency was given a new name, a new focus, a new brand, and a new visual identity. The Minister of Natural Resources was given the power to set all fees and charges, providing more flexibility than occurred when Cabinet set fees. Moos' (2002) evaluation of this implementation of this business model showed the factors responsible for its financial success and political adoption.

In 1996 the Conservative Government of Premier Harris undertook major budget cuts for all government agencies, with the Ministry of Natural Resources losing about 50 per cent of its budget. However, the parks agency lost only about 15 per cent of its budget and was well positioned to move forward by increasing its use of fees and charges to avoid further major service cuts. The financial success of this model after 1995 solidified its acceptance by government, by park users, and by lobby groups.

In both of these situations in British Columbia and Ontario, the approaches used were not closely tied to the dogma of the governing political parties. But there is an ongoing link to the dogma of the role of a government agency in providing services. The differences revolve around the acceptability of a government agency functioning like a business, or the relative acceptability of the outsourcing and the public utility models. Landrum (2005) points out that decisions surrounding the role of entrepreneurism, commercialization, and privatization may be the most controversial aspect of park management.

Another, but contrasting, example of this combination approach is the 58,000-hectare Madikwe Game Reserve in the North West Province of South Africa. This model sees ownership of land by government, resource management by a parastatal, and tourism management by for-profit corporations. Madikwe is a unique, designed game reserve. Starting in 1991 the Parks Board of the Province purchased degraded farmland from private landowners. The Board placed a 150-kilometre-long fence around the area, removed farm structures, and restored a bushveld ecosystem. The introduction of some 8,054 animals of 24 large mammal species into this reserve makes this the largest game reserve restoration project ever attempted. This initial work was financed by a combination of government grants and loans (MDTT, 1997).

Once the restored ecosystem was in place, the North West Park Board sought private-sector operators to build and operate up to 15 ecolodges. Each corporation was given the licence to operate ecotourism on an exclusive portion of the reserve. Each corporation created a tourism infrastructure with its own funds and according to its market objectives. The concessionaires pay yearly fees to the Parks Board based on tourism volume (Davies, 2003). The concessionaires earn their income from tourism fees and charges.

This approach at Madikwe provided to be successful on two fronts, ecological restoration and economic impact. First, the ecological restoration of the savannah faunal structure at a massive scale was accomplished. Second, by 1999 the three existing tourist operations in place at that time created more cash flow than that of the former farms in the area (Davies, pers. comm.). This innovative project in South Africa shows that a combination of public ownership and for-profit management can produce a successful project that fulfils both ecological restoration and financial viability targets.

Innovative approaches within this combination model are under development in China. To service the rapidly developing tourism market and to earn income from tourism to support park management, various local governments, who have responsibility for park management, are entering into long-term arrangements with private tourism companies. The unique aspect of the Chinese approach is that the governments sometimes take an equity position within the tourism companies. For example, in the Wolong Panda Reserve the tourism concessionaire is 45 per cent owned by the park management. The concession contract gives the Wolong Panda Tourism Co. Ltd monopolistic rights to operate all the tourism in the reserve for 50 years. This type of arrangement gains access to private capital monies, increases revenue to the agencies, and enables the agencies to gain increased influence in company management through the board of directors (Su *et al.*, 2006). The potential difficulty with this situation is that the separation between contractor and contractee is blurred, potentially leading to internal conflicts of interest. The 50-year length of concession contract is also a major point of public debate.

This combined public and for-profit private model is the one that is frequently utilized by parks and protected areas. The relative use of government management of tourism services is highly variable. However, a comprehensive analysis of the pros and cons of this approach, according to standard principles of governance, has not been fully undertaken.

Model 6: Public and Non-profit, Private Combination Model

A less common approach involves a public–non-profit, private combination, Model 6 (Figure 3.9). In this situation, all the factors are similar to the public, profit-making model except that management is split between government and non-profit entities rather than government and profit-making companies. This model is not explored by More (2005). The National Park Service of the USA started using this approach in several locales, such as at Muir Woods in California, in the mid-1980s. The concept spread to Canada, first within Parks Canada and, later, various provincial park agencies. An example is Ojibway Provincial Park in Northern Ontario which is owned by the Government of Ontario but is operated by a community association based in Sioux Lookout (Robson, pers. comm.). The Gulf of Georgia Cannery is a National Historical Site owned by Parks Canada and operated by the Gulf of Georgia Cannery Society, an NGO (Parks Canada, 2003). In a unique twist for Canada, Ontario Parks contracted management of Aaron Lake Provincial Park to a municipal government, the City of Dryden.

It is rare for park agencies to assign all aspects of park management to a non-profit entity. However, it is common for park agencies to cooperate with a non-profit entity in some form of shared operation. Park Friends' Groups are common in the UK, the USA, and Canada. These groups often provide education and interpretive services, and, in Canada, sometimes expand into recreation equipment rental, guiding, and food services. Park managers most frequently cooperate with Friends' Groups when the park agency is restricted by law, policy, or practice in providing those services directly. These groups are typically seen by park agencies as fund raisers. There is some conflict arising in those park agencies that are moving from solely being funded by government to partial funding from tourism, for example Parks Canada and Ontario Parks. In these situations the park agencies sometimes see the Park Friends' Groups as competitors for fund raising.

This use of non-profits for service delivery is at a much lower level than the use of profit-making companies, but the use level is increasing slowly. The few examples investigated suggest that it may have a wider utility than is now undertaken.

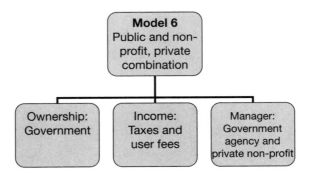

Figure 3.9 Public and Non-profit, Private Combination Model

Model 7: Aboriginal Ownership and Government Management Model

In many locales, the roles of Aboriginal peoples in land ownership and management of parks and protected areas occupy an increasingly important debate. Aboriginal people, people with a prior and historical association with an area of land, often demand land ownership, access to resources, access to financial benefits, and access to park-based employment. Australia has a unique arrangement, whereby several national parks are owned by Aboriginal communities and leased to Commonwealth, State, or Territorial agencies for management (Figure 3.10). Park management is generally undertaken under the direction of a management board. A majority of the members of this board are from the Aboriginal communities. User fees are sometimes charged for admission and use. A portion of the user fees goes to the Aboriginal communities. In addition, the national government provides money for park management and for an annual fee paid to the Aboriginal communities. Aboriginal people have the right to continue the traditional use of any area of the park for hunting and food gathering other than for the purposes of sale. Uluru-Kata Tjuta, Kakadu, and Bonderee National Parks have such arrangements (Worboys *et al.*, 2001). This model is not discussed by More (2005).

In this Australian example, ownership of the land is by Aboriginal people, funding is from both user fees and government grants, and management is by a government agency in cooperation with a local community. This model has been in place in some locales for more than 20 years and is a policy response that recognizes both the public desire for national parks and the resource needs of Aboriginal people and communities. In Australia, opinion about the model is polarized, and the managers indicate it is more difficult to administer a national park where Aboriginal interests are recognized. On the other hand, these difficulties need to be balanced against contemplating what might have resulted in the absence of such an approach (Haynes, pers. comm.). There are concerns among some scientists that the increasing Aboriginal population size could result in exploitation of the natural resources in the parks at unsustainable levels. In 2004, the national government abolished user fees in the national parks which caused concern in the Aboriginal communities about loss of income. The national government agreed to compensate these communities accordingly (Haynes, pers. comm.). Given that the park area often involves several Aboriginal cultures and communities, the relative distribution of responsibilities and benefits is an ongoing and complex discussion.

This Aboriginal and government model often involves various forms of shared management, sometimes called co-management. Parks Canada has several co-management structures with Gwaii Haanas National Park and Kluane National Park being prominent examples. In these two cases the land is owned and managed by government, but the

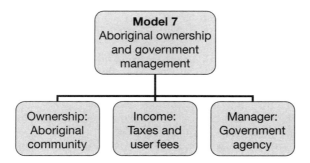

Figure 3.10 Aboriginal Ownership and Government Management Model

management is directed by a management board with Aboriginal involvement. Yamamoto (1993) reported difficulties in governance with Kluane due to the involvement of several different Aboriginal groups and the rigidity of the legal arrangements.

The Aboriginal and government model is expanding and lauded by some. However, it is not yet clear if the involvement of this one powerful stakeholder disenfranchises other stakeholders, such as non-Aboriginal local people or park visitors.

Debates around the appropriateness of models

A scholarly discussion focuses on the role of government in the delivery of services to its constituents (Friedman, 1973; Gormley, 1990; Osborne and Gaebler, 1992; Walsh, 1995). Government's concern with particular services, and not others, is often predicated on the need to ensure the delivery of certain public goods. A public good is one that benefits an entire populace, rather than simply those individuals who partake of the service (Crompton and Lamb, 1986; Walsh, 1995). Savas (2000) summarized other commonly cited traits in providing a typology of four types of goods based on two dimensions: (1) the level of exclusion, and (2) joint versus individual consumption. Public goods are those characterized by joint consumption in that they can be consumed by more than one person at a time, and are also non-exclusive in that they are available to all. In contrast, private goods are those that are consumed individually and exclusively (Savas, 2000). Arguments for public service provision often centre on the need to intervene when the market fails to deliver certain public goods or when certain activities carry such moral significance that their provision must be ensured via the public sector (Walsh, 1995).

Parks, recreation, and tourism (PRT) services are generally considered to be merit goods, which fall along the middle of the public–private spectrum (Glover and Burton, 1998). However, this classification is subject to debate within the field. Some scholars argue that government is justified in adopting private-type goods and methods of service provision in certain situations (Crompton, 1999). Others contest that such services and management models are inappropriate for services such as parks and protected areas (Murdock, 1994; Smale and Reid, 2002; More, 2005). To date, this debate has been largely polemical and has received little empirical examination. In the absence of such, these mixed sentiments have contributed to the wide array of management models that have been adopted for their delivery. Greater understanding is needed on the role of government and the private sector in the provision of PRT services and the inclusion (or exclusion) of private-type services and management models in parks and protected areas.

A second line of research is related to the public–private partnership or networking in the marketing and delivery of public services. In theory, these partnerships are expected to create synergistic dynamics that draw on the strengths and weaknesses of each partner (Rosenau, 1999). Linder (1999) explored six distinct uses of public–private partnership (as management reform, as problem conversion, as moral regeneration, as risk shifting, as restructuring public service, and as power sharing) at the conceptual, ideological, and pragmatic levels.

Frameworks for evaluating the effectiveness of partnerships or inter-organizational networks were proposed at varying levels of analysis, encompassing the community, the network, the organization, and the participant (Provan and Milward, 2001). To date, lessons and experiences of public–private partnerships and their corresponding state-of-the-art research are documented in various policy sectors, such as technology (Stiglitz and Wallsten, 1999), nuclear power (Rosenbaum, 1999), transportation (Dunn, 1999), environmental policy (Kamieniecki et al., 1999), education (Levin, 1999), health (Sparer, 1999), welfare reform (Rom, 1999), criminal justice or the prison system (Lovrich, 1999; Schneider, 1999), and patent policy (Ghere, 2001). By contrast, PRT is a rather weakly documented sector in the public–private partnership research, especially in the planning and management of contracts. This is unfortunate given that parks and protected areas now cover close to 12 per cent of the Earth's surface and the many combinations of public and private operations provide billions of visitor-days of recreation each year.

A related body of research exists on privatization, outsourcing, and contracting out government services (Rehfuss, 1989; Osborne and Gaebler, 1992; Savas, 2000; Cohen, 2001; Peters, 2001). While the research originates from the discipline of public administration, only recently was this line of scholarly research extended to the field of PRT. For example, a comprehensive delivery model of public leisure services was proposed (Glover and Burton, 1998); propositions with research implications for efficiency, effectiveness, and equity in PRT articulated (Glover, 1999a); cases and consequences of partnership discussed (Glover, 1998, 1999b); and processes and procedures for financing or contracting out PRT services outlined (Havitz and Crompton, 1999; Havitz and Glover, 2001). These efforts provide a basis for a further extension and cross-fertilization to address the planning and management issues of PRT contractors.

Evaluations of the models

It is possible to develop an evaluation framework using accepted principles of governance, such as those proposed by Graham *et al.* (2003), combined with the two principal goals of protected areas: (1) the conservation and management of natural resources, and (2) the provision of education and recreation services to the public. The framework is next applied within each of the three categories of investigation: (1) resource ownership; (2) sources of income; and (3) identity of the management body.

Ownership

The alternatives for resource ownership – (1) government, (2) non-profit institution, or (3) for-profit corporation – can be compared according to the two overall goals of conservation and recreation.

As seen in the case studies discussed above, all three alternatives for resource ownership are currently in use. In terms of area covered, government ownership is by far the most popular method. For the effective conservation of very large areas, this might be the only viable option. There are large areas in Aboriginal ownership and management, for example in Brazil, but there is little literature on the long-term effectiveness of this approach for both resource conservation and tourism delivery.

Non-profit ownership is widespread geographically, but typically with small patches of habitat of local significance. NPO ownership of natural land involves organizations categorized as environmental groups or land trusts. They can either be membership or closed organizations, the latter functioning like private corporations. This NPO option is used most frequently in developed countries with an advanced, non-governmental sector. Non-profit land and tourism management is heavily tied to donations of time and money. This also requires a dedicated clientele who live near to the reserve. There are only a few examples of non-profit ownership and operation of globally significant lands.

For-profit ownership of conservation land is a new concept developed in recent years. It is typically done when there is potential for income from tourism. However, some maintain an income stream from various forms of resource extraction. Research shows it can be successful for tourism management, but there is little research available on the long-term viability of conservation efforts.

All three models can be effective in both conservation and recreation, but this success is strongly dependent upon the locale, the institutional structure, and the income available.

Sources of income and management body

The three sources of income can be compared: (1) societal taxes; (2) user fees; or (3) donations, as well as the four alternatives for the management body: (1) a government agency; (2) a parastatal; (3) a non-profit corporation; and (4) a for-profit corporation. It is possible to discuss the effectiveness of each income source and each management body in delivering the two goals of conservation and recreation. A discussion of the combinations can assist in developing an understanding of the advantages and disadvantages of the many possible combinations of sources of income and management body.

Societal taxes can be a large and reliable source of income, and have been for a long time in many countries. Such income is closely tied to the government agency model for management. It has a proven history of providing effective levels of conservation in developed countries. It has mixed success in providing recreation services of high calibre. In some locations, the government tourism services are excellent, based on government grants, but in others they are notably weak. More management research is necessary to fully understand why this approach has such varied levels of success in the delivery of tourism services.

Tax-based grants for conservation and resource management are only available in developed countries. Models based on such income are not viable for developing nations, a fact not always appreciated by scholars and park users based in wealthy countries.

Besides being large and reliable, tax income can pose difficulties. Government budgets must be requested long before expenditure. For operating expenses, requests to government must be made many months or even years before the anticipated time of expenditure. For capital expenditures the requests must be made from a few months to many decades before expenditure. Once the money is granted, it typically has to be spent within a fiscal year, without year-end carry-over. This approach is generally acceptable for resource conservation, but is problematic in regards to recreation provision.

Occasionally governments provide special monies that must be used quickly. This often happens for political reasons, such as an impending election. These funds are welcomed by park agencies, but can be very difficult to handle. Older, established park agencies have contingency plans ready for such opportunities.

Fixed government grants based upon a yearly expenditure plan work best for management issues that are predictable and unchanging. This approach is problematic for unpredictable events and for situations requiring varying levels of funds on short notice. It is, therefore, very challenging for the management of unanticipated and increasing volumes of tourism.

Reliance on government grants turns the management agency into an expenditure body, frequently with scant attention paid to income. It focuses the attention of managers on the chain of command that controls the flow of money. This leads to very long decision structures as bureaucrats manipulate their positions so as to put themselves into the chain of command around financial allocation. For example, the Ontario Provincial Park chain of command in the mid-1980s involved nine levels, leading from field manager to division head. This was a direct consequence of government block funding.

The fixed block funding approach turns the tourist into a problem in two common situations. Unanticipated increases in tourism volume with a fixed park management income, or decreases in government funding with steady tourism volume mean that each new tourist creates stress on the management system. This situation turns every new park visitor into a stress agent and, therefore, by definition, a problem. It is possible that the strong concern expressed by many scholars in the USA and Europe about tourism impacts is the direct result of this management situation.

Block government grants and the associated lack of emphasis on tourist income means that the visitor typically receives low levels of consideration by park managers. Since the tourist is an expenditure agent and not an income agent, the natural reaction of managers is to consider the tourist as a cost problem. This scenario frequently leads to calls for placing caps on use or reducing tourism volumes in order to reduce the problems created by this management structure. It is important to note that the management problem is inherent in the management structure not in tourism per se.

Protected area utilization of tourism user fees reverses the situation dramatically. Under this scenario, each tourist becomes a benefit agent to the management agency. Factors such as length of stay, service quality, and client satisfaction become more important to managers. When Ontario Parks moved in 1996 from an agency funded by government grants to an agency funded by user fees, major changes were made to the agency structure. First, the levels of command shrank in number from nine to three. Second, the field manager's focus moved from the bureaucrats distributing the money, to the park visitors' needs and their associated expenditures. Field staff providing visitor services became much more important to the welfare of the parks and the agency. New service quality measurement was introduced. Many new visitor services were introduced, including better communications, better websites, dramatically improved campsite booking arrangements, new campgrounds, and a new line of park-related merchandise. For the first time, marketing goals were set and achieved (Halpenny, 2007). To ensure financial viability major increases in user fees and new types of fees were introduced. These increases were well accepted by the visitors because of the associated increases in services. Longitudinal studies of park visitor satisfaction generally showed maintenance of good levels of satisfaction with the parks' services and programmes. There was no evidence of an associated loss of biodiversity conservation effectiveness. In fact, an argument could be made that the tourism funding model was more effective in providing biodiversity conservation because in the major budget reductions of 1996 Ontario Parks lost 15 per cent of its budget while the larger Ministry of Natural Resources lost approximately 50 per cent of its budget. The tourism income from parks buffered the impact of the loss of income from government grants.

Texas State Parks are financed by a combination of user fees and government grants. The need to earn income from visitation leads to higher levels of visitor attention. A recent visit to Texas by the author showed dramatic differences in the visitor service quality between Texas State Parks, where visitor income mattered much, and Federal Wildlife Refuges, where visitor income mattered little. The Texas State Parks were much more visitor centric, with new visitor centres, good communication, solid interpretation, and effective resource conservation. By contrast, the Federal Wildlife Refuges showed declining infrastructure and poor interpretation, but apparently acceptable resource conservation.

Donations are only attainable with private, non-profit management models, typically as stand-alone NPOs, occasionally in cooperation with government agencies. Individuals and corporations seldom donate to governments or to for-profit corporations. Donations are closely tied to personal relationships, and are a strength of NPOs. Donation income can be substantial, but is often erratic, high in some periods and low in others. Donations are much more available for capital projects, where status can be attained through grand openings, namings, and other status-earning events. Financial donations are much less available for day-to-day operating activities. Time donations for operating activities can be had in substantial quantities. Donations-based conservation management is widespread in countries with a tradition of conservation education and nature conservation concern. The lands conserved by NPOs are typically small, but can contribute significantly to local conservation. Seldom do NPO lands financed by donations contribute significantly to national or international levels of conservation. There is little published research available on the effectiveness of conservation in non-profit reserves.

The Ontario Provincial Park agency employs a combination model developed where government grants partially fund conservation, and tourist fees fund visitor services (Halpenny 2007). This approach has equity justification. Conservation benefits all of society and, therefore, can reasonably be expected to be paid for by taxes on all of society. Outdoor recreation benefits those who participate and, therefore, reasonably should be paid for by those who benefit.

For-profit conservation efforts are developing in concert with ecotourism. Commonly known as ecolodges, these efforts are developing quickly in some countries, but are virtually non-existent in others. Costa Rica, Belize, Australia, and South Africa have substantial and rapidly increasing ecolodge industries. There is tremendous variability in the size and importance of individual properties. Some are large, on the scale of thousands of hectares, while others are much smaller. In South Africa and Costa Rica the properties are often adjacent to national parks and thus add an important ecological buffer function to these parks. In Belize, the ecolodges are scattered across the country, without any specific geographic relation to protected areas.

The ecolodges along the western border of Kruger National Park in South Africa have a unique resource ownership situation. The land is privately owned. However, the wildlife is owned by the State and is free to move back and forth across the national park land and the private ecolodge land. Therefore, in this situation the resource ownership is split between private companies and public bodies.

Recent research with ecolodges in Belize showed that these financially successful operations had very high customer satisfaction across a range of price bands (Kwan, 2005). The customer satisfaction ratings were much higher than those in Ontario Provincial Parks (Stevcor, 1997) or Tanzania National Parks (Eagles and Wade, 2006). The research suggests that ecolodges can also be financially successful in the provision of nature tourism services (Langholz, 1996; Saunders and Halpenny, 2001). As the gigantic baby boom generation enters retirement, there will likely be a considerable increase in the market for the high-quality nature tourism services available in privately operated ecolodges. There is little research available on the effectiveness of the conservation efforts of ecolodges.

Models seldom used

There are several models that are seldom, if ever, used. The model combining for-profit, private ownership, government funding, and government agency management is not used. In addition, for-profit ownership is not combined with government or non-profit management. Non-profit ownership is not combined with government or for-profit management. Apparently, neither the for-profit nor the non-profit sector is interested in having government as a manager of their land. It appears that these models have too many disadvantages to be useful. However, it is important to note that occasionally sites owned and managed by non-profit bodies are assisted financially by tax breaks or subsidies provided by government, such as the Ontario Conservation Land Tax Incentive Program (OMNR, 2006).

Using good governance criteria for evaluation

Research concerning the overall governance of protected areas is rare (Hannah, 2006). Dearden *et al.* (2005), in a review of 41 countries, found trends towards increased levels of participation by stakeholders in management, and greater use of formal accountability mechanisms. The authors indicated an improvement in protected area governance over the 1990s. A financial trend showed decreases in the relative proportion of budgets coming from governments, with increases from the NPOs and other sources. There was an overall sense that funding did not keep up with the growth of area and responsibilities of the park systems surveyed.

The description of the various models outlined in this chapter suggests that there is a need for a better understanding of the governance of protected areas. An evaluation of the models can be structured using the key aspects of governance. Using criteria developed by the United Nations Development Programme, Graham *et al.* (2003) suggest that good governance can be understood through the use of a set of nine major characteristics:

- public participation;
- the application of the rule of law;
- transparency in decisions;
- responsiveness;
- consensus-oriented decisions;
- equity and inclusiveness;
- effectiveness;
- efficiency; and
- accountability.

Graham *et al.* (2003) further suggest that these nine principles for good governance can be grouped under five categories for consideration in parks and protected areas (Table 3.1):

- legitimacy and voice;
- direction;
- performance;
- accountability; and
- fairness.

The World Parks Congress (2003: 40) accepted these five categories and recommended that they serve as a basis for assessing protected area governance.

Glover (1998) suggested that three of these good governance criteria – efficiency, effectiveness, and equity – are particularly important in understanding PRT services.

Efficiency is the capability of acting or producing effectively with a minimum amount or quantity of waste, expense, or unnecessary effort. In general, efficiency is a measurable concept that can be quantitatively determined by the ratio of output to maximal possible output, or the ratio of output compared to input.

Effectiveness can be defined as how well plans are carried out. This involves leadership by political officials and managers. It requires the delegation of authority to members of an organization. The organization must be capable of managing conflict and motivating employees and volunteers. Financial return on investment is a common measure of management effectiveness. Effectiveness is a vague, non-quantitative concept, mainly concerned with achieving objectives. Effectiveness is harder to measure than efficiency.

Equity is the name given to the portion of the legal system in countries following the English common-law tradition, which resolves disputes between persons by resorting to principles

Table 3.1 Governance principles

Combined categories	Basic governance principles
Legitimacy and voice	• Public participation • Consensus orientation
Direction	• Strategic vision, including human development and historical, cultural, and social complexities
Performance	• Responsiveness to stakeholders • Effectiveness and efficiency
Accountability	• Accountability to the public and stakeholders • Transparency
Fairness	• Equity • Rule of law

of conscience, fairness, and justness. An aspect of equity is inclusiveness, which is the equal ability of all people to participate in and benefit from an activity.

An analysis of the protected area management models using these principles of governance could be useful in providing a better understanding of the comparative differences of the many approaches. Hannah (2006) developed a framework for governance analysis using Graham et al.'s (2003) principles of: (1) legitimacy and voice, (2) direction, (3) performance, (4) accountability, and (5) fairness. Hannah applied this framework to private protected areas in Canada. She concluded that the private protected areas studied generally showed good governance, with different levels of competence in each of the five categories. Hannah's work shows that the framework developed by Graham et al. (2003) can be used for an analysis for private protected area governance, and might be appropriate for a more all-encompassing application to the full range of management models outlined in this chapter. Such work needs to done and should be an area of increased research emphasis.

Summary

This analysis reveals many governance models currently in use in protected areas. The varying financial status, political propensities, and histories in different countries have led to a variety of approaches to governance. Oddly, a full analysis of the various approaches using generally acceptable good governance criteria is lacking.

There is no one, universal approach that is suitable in all situations. The design of the governance and management structure must be appropriate to the social and political systems at both a national and local level. The success of site-level management is ultimately dependent upon the national, political, social, and legal structures of a country. The allocation of the costs and benefits to each one of a large number of stakeholders is an important component.

Much of the scholarly comment on management models concentrates on only a few of the governance principles, most specifically equity of access. Frequently, efficiency is ignored. Only recently has a full consideration of effectiveness been considered (Hockings et al., 2000). A complete evaluation of each of the models would involve the application of all of the principles of good governance with particular emphasis on effectiveness, efficiency, and equity. However, fairness, responsiveness, and the rule of law must not be forgotten. Such an evaluation should be undertaken for the wide range of models now being used for the delivery of PRT services. This chapter is an initial step towards this end.

Protected area management is in a period of dynamic experimentation. It is important for this experimentation to continue. Further research and debate about the appropriateness of various governance approaches is needed. This is critical as the long period of protected area establishment is probably nearing an end, as suitable areas become scarce, human population pressures continue to mount, and competition for land increases. As the grand period of significant protected area establishment is nearing an end, the much longer period of management stretches before all of us into the foreseeable future. The world must now refine governance and site management approaches to ensure that both effective resource protection and quality visitor experiences continue into the future.

Acknowledgements

Ms Anne Ross, Ms Grace Bandoh and Dr Troy Glover provided very useful comments on this chapter. Dr Mark Havitz provided insight into aspects of this work. Mr Andy Kaczynski and Mr Honggen Xiao undertook valuable background research. Doug Barrett provided up-to-date data on Ontario Parks. Mohsin Farooque gave insightful critical comment. Dan Su contributed important background information on the rapidly developing institutional arrangements in China. Richard Davies provided information on the park and tourism developments in the North West Province of South Africa. Chris Haynes gave up-to-date information on Australia. Bob Ditton commented on funding issues in Texas State Parks. Bill Stewart provided comments on many aspects of the work.

Literature cited

Alderman, C.L. (1990). *A Study of the Role of Privately Owned Lands Used for Nature Tourism, Education and Conservation*. Washington, DC: Conservation International.

Baldin, E. (2003). *Visitor Statistics for Conservation Authorities in Ontario: current status and methods*. Paper presented to the Latornell Conservation Symposium.

BC Parks (2006). *BC Parks Attendance Review, 1999–2004*. Ministry of Water, Land and Air Protection. Retrieved 28 January 2006 from www.env.gov.bc.ca/bcparks/facts/attendance_overview_report.pdf.

Bowman, M. (2003). Personal communication.

Bruce, C. (2001). Contracting out at Parks Canada: employee takeovers. In T.L. Anderson and A. James (eds), *The Politics and Economics of Park Management*, pp.107–125. New York: Rowman & Littlefield Publishers.

Cohen, S. (2001). A strategic framework for devolving responsibility and functions from government to the private sector. *Public Administration Review* 61 (4), 432–440.

Crompton, J.L. (1999). *Financing and Acquiring Park and Recreation Resources*. Champaign, IL: Human Kinetics.

—— and Lamb, C.W. (1986). *Marketing Government and Social Services*. New York: John Wiley & Sons.

Davies, R. (2003). *Madikwe Game Reserve – A Decade of Progress*. Rustenburg, South Africa: North West Parks Board.

Dearden, P., Bennett, M., and Johnston, J. (2005). Trends in global protected area governance, 1991–2002. *Environmental Management* 36 (1), 89–100.

Dunn, Jr, J.A. (1999). Transportation: policy level partnerships and project-based partnerships. *American Behavioral Scientist* 43 (1), 92–106.

Eagles, P.F.J. and McCool, S.F. (2002). *Tourism in National Parks and Protected Areas: planning and management*. England: CABI.

—— and Wade, D. (2006). Tourism in Tanzania: Serengeti National Park. *Bois et Forêts des Tropiques* 290 (4), 73–80.

——, McCool, S.F., and Haynes, C. (2002). *Sustainable Tourism in Protected Areas: guidelines for planning and management*. United Nations Environment Programme, World Tourism Organization and World Conservation Union (also available at: http://iucn.org/themes/wcpa/pubs/pdfs/tourism_guidelines.pdf).

Friedman, D. (1973). *The Machinery of Freedom*. New York: Harper and Row.

Ghere, R.K. (2001). Probing the strategic intricacies of public–private partnership: the patent as a comparative reference. *Public Administration Review* 61 (4), 441–451.

Glover, T.D. (1998). Reinventing local government: consequences of adopting a business model to deliver public leisure services. *Journal of Applied Recreation Research* 23 (4), 339–366.

—— (1999a). Propositions addressing the privatization of public leisure services: implications for efficiency, effectiveness, and equity. *Journal of Park and Recreation Administration* 17 (2), 1–27.

—— (1999b). Municipal park and recreation agencies unite! A single case analysis of an inter-municipal partnership. *Journal of Park and Recreation Administration* 17 (1), 73–90.

—— and Burton, T.L. (1998). A model of alternative forms of public leisure services delivery. In M.F. Collins and I.S. Cooper (eds), *Leisure Management: issues and applications*, pp. 139–155. England: CABI.

Gormley, W.T. (ed.) (1990). *Privatization and its Alternatives*. Madison, WI: University of Wisconsin Press.

Graham, J., Amos, B., and Plumptre, T. (2003). *Governance Principles for Protected Areas in the 21st Century*. Ottawa, ON: Institute on Governance. Available at: www.iog.ca/publications/pa_governance2.pdf.

Halpenny, E. (2007). Financing parks through marketing: a case study of Ontario Parks. In R. Bushell and P.F.J. Eagles (eds), *Innovations in Park Tourism*, pp. 277–300. England: CABI.

Hannah, L. (2006). *Governance of Private Protected Areas in Canada: advancing the public interest?* Unpublished Ph.D. thesis, Department of Geography, University of Victoria, Victoria, British Columbia, Canada.

Harper, M. (2000). *Public Services through Private Enterprise: micro-privatization for improved delivery.* London, UK: Intermediate Technology Publications.

Havitz, M.E and Crompton, J.L. (1999). Contracting out services. In J. Crompton, *Financing and Acquiring Park and Recreation Resources*, pp. 227–261. Champaign, IL: Human Kinetics.

—— and Glover, T.D. (2001). *Financing and Acquiring Park and Recreation Resources: a Canadian supplement.* Champaign, IL: Human Kinetics. Available online: www.humankinetics.com/havitz/Table%20of%20Contents.htm.

Haynes, C. (2005). Personal communication.

Hockings, M., Stolton, S., and Dudley, N. (2000). *Evaluating Effectiveness: a framework for assessing the management of protected areas.* Gland, Switzerland: IUCN – World Conservation Union.

——, ——, Leverington, F., Dudley, N., and Courrau, J. (2006). *Evaluating Effectiveness: a framework for assessing management effectiveness of protected areas.* 2nd edn. Best Practice Protected Area Guidelines Series No. 14. Gland, Switzerland and Cambridge, UK: IUCN Publications.

Kamieniecki, S., Shafie, D., and Silvers, J. (1999). Forming partnerships in environmental policy: the business of emissions trading in clean air management. *American Behavioral Scientist* 43 (1), 107–123.

Kwan, P. (2005). *Ecolodge Patrons' Expectation and Perception of Performance.* Unpublished M.A. thesis, Department of Recreation and Leisure Studies, University of Waterloo, Waterloo, Ontario.

Landrum, N.C. (2005). Entrepreneurism in America's State Parks. *The George Wright Forum* 22 (2), 26–32.

Langholz, J. (1996). Economics, objectives and success of private nature reserves in sub-Saharan Africa and Latin America. *Conservation Biology* 10 (1), 271–280.

Levin, H.M. (1999). The Public-Private Nexus in Education. *American Behavioral Scientist* 43 (1), 124–137.

Linder, S.H. (1999). Coming to terms with the public-private partnership. *American Behavioral Scientist* 43 (1), 35–51.

Loon, R., Harper, I., and Shorten, P. (2007). Sabi Sabi: a model for effective ecotourism, conservation and community involvement. In R. Bushell and P.F.J. Eagles (eds), *Innovations in Park Tourism*, 264–276. England: CABI.

Lovrich, Jr, N.P. (1999). Policy partnering between the public and the not-for-profit private sectors: a key lever or a dire warning of difficulty ahead? *American Behavioral Scientist* 43 (1), 177–191.

MDTT (Madikwe Development Task Team) (1997). *The Madikwe Game Reserve Management Plan.* Rustenburg, South Africa: North West Parks Board.

Moos, R. (2002). Ontario Parks – a successful business operating model. *Parks* 12 (1), 17–25.

More, T. (2005). From public to private: five concepts of park management and their consequences. *The George Wright Forum* 22 (2), 12–20.

Murdock, G. (1994). New times/hard times: leisure, participation and the common good. *Leisure Studies* 13, 239–248.

Ontario Ministry of Natural Resources (2006). *Ontario Land Tax Incentive Program.* Retrieved 18 November 2006 from www.mnr.gov.on.ca/MNR/cltip/.

Ontario Parks (1996). *The Ontario Parks Business Plan.* Peterborough, ON: Ministry of Natural Resources.

—— (2005). *Current Regulation – Detailed Summary.* Unpublished data from the Ontario Provincial Parks Database.

Osborne, D. and Gaebler, T. (1992). *Reinventing Government: how the entrepreneurial spirit is transforming the public sector.* New York: Addison-Wesley.

Osland, G. and Mackay, R. (2004). Ecolodge Performance Goals and Evaluations. *Journal of Ecotourism* 3 (2), 109–128.

Parks Canada (2003). Parks Canada Guiding Principles and Operational Policies. Available at: www.parkscanada.gc.ca/docs/pc/poli/princip/part3/part3e_e.asp.

Peters, B.G. (2001). *The Future of Governing* (2nd rev. edn). Lawrence, KS: University Press of Kansas.

Provan, K.G. and Milward, H.B. (2001). Do networks really work? A framework for evaluating public-sector organizational networks. *Public Administration Review* 61 (4), 414–423.

Rehfuss, J.A. (1989). *Contracting Out in Government: a guide to working with outside contractors to supply public services*. San Francisco, CA: Jossey-Bass Publishers.

Rom, M.C. (1999). From welfare state to opportunity: public-private partnerships in welfare reform. *American Behavioral Scientist* 43 (1), 155–176.

Rosenau, P.V. (1999). Introduction: the strengths and weaknesses of public-private policy partnerships. *American Behavioral Scientist* 43 (1), 10–34.

Rosenbaum, W.A. (1999). The good lessons of bad experience: rethinking the future of commercial nuclear power. *American Behavioral Scientist* 43 (1), 74–91.

Royal Society for the Protection of Birds (RSPB) (2006). *Reserves*. Retrieved 27 January 2006, from www.rspb.org.uk/reserves/.

SANParks (2006). *About SANParks Brief History*. Retrieved 31 May 2006 from www.sanparks.org/about/history.php.

Saunders, E.G. and Halpenny, E. (2001). *The Business of Ecolodges: a survey of economics and finance*. Burlington, VT: The International Ecotourism Society.

Savas, E.S. (2000). *Privatization: Public and Private Partnerships*. Chatham, NJ: Chatham House Publishers.

Schneider, A.L. (1999). Public-private partnerships in the US prison system. *American Behavioral Scientist* 43 (1), 192–208.

Segal, G.F. (2005). Competitive sourcing in our national parks. *The George Wright Forum* 22 (2), 43–50.

Smale, B.J.A. and Reid, D.G. (2002). Public policy on recreation and leisure in urban Canada. In E.P. Fowler and D. Siegel (eds), *Urban Policy Issues: Canadian perspectives* (2nd edn), pp. 172–193. Toronto, ON: Oxford University Press.

Sparer, M.S. (1999). Myths and misunderstandings: health policy, the devolution revolution, and the push for privatization. *American Behavioral Scientist* 43 (1), 138–154.

Stevcor (1997). *The 1996 Ontario Camper Survey: an analysis of the facilities and services*. Unpublished consultant report to Ontario Parks.

Stiglitz, J.E. and Wallsten, S.J. (1999). Public-private technology partnerships: promises and pitfalls. *American Behavioral Scientist* 43 (1), 52–73.

Su, D., Wall, G., and Eagles, P.F.J. (2006). Emerging governance approaches for tourism in the protected areas of China. *Environmental Management* 39 (6), 749–759.

TANAPA (2001). *National Parks Visitor Use Statistics*. Unpublished internal government report.

—— (2005). *Regulation and Park Fees*. Retrieved 14 December 2005, from: www.tanzaniaparks.com/.

Tropical Science Center (TSC) (2005). *A Brief History of the Monteverde Cloud Forest Reserve*. Retrieved 28 January 2005, from: www.cct.or.cr/en/historia_mtv.htm.

Van Sickle, K. and Eagles, P.F.J. 1998. User fees and pricing policies in Canadian senior park agencies. *Tourism Management* 19 (3), 225–235.

Wade, B. (2005). A new tragedy for the commons: the threat of privatization to national parks (and other public lands). *The George Wright Forum* 22 (2), 61–67.

Walsh, K. (1995). *Public Services and Market Mechanisms: competition, contracting and the new public management*. New York: St Martin's Press.

Wilkinson, P. (2003). Ecological integrity, visitor use, and marketing of Canada's national parks. *Journal of Park and Recreation Administration* 21 (2), 63–83.

Worboys, G., Lockwood, M., and DeLacy, T. (2001). *Protected Area Management: principles and practice*. Oxford, UK: Oxford University Press.

World Parks Congress (2003). *Private Protected Area Action Plan*. WPC Governance Stream.

Yamamoto, W. 1993. *The Nature and Evolution of Cooperative Management at Kluane National Park Reserve: a case study*. M.A. thesis, Department of Recreation and Leisure Studies, University of Waterloo, Waterloo, Ontario.

Chapter 4
Information technology and the protection of biodiversity in protected areas

Michael S. Quinn and Shelley M. Alexander

> Environmental degradation and habitat loss continue to accelerate. Solutions may be found to reverse this trend, but only with comprehensive data, information and knowledge on the conservation and sustainable use of biodiversity. Data access and knowledge sharing are not simple tasks, however. Difficulties abound. Much of the data, information and knowledge conservationists require is fragmented, difficult to find, or simply not accessible to the conservation community. This challenge is magnified in many developing countries where the consequences of the ever-widening 'digital divide' impede conservation efforts at national, regional, and global levels.
>
> IUCN, 2005: 1

Introduction

We live in an age of information (Wade, 2006). The production, storage, and sharing of information (much of it digital) is a significant facet of economic, cultural, and scientific affairs at scales from local to global. Societal commentators and analysts suggest that we are experiencing a *digital revolution*, an *information explosion*, and we are, as individuals, highly susceptible to *information anxiety or overload* (Biggs, 1989; Holtham and Courtney, 1999). This is a fact that anyone who must contend with their daily e-mail, phone calls, and information requests will have no trouble relating to (Weber, 2005). 'One of the great ironies of the information age is that as the technology of delivering information becomes more sophisticated, the possibility that we can process it all becomes more remote' (Wurman, 1989: 294). The amount of information being produced globally is both astounding and increasing. Lyman and Varian (2003) estimated that 5 exabytes (5,000,000,000,000,000,000 bytes or 5×10^{18} bytes if stored in digital form) of new information was created and stored in 2002 and this value increased by 30 per cent per year between 1999 and 2002. Five exabytes of information equates to approximately 800 megabytes of information for every one of the 6.3 billion people on the planet in 2002; an amount of information that would require ten metres of bookshelf space per person if the information was stored on paper.

It is not only the volume of information being generated and stored; it is the transmission or flow of information that contributes to our information management challenges (Wurman, 1989; Wade, 2006). It has never been easier to access and share information. Concomitantly, it has never been harder to control and manage information. The internet, wireless communication, and a host of emerging digital media make data proliferation and exchange almost effortless. Rapid changes in information and communication technology are making more information available to more people at ever-increasing rates. Filtering and selecting the 'right' information has become a primary task of modern managers, including those of protected areas.

Although there is a great deal of academic discourse and debate around theories of the information society (Fuller, 2005; Webster, 2005), it is clear that information technology has become a driving force in the area of biodiversity conservation and the establishment and management of protected areas (ICEM, 2003; Wilson, 2003; Henry and Armstrong, 2004). In the United States National Park Service, for example, before any action with the potential for adverse effects can be taken, the law requires that adequate information be developed, utilized, and clearly documented in all decisions (Miller, 2001). And, while it may not be as explicitly mandated in other jurisdictions, protected area managers around the globe recognize the value of marshalling the best available information to support decision-making in the complex milieu of parks and protected spaces (see Boxes 4.1 and 4.2).

The availability of data and information is both a blessing and a curse for protected areas managers. On one hand, information represents the potential for greater understanding and decision support. On the other hand, the sheer volume of data available or required, and the systems needed to store, analyse, interpret, and employ the information, create management and financial challenges of their own. The purpose of this chapter is to explore some of the foundational concepts and current trends surrounding the management of protected areas in the information age.

The kind of data and information that we discuss in this chapter is specific to decision support in achieving protected area goals and objectives as they pertain to wildlife, reserve design, and monitoring. We focus primarily on ecological spatial data and the ability of protected areas managers to store, analyse, and integrate those data over a variety of spatial and temporal scales. We recognize the importance of other types of data (e.g. social, economic, political) and applications within protected areas, but these are beyond the scope of the chapter.

Box 4.1 Case study – The Banff-Bow Valley Study and information technology

Banff National Park (BNP) is an international icon for tourism with spectacular mountain scenery, world-class recreation opportunities and a wealth of biodiversity. Since its establishment as Canada's first national park in 1885, visitors from around the world have been attracted to the wonders of BNP in ever-increasing numbers; from 70,000 annual visits in 1912 to approximately 5 million today. Coincident with the increase in visitation has been the proliferation of facilities and services, most of which are concentrated around the park townsites of Banff and Lake Louise in the Bow River Valley. The valley bottom, which has been so attractive for development, comprises a unique montane ecoregion (only 4.2 per cent of the watershed) characterized by relatively warm temperatures, high biodiversity, and prime wildlife habitat. By 1995, the Bow Valley in BNP contained 5,600 hotel rooms, 60 restaurants, 175 specialty shops, and a permanent resident population of over 9,000. In addition, the valley was bisected by two highways and a railway as well as possessing hundreds of kilometres of trails, three downhill ski areas, and a golf course. In total, the footprint of the developed infrastructure covered approximately 20 per cent of the small ecoregion with an effect on a much greater area.

In 1994, concern over the effects of visitation, infrastructure and management of BNP, along with a strengthening of the Canadian National Parks Act, resulted in a call for a comprehensive, two-year study to assess the cumulative environmental effects of development and use in the Bow River watershed within BNP. The Banff-Bow Valley Study (BBVS) was established to:

- develop a vision and goals for the Banff-Bow Valley that will integrate ecological, social, and economic values;
- complete a comprehensive analysis of existing information, and provide direction for future collection and analysis of data to achieve ongoing goals; and
- provide direction on the management of human use and development in a manner that will maintain ecological values and provide sustainable tourism.

The BBVS was directed by a five-member, multidisciplinary task force, and supported by a dedicated secretariat for administrative, professional, technical, and research coordination. Science was coordinated by a technical committee and peer-reviewed by a multidisciplinary team of national experts. The study also included an extensive public involvement process and a comprehensive round table representing the diverse interests in the Bow Valley.

The details and findings of the Banff-Bow Valley Study are beyond the scope of the current discussion, but readers are encouraged to review the final report, a landmark document in protected areas management (Banff-Bow Valley Study, 1996). What is of relevance here is the management and synthesis of information that was conducted for the study. Although the study did commission the collection of some new data, primarily on economic and human-use issues, the combination of a short timeframe and a wealth of existing data (Pacas et al., 1996) resulted in an effort to utilize existing information. The use of spatial data and geographic information systems (GIS) mapping products was essential to communicating current conditions, study results, and future scenarios to public and professional study participants. It is difficult to imagine how the study could have been completed without the efficient and effective use of information technologies.

One of key modelling efforts in the Banff-Bow Valley Study was the *Ecological Outlook Project* (Green et al., 1996) which attempted to evaluate the cumulative environmental effects of human use in the Bow Valley and to predict how current behaviour, trends, and decisions will shape its future. Spatial databases, GIS, and systems modelling were employed to assess effects of key stressors (socio-economic and biophysical) on a suite of ecological indicators (e.g. grizzly bear, wolf, vegetation, aquatic systems, elk) from 1950 into the foreseeable future. The cumulative effects model was a dynamic simulation model developed using STELLA software. This is a 'stock-and-flow' modelling approach that is ideally suited to modelling the relationships between components of interest. The final model was used to predict the ecological and socio-economic effects of future human-use scenarios (–0.5 per cent, 1 per cent, 3 per cent, and 6 per cent) recommended by the round table. The combination of ecological and socio-economic indicators allowed for meaningful discussions of trade-offs among different management alternatives. Although there was a significant amount of ecological and socio-economic data available for the modelling exercise, the process also identified key information gaps and made recommendations for future research priorities.

In the end, the BBVS made more than 500 specific recommendations concerning the future of the Banff-Bow Valley. The efforts to conduct the study included extensive use of information technology, spatial analysis, and scenario modelling. Furthermore, the work provides a foundation for the continued use and management of information technology to support decision-making in BNP. The BBVS directly influenced the development of a new management plan for BNP and many, although certainly not all, of the recommendations have been implemented (see Box 4.2 for an example).

Box 4.2 Case study – Wolf recovery and spatial analysis in Banff National Park – a follow-up to the Banff-Bow Valley Study

Wolves were selected as an environmental indicator for cumulative effects modelling in the Banff-Bow Valley Study (BBVS; see Box 4.1). In particular, wolves were selected as an indicator of permeability for wildlife movement in the montane ecoregion. An empirically derived habitat/movement model was developed based on the results of radio-telemetry, winter tracking, and GIS analysis of habitat (Paquet et al., 1996). The model indicated that the presence of human infrastructure was acting as a barrier to effective wildlife movement through a key travel corridor near Banff townsite, the *Cascade Corridor*. The corridor is a natural 'pinch point' in the Bow Valley that links three important valleys containing critical wildlife habitat. The corridor is approximately 6 kilometres long and ranges from 350 metres to 1.5 kilometres in width. Therefore, one of the recommendations of the BBVS was the removal of all built facilities along the lower slopes and the valley floor near Cascade Mountain, including horse corrals, a fenced bison paddock, an airstrip, and a military cadet camp.

In 1997, Parks Canada removed the bison paddock, barns, and horse corrals and closed the airstrip to all but emergency traffic, significantly reducing the level of human activity and the infrastructure footprint (the cadet camp was subsequently removed as well). Wolves from packs known to utilize habitat on either side of the Cascade Corridor were captured and fitted with VHF radio collars beginning in 1989. Following the removal of human facilities, wolves responded by using the Cascade Corridor significantly more than before 1997 (Duke et al., 2001). In fact, in the ten years of radio-collar monitoring, packs of wolves were only recorded moving through the corridor in the two years following restoration.

The restoration of wolves to the Cascade Corridor in Banff National Park is an exemplary case of applied research and information technology being used to make a cogent case for a management intervention. The removal of popular human-use facilities was a difficult decision for park managers, but the data and analyses were compelling enough to convince them that the park mandate would best be served by embarking on a restoration project. The cumulative effects analysis conducted as part of the BBVS provided further support for action. Subsequent monitoring of radio-collared animals and GIS analysis demonstrated the success of the intervention. This project, which was really a landscape-scale experiment, demonstrates the value of habitat/movement models in determining travel linkages and in understanding the relationships between human use and wildlife movement. Moreover, it demonstrates the use of information technology in the protected area management process.

The first section of the chapter provides an overview of key concepts necessary to understand the collection and use of spatial data in protected areas management and planning. Next we outline some of the current analytical methods employed to assess wildlife–environment relationships and delineate important habitat for reserve design, which are both fundamental to monitoring protected areas over time. The third section provides a brief overview of four critical protected area topics with recent advancements in information technology application: global biodiversity and protected area databases, protected areas system design and evaluation, ecological monitoring, and public engagement through GIS. Finally, we provide some conclusions with respect to the role of information technology in protected areas biodiversity conservation.

Key principles in spatial information technology

In protected areas management, spatial information technology for biodiversity includes databases of the species or phenomena of interest, remote sensing (RS), global positioning systems (GPS), and GIS. GIS is commonly used as a synthesizing tool, characterized by:

- the ability to integrate data from different sources, at different scales, and using more than one mode of representation;
- the ability to store and analyse an unlimited number of attributes of each feature;
- the ability to store and analyse spatial relationships between features; and
- an emphasis on analytical functionality of the software.

Effective and meaningful application of GIS in protected areas management is predicated on having 'good' data. Data quality is a function of: data acquisition, data accuracy, and data validation. Deficiencies in any one of the data processes can have negative consequences for the reliability of the final GIS product. The comforting absolutes of colourful GIS map outputs often mask significant problems and/or invalid assumptions concerning source data. Protected area planners, policy makers, and managers should expect that uncertainties associated with data and resultant models are clearly quantified and articulated (Fieberg and Jenkins, 2005; Prato, 2005; Moilanen *et al.*, 2006). Responsibility lies both with GIS technical staff to be transparent about data and model quality (e.g. clear metadata and explicit statements regarding levels of uncertainty) and users of GIS output to understand the limitations of final models. The following sections identify some of the key data issues that should be clearly articulated by GIS technicians and understood by protected areas managers. Failure to embrace these issues can lead to incorrect conclusions and inappropriate decisions.

Data availability and acquisition

Despite the vast amount of data available, one of the most challenging aspects of conducting spatial analysis in protected areas is acquiring reliable and compatible spatial data. Data acquisition is constrained primarily by cost and secondarily by its format or the ease with which the raw digital data can be interchanged with the GIS or other analytical software. Data acquisition can be particularly challenging in jurisdictions where much of the natural resource data are proprietary and costly to acquire. Furthermore, data are often fragmented with individual jurisdictions or departments holding different datasets with differing quality standards. For example, a digital elevation model (DEM) is central in most analyses conducted by, or for, protected areas managers. Elevation is one of many useful variables that may be developed from a DEM; others include, but are not limited to, slope, aspect, and terrain complexity. These geographic features are central in determining species–environment relationships and distribution, delineation of watersheds, and construction of hydrological maps. Although the utility of DEMs is considerable, their use is limited by locating which organization, jurisdiction, or government agency has acquired the data, which of these has the best resolution and coverage for the project needs, and then securing permissions to use the data. These challenges are becoming more significant as protected areas managers adopt an ecosystem approach that transcends their jurisdictional boundaries and, concomitantly, their data requirements. For example, ecoregional management for Waterton/Glacier International Peace Park in western North American transcends a national boundary, a state boundary, two provincial boundaries and more than 20 management jurisdictions, all with unique datasets and data standards. Furthermore, whereas these issues are challenging when the data exchange needs to happen between government departments, they are even more problematic for academic or private analysts who are working independently of government or across jurisdictional boundaries on behalf of protected areas.

Assuming data can be acquired, their usefulness in decision-making will be determined by the spatial data dimensions and the method of acquisition. Spatial data dimensions include: space, time, and theme. Each dimension has a bearing on whether data are: applicable to the location and problem, reliable for analysis, and useful to determine the confidence placed on final GIS-based management prescriptions. Method of acquisition refers to the way in which the original data were collected. Evaluation of GIS outputs requires managers to have knowledge of the relevant facets of spatial data dimensions and methods of acquisition. This, in turn, requires that the relevant information to evaluate GIS outputs is available. Information about data (metadata) is an essential component of data quality and should always be stored with raw data.

The first dimension of spatial data is their explicit link to a geographic location on the Earth. This process of linking a spatial entity to a geographic coordinate is called 'georeferencing' (Longley *et al.*, 2001; DeMers, 2005). Different nations and international agencies use different datums (starting points for spatial grid systems) as the basic coordinate systems in GIS. As a consequence, the World Geodetic System of 1984 (WGS84) was developed, and remains the accepted standard against which all conversions can be made across datums. A mismatch in the true and applied datum can lead to enormous error in the spatial database and the results of any spatial analysis. Despite knowledge of its importance to management, data often are not accompanied with appropriate metadata files that include datum information.

The temporal dimension specifies when the data were collected and this information should always be included in the metadata file. Because many spatial data are acquired using RS approaches, and subsequently used to determine species–environment relationships, it is important to match the time of data acquisition among multiple datasets as closely as possible. When imagery from multiple years must be combined, dry or summer season images often are preferred to minimize the influences of moisture and changes in leaf cover (Mas, 1999). Many wildlife data are collected throughout the year and across various seasons, which can make it challenging to draw inferences about species–environment relationships using RS imagery. Perhaps a greater temporal issue faced by protected areas managers is the difficulty of maintaining *current* spatial data. All spatial data are static and represent one snapshot in time. Yet, data that are critical to protected areas management (e.g. vegetation cover) are constantly changing. Updating spatial datasets can be time consuming and costly. In addition, while databases become obsolete from a temporal perspective, so too do the technologies (e.g. Landsat is now plagued with a digital scan line problem). Decisions surrounding updating imagery must consider the relevance of staying with the same type of imagery as a means of standardizing across time (if the objective is to measure change) – or introducing bias in comparisons across time by moving to alternative, newer, and potentially better, alternatives. Again, there are pros and cons associated with any choice and limited sources to guide such a decision. We recommend that protected area managers develop a strategic and long-term view to data acquisition that is consistent with their budgets and research needs.

The third and final data dimension in spatial analysis is the theme. This is the non-spatial attribute information that describes the phenomena of interest. These data arise from a number of sources (e.g. vegetation data might come from manual interpretation of aerial photos or from a classification of an RS satellite image). Each acquisition and classification technique has inherent strengths and weaknesses related to the equipment used to derive the image and its applicability to the problem at hand. As illustration, we describe RS data, which are now routinely employed in protected area design, management, and species conservation (Alexander *et al.*, 2005). RS data can be acquired through a variety of platforms (e.g. aircraft and satellite), a range of resolutions, and a variety of electromagnetic energy forms (Lillesand and Kiefer, 2000). The advantage of most RS imagery is the ability to rapidly acquire the necessary spatial data and to acquire them repeatedly. The disadvantage of RS data is that they are often expensive to acquire and can require high levels of expertise to process for analysis. Moreover, RS data can be compromised by cloud cover, moisture, and spectral characteristics of vegetation (Lillesand and Kiefer, 2000). Seasonal, monthly, and daily variability in moisture influences spectral signatures, which makes time-series comparison of images collected over weeks, months, or years less reliable (Foody and Hill, 1996; Mayaux and Lambin, 1997; Mas, 1999). Classification schemes used to create thematic data from RS images also can be plagued with problems (Jensen, 1996).

Data accuracy and validation

Deriving useful and accurate ecological data and models from RS and GIS is not without challenges. The limitations related to data space, time, and theme, and the inherent problems with image processing, underscore the need to quantify the accuracy of images. Accuracy assessment and validation are critical, but often ignored, steps in GIS applications for protected areas (Pearce and Ferrier, 2000). Testing a model with independent data is unquestionably the best way to determine accuracy (Verbyla and Litvaitis, 1989; Fielding and Bell, 1997; Pearce and Ferrier, 2000; Fielding, 2002). However, testing with independent data is not always possible and there remains substantial debate over the optimal method for verifying management models developed using ecological spatial data (and GIS).

Several model validation methods exist that do not require independent data, including cross-validation, jackknife, and bootstrap methods (Fielding and Bell, 1997). Cross-validation requires the dataset be divided into two – one to construct and the other to test. Although straightforward, this method results in loss of degrees of freedom which reduces model significance (Verbyla and Litvaitis, 1989). It also is not suitable for analyses based on small sample size. Jackknife validation is much more precise. Here, one data record is withheld and a model is then compiled with the remaining (n – 1) dataset, and the withheld record is used to test the model (Fielding and Bell, 1997). Bootstrapping is similar to jackknifing but sampling is undertaken with replacement and requires greater computation time (Verbyla and Litvaitis, 1989). There has been no definitive answer to the question of which error test is optimal under which circumstances. However, we can say with certainty that the approach used will be limited by the available dataset, the context, and the technical competence of the GIS analyst.

There remains a gap between the scientist's ability to determine accuracy and to communicate it meaningfully to managers. This is a critical gap to bridge. An incorrect model or an error assessment that is too lenient or misunderstood could result in management decisions that are detrimental to a species. This is particularly true if a model predicts absence when a species is actually present, or documents a false accuracy, and management follows the prescriptions of the model (Fielding, 2002).

GIS and species-based protected area management

Biodiversity management in protected areas is often focused on key threats to wildlife viability, including habitat destruction, habitat fragmentation, over-harvest, introduction of invasive or exotic species, environmental variation and catastrophes, demographic stochasticity, and the potential for genetic drift or inbreeding (Alexander et al., 2005). The development of information technologies, specifically GIS and RS, and Web-based mapping have been critical to advancing wildlife management practice. GIS and RS allow users to assess phenomena and impacts at a variety of scales from landscape to global. Managers are able to recognize and account for multiple scales of ecological organization in analysis and decision-making. Moreover, GIS provides a quantitative basis for decision-making and has grown popular for a variety of reasons, including: (1) landscape data are easier to obtain as the cost declines (Tischendorf, 2001); (2) it is a straightforward platform to analyse data across ecological scales (Luoto et al., 2001); and (3) specialized software has been coupled with GIS and provides a means to conduct advanced spatial analyses.

GIS-based habitat and species–environment analyses have become cornerstones in management and conservation efforts. Here, the greatest strengths of GIS are the capacity to rapidly identify habitat associations and relationships among species and to generalize predictive models for individual species and guilds to other areas where data are lacking. A multitude of studies demonstrate the usefulness of GIS and RS for many taxa, including reptiles, insects, birds, and mammals (Debinski et al., 2002; Fleishman et al., 2002; Lee et al., 2006; Alexander et al., 2006). These techniques have become substantially more sophisticated over the past ten years (Alexander et al., 2005) and now range in complexity from simple additive/subtractive models, to multivariate analyses using high-level spatial statistics. The highly specialized nature of the statistical modelling approaches, coupled with the sheer volume of approaches that can be used to analyse spatial information, can be difficult for a spatial analyst to navigate; it can pose a significant road block to managers, who simply might not have the time to examine the alternatives. In order to illustrate the complexity of the underlying assumptions, we review some of the most commonly applied techniques.

Species presence (occurrence) data are perhaps the most widely used in parks management (e.g. telemetry locations of a grizzly bear). Among other questions, these data have been used to understand: critical habitat for a given species (Clark et al., 1993; Mace et al., 1999), the distribution of a species in a protected area (Alexander et al., 2005), the interactions of species in space and time (Carroll, 2005; Alexander et al., 2006), the effect of roads and human activity on species movement (Gibeau et al., 2002; Alexander et al., 2004), and the value of an umbrella species approach (Carroll et al., 2001). To address such topics, occurrence data have been analysed with a range of empirically based statistical techniques, such as multivariate regression analysis, discriminant function analysis, Mahalanobis distance, and information criteria, among others (Clark et al., 1993; Mace et al., 1996; Dettmers et al., 2002; Wright and Fielding, 2002). Below, we provide a synopsis of the more common statistical approaches in wildlife conservation and parks management as these are critical to the use and understanding of spatial analysis. Understanding the basic assumptions of these approaches is important to understanding the limits of IT approaches in protected area design and management.

Multivariate linear regression analysis was once widely used for modelling species–habitat relationships. However, linear regression has proven less appropriate for modelling these relationships because the assumptions of this parametric statistical approach tend to be violated. In addition, wildlife surveys in protected areas tend to result in databases of species presence locations (e.g. telemetry data points). Combined with an estimation of the absence of survey animals, the available species data are binary and not suited to linear regression. An alternative that is almost ubiquitously used is logistic regression.

Logistic regression has emerged as the most extensively used method for predicting species–environment relationships (Mladenoff et al., 1995, 1999; Mace et al., 1996; Boyce and McDonald, 1999; Carroll et al., 2001; Glenz et al., 2001; Crooks, 2002; Tobalske, 2002; Dettmers et al., 2002; Alexander et al., 2005). Logistic regression has advantages over other techniques. In addition to accommodating dichotomous dependent data it allows the use of categorical independent data, which is not possible in techniques such as linear regression and discriminant analysis (Hosmer and Lemeshow, 2000). The inclusion of categorical independent data is important because many data sources of interest in wildife conservation are categorical (e.g. landcover classes, vegetation composition, soil structure). Logistic regression has further advantages in that it is more tolerant of violations of the assumptions of linear regression or discriminant function analysis (Hosmer and Lemeshow, 2000).

An alternative statistic, Mahalanobis distance, has recently come into favour for GIS-based habitat analysis. Mahalanobis distance does not require the creation of absence points, but measures dissimilarity between 'ideal' mean conditions (usually based on the literature) and the conditions of the observed presence data (Clark et al., 1993). Here, any location in the study area is described by its distance from the optimum 'occupied' habitat as predicted by alternative literature (Rotenberry et al., 2002).

Other popular alternatives to regression-based approaches are discriminant function analysis (DFA) and principal components analysis (PCA). These are used to predict class membership based on the characteristics of independent variables. In particular, DFA generates a function that is based on linear combinations of the predictor variables and provides the best discrimination between the groups. Similarly, PCA determines associations among dependent variables (for instance, different species), and categorizes these as components based on certain spatial attributes. These functions can be used to classify unknown group memberships. Examples of their use include studies of bird species distribution (Dettmers et al., 2002), badger habitat associations (Wright and Fielding, 2002), and multi-species assemblages by habitat attributes (Alexander et al., 2006).

Notwithstanding the usefulness of the above approaches, there has been a dramatic shift in all natural areas and wildlife analysis away from the traditional statistical approaches towards information theoretic approaches (such as Akaike's Information Criterion (AIC) or Bayesian Information Criterion (BIC)). This is an important evolution in the analysis that underscores parks management with respect to species and the proper evaluation of habitat requirements or parks design. Because these modelling approaches are germane to current planning at a species or community level, we provide detail of the rationale for their use. We believe this merits inclusion in our discussion of IT because understanding why these models are used and the basic assumptions of the approaches provides the foundation to understand their usefulness or limitations within real-world applications.

The core argument for using these alternative approaches is that the central assumptions of classical (frequentist) statistics are violated in ecology: true randomization is difficult, replication is often small, misidentified, or non-existent, and ecological experiments rarely are repeated independently (Ellison, 1996; Anderson et al., 2000; Wade, 2000). Another central assumption of both the AIC and BIC approaches is that testing null-hypotheses adds little to scientific understanding: we expect to accept the test hypothesis and reject the statistical null-hypothesis (Ellison, 1996). Anderson et al. (2000) noted that nearly all null-hypotheses are false on a priori grounds. Likewise, Ellison (1996) and Anderson et al. (2000) both argued that a critical problem with the use of p-values (central to classical statistics) is the inability to discern what variable has statistical rather than biological importance (is $p < 0.05$ biologically significant and $p = 0.051$ not?). Information theoretic approaches (AIC or BIC) circumvent these noted problems by providing a framework to analyse multiple working hypotheses (Anderson et al., 2000). In the case of AIC, the measure of comparison is not a p-value, but a calculation that determines a relative rank, based on the log-likelihood value of a model and the number of variables used to develop the model (Anderson et al., 2000). The objective with information theoretic approaches is to find the model with the best fit (log-likelihood) that is achieved using the fewest number of variables.

Spatial information technology in protected areas practice

Global protected areas information

Advances in IT and in international cooperation have resulted in initiatives to compile and analyse data on protected areas from around the globe. The most comprehensive of these is the World Database on Protected Areas (WDPA) which was publicly launched in 2003 at the Fifth World Parks Congress. Global lists of protected areas have been compiled periodically since 1962, but the WDPA represents a quantum leap in the flexibility, accessibility, and transparency of world protected areas data. The vision for the WDPA is

> [a] widely [and freely] available, accurate and up-to-date World Database on Protected Areas that is accepted as a world standard by all stakeholders (government, intergovernmental and non-government), providing the essential link to information from multiple sources on protected areas and contributing to effective resolution of protected area planning and management issues at global, regional and national levels.
>
> (Chape et al., 2003: 6)

The WDPA is managed by the United Nations Environment Programme – World Conservation Monitoring Centre (UNEP-WCMC) in partnership with the World Conservation Union (IUCN), the World Commission on Protected Areas (WCPA), and the World Database on Protected Areas Consortium. Data are provided through agreements with participating national authorities. The WDPA currently (2006) houses data from 233 countries representing 107,107 protected areas covering 169.5 million square kilometres of the Earth's surface (Table 4.1). The WDPA is updated on a regular basis and is publicly available via the Web as well as via an annual compact disc compilation. The WDPA is comprised of spatial data (boundaries of protected areas or points where boundary data are unavailable) and a relation database of non-spatial management, status, and environment data. On-line tools allow users to produce data summaries and maps.

The WDPA is a central source for tracking global, regional, and national targets related to sustainability and biodiversity conservation. For example, Indicator 26 (protected area coverage) under Millennium Development Goal 7 (ensure environmental sustainability), Target 9 (integrate the principles of sustainable development into country policies and programmes and reverse the loss of environmental resources) is assessed using WDPA data (United Nations, 2005). The WDPA also provides essential data and analysis for tracking progress towards the 2010 Biodiversity Target under the Convention on Biological Diversity, as well as key biodiversity indicators from the United Nations Commission on Sustainable Development and the Millennium Ecosystem Assessment. Other uses of the WDPA include emergency oil spill response and contingency planning.

Although the WDPA is an invaluable source of data and analyses on global protected areas, there is a need to expand the nature of the information beyond the geographic measures of coverage and areal extent to encompass the effectiveness of management and protection (Hockings et al., 2000). Recent reviews of protected areas management have revealed that designation does not necessarily result in effective biodiversity conservation (Margules and Pressey, 2000; Pressey et al., 2002; Rodrigues et al., 2004). To evaluate fully progress toward global biodiversity, protection targets will require metrics to assess: (1) effectiveness of coverage: how much and what biodiversity is included within protected areas, and (2) effectiveness in achieving conservation objectives. 'The challenge is to define a standard methodology and apply it consistently in countries so that meaningful results can be derived' (Chape et al., 2005: 454). Rising to this challenge represents the next area of development for the WDPA.

The storage and distribution of spatial data and information are also being facilitated through the creation of Web-based portals for protected areas practitioners. Regional consortia are emerging to create more efficient information sharing and capacity building for managing protected areas in a greater landscape context. One of the best examples is NATURE-GIS, a pan-European network for protected areas, nature preservation, and geographic information (Figure 4.1). NATURE-GIS provides a network for a suite of existing protected areas and GIS organizations and helps to create the productive interchange

Table 4.1 Summary of global protected areas from the World Database on Protected Areas (2006)

Region	Number of protected areas	Total surface area (km²)	Total area protected (km²)	% of source information sourced from					% of surface area protected
				National agency	IGO/NGO	Other named source	No source listed	
Antarctic	122	14,024,488.00	70,323.28	23.77	0.00	4.10	72.13	0.50
Australia/New Zealand	9,549	8,961,660.00	1,511,992.14	99.98	0.00	0.00	0.02	16.87
Brazil	1,287	8,765,502.00	1,638,866.92	25.10	1.40	30.15	43.36	18.70
Caribbean	958	810,199.00	66,209.68	35.91	7.10	7.52	49.48	8.17
Central America	781	704,519.00	158,192.58	55.95	1.66	12.42	29.96	22.45
East Asia	3,265	12,615,201.00	1,764,647.53	69.89	2.08	1.75	26.28	13.99
Eastern and Southern Africa	4,060	12,066,365.00	1,696,304.34	41.70	1.70	0.86	55.76	14.06
Europe	46,194	6,344,292.00	785,012.02	100.00	0.00	0.00	0.00	12.37
North Africa and Middle East	1,325	13,485,380.00	1,320,411.38	79.70	3.40	2.87	14.04	9.79
North America	13,447	27,509,433.00	4,459,304.69	31.73	0.29	1.19	66.79	16.21
North Eurasia	17,724	23,488,200.00	1,816,987.09	5.47	0.60	0.17	93.76	7.74
Pacific	430	3,571,070.00	67,502.21	23.49	1.40	82.33	7.21	1.89
South America	1,456	10,118,418.00	1,955,420.15	51.17	12.36	2.13	35.16	19.33
South Asia	1,216	4,909,391.00	320,635.23	57.89	14.31	0.74	27.06	6.53
South East Asia	2,689	9,084,688.00	867,186.37	47.79	3.68	0.52	48.94	9.55
Western and Central Africa	2,604	13,081,820.00	1,131,153.18	14.32	3.80	3.46	78.42	8.65
Total	107,107	169,540,626.00	19,630,148.78	65.68	0.92	1.29	32.21	11.58

IT and the protection of biodiversity

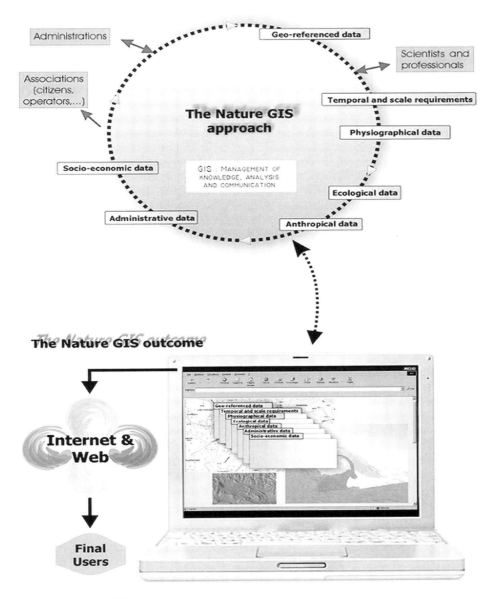

Figure 4.1 NATURE-GIS

between them. Other examples of Web portals for protected area data and information sharing are included in a list at the end of this chapter.

Protected areas system assessment and design

Protected areas serve a vital role in providing *in situ* conservation of biodiversity and the ecological processes that maintain it. 'A good network of protected areas forms perhaps the pinnacle of a nation's effort to protect biodiversity, ensuring that the most valuable sites and representative populations of important species are conserved in a variety of ways' (Vreugdenhil et al., 2003: 7). Global commitments (e.g. Convention on Biological Diversity) recognize the need for completing comprehensive protected area systems that optimize biodiversity protection in concert with *ex situ* management practices. The ability of protected areas systems to achieve biodiversity protection goals is a function of: (1) representation, the extent to which they capture the full extent of native biodiversity, and (2) persistence, their effectiveness in supporting the long-term survival of species (Cowling et al., 1999; Margules and Pressey, 2000; Salomona et al., 2006). The phases required for reserve system design include: (1) mapping biodiversity, (2) identification of candidate sites with significant and complementary biodiversity value, (3) selection of sites to achieve predetermined objectives for biodiversity, (4) establishing protected area boundaries, (5) developing and implementing monitoring and management strategies to ensure the effectiveness of biodiversity protection, and (6) integrating and coordinating with land use planning processes outside of protected areas (Possingham et al., 2000; Pierce et al., 2005; Salomona et al., 2006).

The advancement of remote sensing, GIS, spatial analysis techniques, and optimization algorithms have greatly benefited the field of conservation area design. In addition, the growth of conservation biology and landscape ecology has provided the theoretical frameworks and interdisciplinary approaches to better employ the technology. In order to optimize the biodiversity value of the limited area that will be dedicated to formal protection, all six of the above steps are facilitated greatly through the use of spatial analyses and the technologies that support them (for reviews, see: Pressey et al., 1997; Bibby, 1998; Williams, 1998; Balmford, 2002; Vreugdenhil et al., 2003; Bonn and Gaston, 2005). For example, to achieve protected area conservation efficiency through the principle of complementarity (the lowest number of sites with the highest combined biodiversity coverage) requires the use of complex spatial algorithms best managed within a GIS. Additionally, for jurisdictions with existing protected areas that might not have been originally established with ecological representation criteria, spatial gap analyses are required to assess current conditions and identify opportunities to complete the system (Davis et al., 1990; Scott et al., 1993; Jennings, 2000).

A variety of spatial analytical approaches and reserve design algorithms have been developed to aid in the process of protected area system design and gap analysis (Williams, 1998). In general, these have evolved from being species-focused and static towards more holistic and dynamic (Rodrigues et al., 2000). Coarse-filter, landscape-scale approaches based on ecological systems and enduring features are thought to be the most comprehensive in capturing the majority of biodiversity elements, meeting the needs of species with large area requirements, and encompassing critical ecological processes (Noss, 1992). Furthermore, a focus on the system scale provides greater opportunity for adaptation in the dynamic and uncertain world of land use and climate change (Coulston and Riitters, 2005). However, approaches to identifying spatial priorities for biodiversity conservation will necessarily encompass multiple spatial scales and levels of biological organization. Factors such as zones of high endemism, unique genetic diversity, endangered species, and other fine-scale features will always be required to fully meet biodiversity conservation targets. A comprehensive listing and review of IT approaches for protected area system design is beyond the scope of this chapter, but following is a range of examples that reflect recent large-scale applications.

1 MICOSYS (Minimum Conservation System), a GIS-supported, database program developed for the World Bank for application in Costa Rica and which has now been applied in Belize, Honduras, Nicaragua, and Panama (Vreugdenhil and House, 2002; Vreugdenhil et al., 2003). MICOSYS compares areas on the basis of representation of ecosystems, species of special concern, and socio-economic and cultural variables. The system is transparent in its process (i.e. clear analytical steps) and is designed for ease of application by practitioners with little

technical GIS knowledge. It is the oldest and most widely used comprehensive protected areas system analysis tool.

2 BIMS (Biological Information Management System), a system developed by the Asian Bureau of Conservation (ABC), which integrates GIS coverages with database files to allow monitoring of the status of individual species, habitat types, and protected areas. BIMS (formerly MASS = MacKinnon-Ali Software Systems) also contains a number of analytical tools for evaluating species conservation status and gaps in the protected area system of a given country based on the remaining habitat types. It has been applied across the Indo-Malayan realm whereby conservation effectiveness was evaluated through a set of indices, including the development of priorities for further action (MacKinnon, 1996).

3 BioRap, an approach that utilizes database, GIS, and heuristic analysis tools and includes a heavy reliance on individual species field data to assess biodiversity. BioRap includes both biotic and abiotic surrogates for biodiversity and employs algorithms to integrate economic trade-offs in site selection. The approach has been applied successfully in Papua New Guinea (Nix et al., 2000; Faith et al., 2001).

4 SITES, a spatial decision support tool for ecoregional planning developed for The Nature Conservancy by the University of California in Santa Barbara (Andelman et al., 1999, Groves et al., 2002). The development of SITES built on the site selection model SPEXAN (Spatially Explicit Annealing) at the University of Adelaide, in Australia. SITES employs a pair of heuristic procedures, the 'greedy heuristic' and 'simulated annealing', within a GIS (ArcGIS) interface to optimize the selection of a conservation portfolio (reserve system). The former is a stepwise, iterative procedure that accumulates one site at a time, choosing the 'best' site at each step (as determined by user criteria), until the identified system goals have been met. The latter evaluates alternative complete reserve systems at each step, and compares a very large number of alternative reserve systems to identify a good solution. The programme attempts to minimize cost while meeting goals for representation and spatial configuration. SITES has been applied extensively in US ecoregional planning exercises by The Nature Conservancy.

5 F-TRAC (Florida Forever Tool for Efficient Resource Acquisition and Conservation), a systematic reserve design analysis based on a simulated annealing site-selection algorithm using Marxan software (Oetting et al., 2006). F-TRAC takes the dynamic and uncertain nature of landscape change into account and is being conducted to guide a ten-year, $3 billion state programme of land acquisition for conservation.

6 WWF-Canada Assessment of Representation (AoR) Analyst, an automated GIS application developed as an extension to ArcGIS that provides the capability to assess enduring feature representation (landform and climate as spatial unit surrogates for biodiversity) by protected areas or protected area candidate sites (Iacobelli et al., 2006). Representation is measured according to several conservation criteria that include size requirements to maintain viable populations of native species and sustain ecological processes, environmental gradients (e.g. elevation), important habitat types, habitat quality, and adjacency.

Selection of the appropriate technology and approach to assess biodiversity representation and effectiveness of protected area systems will depend on a host of criteria that include: programme goals and objectives, budget, expertise, quality and quantity of existing data, and how the system coordinates with other spatial analyses within the jurisdiction. The past decade has been witness to a suite of new tools that are readily available for adoption and adaptation by protected area systems managers. Many of these systems have been developed as extensions that can easily be added to existing GIS platforms.

One of the most important considerations in applying protected area system design is to ensure that it is conducted in concert with land use planning for the majority of the landscape that exists outside of protected areas. The importance of landscape connectivity to maintain metapopulation structure (Hanski and Ovaskainen, 2000) and landscape ecological function (Briers, 2002) is recognized as an essential element of protected area design. There are many approaches to assessing landscape connectivity (Tischendorf and Fahrig, 2000), and research approaches have generally focused on tracking animal movement, simulating virtual species movements, or developing connectivity indices from landscape ecology metrics. More recently, the application of graph theory to assess the trade-offs between total area protected and connectivity within protected area systems has proven to be a viable approach (Williams, 1998; Bunn et al., 2000; Urban and Keitt, 2001; Rothley and Rae, 2005). Graph theory approaches offer the benefit of incorporating differing dispersal capabilities of focal species without the data requirements of simulation models. Linking protected area design exercises to comprehensive landscape conservation planning will greatly enhance the conservation of biodiversity within and beyond protected areas. The models and advances briefly discussed above are beginning to bridge the gap between conservation area design theory and land use planning practice (Prendergast et al., 1999).

Monitoring and adaptive management

Protected areas provide society with a broad array of values including the provision of ecological benchmarks. If we are to learn about the effects of our management actions within protected areas, as well as compare outcomes and management effectiveness of activities outside protected areas, we need a commitment to the collection of long-term, strategic monitoring data. Adaptation and improvement are limited by our ability to learn from past action. As Spellerberg (2005) cogently observes, one would not consider purchasing a car without a speedometer, petrol gauge, and warning lights. Yet, with vastly more complicated ecological systems we are often lacking even the most basic indicators to monitor how human activities affect change.

The role of emerging technologies for the collection, storage, retrieval, and analysis of data has played a central role in improving monitoring programmes (Stafford, 1993). The use of remote sensing and GIS in particular has become ubiquitous in all aspects of protected areas monitoring. If there remains a weakness in the overall practice of monitoring, it is in the explicit feedback of information for continuous improvement through adaptive management. Collection of data cannot be an end in itself, but must be clearly linked to decision-making. Adaptive management involves using biodiversity information to set management goals, evaluate the results of monitoring and assessment programmes, and, based on the results, change management actions as warranted.

GIS and spatial analysis for monitoring protected areas gained momentum in the mid-1980s through the development of comprehensive biodiversity monitoring programmes and the adoption of satellite imagery for landscape change analysis. The former is best exemplified by the UNESCO Man and the Biosphere Program (MAB) joining forces with the Smithsonian Institution (SI) in 1986 to form the SI/MAB Biodiversity Program which includes a global network of long-term biodiversity monitoring sites within protected areas (Smithsonian National Zoological Park, 2006). The SI/MAB initiative along with Biosphere Reserve Integrated Monitoring Programme (BRIM) also developed a framework, standardized protocols, and software (BioMon; Comiskey, 1995) for the consistent storage and management of biodiversity data. BRIM explicitly recognizes the value of GIS in integrated monitoring (the measurement of related variables in different biotic and abiotic compartments and coordinated in space and time to provide a comprehensive picture of the system under study) and is exploring the potential of developing a Web-based GIS for biosphere reserves (BRIM, 2001).

Early examples of protected areas employing the monitoring capability from satellite imagery include the use of Landsat MSS (band 6) to examine the regional integrity of protected areas in Botswana (Vujakovic, 1987) and Ward et al.'s (1989) use of Landsat TM to monitor vegetation community change in UK national parks. Today a variety of remotely sensed data are combined with multi-scalar ecosystem and biodiversity monitoring programmes at protected areas around the world. The spatial resolution of multi-spectral and panchromatic imagery has improved tremendously, as has the technical capacity of conservation biologists and landscape ecologists. Data sharing through information portals and multilateral agreements with data providers are becoming the norm in monitoring and managing protected areas within their greater regional landscapes. Combining technologies to better optimize their potential is, perhaps, where the greatest progress is occurring. For example, the use of GPS collars on wildlife, combined with detailed habitat classification derived from remotely sensed imagery is providing researchers and managers with the tools to carefully track the effectiveness of management activities and adjust their actions accordingly. Dibb and Quinn (in press) demonstrate the value of this approach through the deployment of GPS collars on bighorn sheep to monitor and guide restoration fires as part of the historical natural disturbance regime in Kootenay National Park, Canada. Another example of the innovative combination of technologies for protected areas monitoring is the use of GIS to combine Landsat TM imagery, aerial photography, aerial video, and a digital bathymetric model, to assess and to map submerged habitats for Alacranes Reef National Park, Yucatan, Mexico (Bello-Pineda et al., 2005). Only a decade ago, these applications would not have been feasible. Used appropriately and linked directly to planning and management, integrated GIS-based monitoring will continue to provide protected areas with ever-increasing value.

Web-based GIS and the public

The technological provision of spatial information to visitors is an emerging trend in protected areas visitor management. Ubiquitous access to the Web along with the evolution of Web-based GIS and mapping software are creating new venues for interacting with park users. Visitors can now explore the parks through interactive maps to help plan their trip activities and movements. For many people, the Web is now the first source of information in making their travel plans. Protected area managers and communication personnel are embracing the rich opportunities offered by this media to deliver pre-trip information and interpretation. Computer terminals in park visitor centres can also help to provide this information on-site.

An emerging example of this information technology is the US National Park Service Interactive Map Center (IMC). The IMC is publicly accessible on-line (http://maps.nps.gov) and includes a park locator map where users can search for parks by interest topics, activities, park type, cultural heritage, or state, and a park atlas where users can manipulate basic GIS layers to explore park features such as trails and facilities. The IMC is in the early stages of development and the plans are for expanded functionality in the future. The US National Park Service also allows the public to download many of the base GIS layers for further exploration and analysis. Likewise, the UK National Parks Portal contains a publicly accessible GIS site containing individual Web-based park maps with interactive layers that allow users to produce printed maps to enhance their visit (www.nationalparks.gov.uk/index/p_maps.htm). Many other protected area systems are creating sites for visitor access, and we expect this trend to expand rapidly in the next decade.

Web-based GIS is not only a mechanism for delivering information to the public, it also has tremendous potential as a means of receiving data from the public. The interface with the public provides limitless opportunities to go beyond the unidirectional process of serving out data towards two-way communication with park visitors and the collection of new data. Protected areas managers can deliver visitor surveys to visitors both on-site and when they return home. Social science instruments can be administered to collect information on the spatial patterns of visitor use and can be linked to GIS maps of park facilities to collect spatially explicit data. Another emerging application is the use of Web-based GIS as an interface for engaging citizens in observational data collection. Park visitors can interact with on-line GIS to input records of wildlife and plant observations. Simple data collection forms can be created as 'pop-ups' and linked to spatial locations selected on park maps. This approach is in the early stages of development, but shows great promise for citizen science applications (Lee et al., 2006).

Conclusion

Managing protected areas in the information age is both a blessing and a curse. A paradox of our time is that the exponential proliferation of human-generated information is concurrent with a spasm of information loss through the erosion of biodiversity – millions of years' worth of information amassed by life adapting to the planet. A central challenge for protected areas managers is to manage the collection, storage, retrieval, and utilization of the former in protection of the latter. Recent advances in the fields of remote sensing, geographic information systems, and spatial analysis engender tremendous promise for their application to protected areas and beyond. The use of spatial data technology is providing us with a global assessment of progress toward achieving biodiversity protection goals and commitments. Information management frameworks and common data standards are facilitating data-sharing, and the ubiquitous presence of the Web is making it easier for managers in developed and developing countries to access and employ the information they require. Coupling the use of fine-scale information collected through conventional ecological fieldwork with the analytical capabilities of GIS at larger spatial scales is currently a topic of rapid intellectual and methodological advancement. Emerging information technologies to better engage the public in understanding the role and value of protected areas, as well as involve them in contributing to monitoring and management efforts, are changing the public's interface with protected areas.

This chapter focused primarily on multi-spatial issues of ecology, but many of the same principles are transferable to other types of data and information that must be integrated for a truly ecosystem-based approach to protected area management that encompasses the full breadth of our biosocial systems. Protected areas are natural laboratories where such approaches can be developed, tested, adapted, and transferred to contexts beyond the boundaries. The development of information technologies has been a boon to protected area establishment and management. A significantly valuable role that protected areas can now play in conserving global biodiversity is the continued use and advancement of these technologies and methods towards more unified and integrated landscape management.

Literature cited

Alexander, S. (1997). A GIS decision support system for resolving land use conflicts. M.Sc. thesis, Department of Geography, University of Calgary, Alberta.

——, Logan, T.B., and Paquet, P.C. (2006). Spatio-temporal co-occurrence of cougars (*Felis concolor*), wolves (*Canis lupus*) and their prey during winter: a comparison of two analytical methods. *Journal of Biogeography* 33, 2001–2012.

——, Waters, N., and Paquet, P.C. (2004). An uncertainty-based spatial decision support model for restoring landscape connectivity in road fragmented environments. In G. Clarke and J. Stillwell (eds), *Applied GIS and Spatial Modelling*, pp. 233–235. Leeds, UK: University of Leeds.

——, Paquet, P.C., Logan, T.B., and Saher, D.J. (2005). The efficacy of track data in resource selection models for wolves in the Rocky Mountains of Canada. *Wildlife Society Bulletin* 33, 1–9.

Andelman, S., Ball, I., Davis, F., and Stoms, D. (1999). *SITES V 1.0: an analytical toolbox for designing ecoregional conservation portfolios. A manual prepared for The Nature Conservancy.* Santa Barbarba, CA: University of California at Santa Barbara.

Anderson, D.R., Burnham, K.P., and Thompson, W.L. (2000). Null hypothesis testing: problems, prevalence, and an alternative. *Journal of Wildlife Management* 64, 912–923.

Balmford, A. (2002). Selecting sites for conservation. In K. Norris and D.J. Pain (eds), *Conserving Bird Biodiversity: general principles and their application*, pp. 74–104. Cambridge: Cambridge University Press.

Banff-Bow Valley Study (1996). *Banff-Bow Valley: at the crossroads*. Technical report of the Banff-Bow Valley Task Force (R. Page, S. Bayley, J.D. Cook, J.E. Green, and B. Ritchie). Prepared for the Honourable Sheila Copps. Ottawa, ON: Department of Canadian Heritage.

Bartel, A. (2000). Analysis of landscape pattern: towards a 'top down' indicator for evaluation of landuse. *Ecological Modelling* 130, 87–94.

Bello-Pineda, J., Liceaga-Correa, M.A., Hernandez-Nunez, H., and Ponce-Hernandez, R. (2005). Using aerial video to train the supervised classification of Landsat TM imagery for coral reef habitats mapping. *Environmental Monitoring and Assessment* 105, 145–164.

Bibby, J.C. (1998). Selecting areas for conservation. In W.J. Sutherland (ed.), *Conservation Science and Action*, pp. 176–201. Oxford: Blackwell Science.

Biggs, M. (1989). Information overload and information seekers: what we know about them, what to do about them. *The Reference Librarian* 25/26, 411–429.

Bonn, A. and Gaston, K.J. (2005). Capturing biodiversity: selecting priority areas for conservation using different criteria. *Biodiversity and Conservation* 14, 1083–1100.

Boyce, M.S. and McDonald, L.L. (1999). Relating populations to habitats using resource selection functions. *Trends in Ecology and Evolution* 14, 268–272.

Briers, R.A. (2002). Incorporating connectivity into reserve selection procedures. *Biological Conservation* 103, 77–83.

BRIM (Biosphere Reserve Integrated Monitoring Programme) (2001). *Preliminary feasibility study on the possibility of implementing GIS within BRIM*. Retrieved on 22 March 2006 from www.fao.org/gtos/doc/BRIM-gis.pdf.

Bunn, A.G., Urban, D.L., and and Keitt, T.H. (2000). Landscape connectivity: a conservation application of graph theory. *Journal of Environmental Management* 59, 265–278.

Carroll, C. (2005). *Carnivore Restoration in the Northeastern US and Southeastern Canada: a regional-scale analysis of habitat and population viability for wolf, lynx, and marten*. Richmond, VT: Wildlands Project.

——, Noss, R.F., and Paquet, P.C. (2001). Carnivores as focal species for conservation planning in the Rocky Mountain region. *Ecological Applications* 11, 961–980.

Chape, S., Harrison, J., Spalding, M., and Lysenko. I. (2005). Measuring the extent and effectiveness of protected areas as an indicator for meeting global biodiversity targets. *Philosophical Transactions of the Royal Society B* 360, 443–455.

——, Blyth, S., Fish, L., Fox, P., and Spalding, M. (compilers) (2003). *2003 United Nations List of Protected Areas*. Cambridge, UK: IUCN and UNEP-WCMC.

Clark, J.D., Dunn, J.E., and Smith, K.G. (1993). A multivariate model of female black bear habitat use for a Geographic Information System. *Journal of Wildlife Management* 57 (3), 519–526.

Clevenger, A.P., Wierzchowski, J., Chruszcz, B., and Gunson, K. (2002). GIS-generated, expert-based models for identifying wildlife habitat linkages and planning mitigation passages. *Conservation Biology* 16 (2), 503–514.

Comiskey, J.A. (1995). *BioMon: Biodiversity Monitoring Database Ver. 2.01*. Washington, DC: MAB, Smithsonian Institution.

Coulston, J.W. and Riitters, K.H. (2005). Preserving biodiversity under current and future climates: a case study. *Global Ecology and Biogeography* 14, 31–38.

Cowling, R.M., Pressey, R.L., Lombard, A.T., Desmet, P.G., and Ellis, A.G. (1999). From representation to persistence: requirements for a sustainable system of conservation areas in the species-rich Mediterranean-climate desert of southern Africa. *Diversity and Distributions* 5, 51–71.

Crooks, K.R. (2002) Relative sensitivities of mammalian carnivores to habitat fragmentation. *Conservation Biology* 16, 488–502.

Davis, F.W., Stoms, D.M., Estes, J.E., Scepan, J., and Scott, J.M. (1990). An information systems approach to the preservation of biological diversity. *International Journal of Geographical Systems* 4 (1), 55–78.

Debinski, D.M., Jakubauskas, M.E., Kindscher, K., Saveraid, E.H., and Borgognone, M.G. (2002). Predicting meadow communities and species occurrences in the Greater Yellowstone ecosystem. In J.M. Scott, P.J. Heglund, M.L. Morrison, J.B. Haufler, M.G. Raphael, W.A. Wall, and F.B. Samson (eds), *Predicting Species Occurrences: issues of accuracy and scale*, pp. 499–506. Washington, DC: Island Press.

DeMers, M.N. (2005). *Fundamentals of Geographic Information Systems.* New York: John Wiley & Sons.

Dettmers, R., Buehler, D.A., and Bartlett, J.B. (2002). A test and comparison of wildlife habitat modeling techniques for predicting bird occurrence at a regional scale. In J.M. Scott, P.J. Heglund, M.L. Morrison, J.B. Haufler, M.G. Raphael, W.A. Wall, and F.B. Samson (eds), *Predicting Species Occurrences: issues of accuracy and scale*, pp. 607–615. Washington, DC: Island Press.

Dibb, A. and Quinn, M.S. (in press). Response of bighorn sheep to restoration of winter range. *Biennial Symposium of the Northern Wild Sheep and Goat Council.*

Dragicevic, S. and Marceau, D.J. (2000). A fuzzy set approach for modelling time in GIS. *International Journal of Geographical Information Science* 14 (3), 224–245.

Duke, D.L., Hebblewhite, M., Paquet, P.C., Callaghan, C., and Percy, M. (2001). Restoring a large-carnivore corridor in Banff National Park. In D.S. Maehr, R.F. Noss, and J.L. Larkin (eds), *Large Mammal Restoration: ecological and sociological challenges in the 21st century*, pp. 261–276. Washington, DC: Island Press.

Eastman, J.R. (1999). *Idrisi 32. Guide to GIS and Image Processing, Vol. 1.* Wocester, MA: Clark Labs, Clark University.

Ellison, A.M. (1996). An introduction to Bayesian inference for ecological research and environmental decision-making. *Ecological Applications* 6 (4), 1036–1046.

Faith, D.P., Nix, H.A., Margules, C.R., Hutchinson, M.F., Walker, P.A., *et al.* (2001). The BioRap Biodiversity Assessment and Planning Study for Papua New Guinea. *Pacific Conservation Biology* 6 (4), 279–289.

Fieberg, J. and Jenkins, K.J. (2005). Assessing uncertainty in ecological systems using global sensitivity analyses: a case example of simulated wolf reintroduction effects on elk. *Ecological Modelling* 187, 259–280.

Fielding, A.H. (2002). What are the appropriate characteristics of an accuracy measure? In J.M. Scott, P.J. Heglund, M.L. Morrison, J.B. Haufler, M.G. Raphael, W.A. Wall, and F.B. Samson (eds), *Predicting Species Occurrences: issues of accuracy and scale*, pp. 271–280. Washington, DC: Island Press.

—— and Bell, J.F. (1997). A review of methods for the assessment of prediction errors in conservation presence/absence models. *Environmental Conservation* 24, 38–49.

Fleishman, E., Ray, C., Sjögren-Gulve, P., Boggs, C.L., and Murphy, D.D. (2002). Assessing the roles of patch quality, area, and isolation in predicting metapopulation dynamics. *Conservation Biology* 16 (3), 706–713.

Foody, G.M. and Hill, R.A. (1996). Classification of tropical forest classes from Landsat TM data. *International Journal of Remote Sensing* 17 (12), 2353–2367.

Fuller, S. (2005). Another sense of the information age. *Information, Communication & Society* 8 (4), 459–463.

Garshelis, D.C. (2000). Delusions in habitat evaluation: measuring use, selection and importance. In L. Boitani and T.K. Fuller (eds), *Research Techniques in Animal Ecology: controversies and consequences*, pp. 111–164. New York: Columbia University Press.

Gibeau, M.L., Clevenger, A.P., Herrero, S., and Wierzchowski, J. (2002). Grizzly bear response to human development and activities in the Bow River Watershed, Alberta, Canada. *Biological Conservation* 103, 227–236.

Glenz, C., Massolo, A., Kuonen, D., and Schlaepfer, R. (2001). A wolf habitat suitability prediction study in Valais (Switzerland). *Landscape and Urban Planning* 55, 55–65.

Green, J., Pacas, C., Bayley, S., and Cornwall, L. (eds) (1996). *Ecological Outlooks Project: a cumulative effects assessment and futures outlook of the Banff Bow Valley.* Prepared for the Banff Bow Valley Study. Ottawa, ON: Department of Canadian Heritage.

Groves, C.R., Jensen, D.B., Valutis, L.L., Redford, K.H., Shaffer, M.L., *et al.* (2002). Planning for biodiversity conservation: putting conservation science into practice. *BioScience* 52, 499–512.

Hanski, I. and Ovaskainen, O. (2000). The metapopulation of a fragmented landscape. *Nature* 404, 755–758.

Henry, M. and Armstrong, L. (eds) (2004). *Mapping the Future of America's National Parks: stewardship through geographic information systems.* Redlands, CA: ESRI Press.

Hockings, M., Stolton, S., and Dudley, N. (2000). *Evaluating Effectiveness: a framework for assessing the management of protected areas*. Gland, Switzerland and Cambridge, UK: IUCN.

Holtham, C. and Courtney, N. (1999). Perspectives on information overload. *Aslib Proceedings* 51 (8), 249–256.

Hosmer, D.W. and Lemeshow, S. (2000). *Applied Logistic Regression*. New York: Wiley.

Iacobelli, A., Alidina, H., Blasutti, A., Anderson, C., and Kavanagh, K. (2006). *A Landscape-based Protected Areas Gap Analysis and GIS Tool for Conservation Planning*. Toronto: World Wildlife Fund Canada.

ICEM (International Centre for Environmental Management) (2003). Information technology and protected areas. In *Lessons Learned from Global Experience. Review of protected areas and development in the Lower Mekong River Region*, pp. 102–114. Indooroopilly, Queensland, Australia: ICEM.

IUCN (World Conservation Union) (2005). *Conservation Commons*. Retrieved 2 June 2006 from www.conservationcommons.org/section.php?section=common&langue=en.

Jennings, M.D. (2000). Gap analysis: concepts, methods, and recent results. *Landscape Ecology* 15, 5–20.

Jensen, J.R. (1996). *Introductory Digital Image Processing: a remote sensing perspective*. Englewood Cliffs, NJ: Prentice-Hall.

Landres, P., Spildie, D.R., and Queen, L.P. (2001). *GIS Applications to Wilderness Management: potential uses and limitations*. Gen. Tech. Rep. RMRS-GTR-80. Fort Collins, CO: US Department of Agriculture, Forest Service, Rocky Mountain Research Station.

Lee, T., Quinn, M., and Duke, D.L. (2006). Citizens, science, highways, and wildlife: using a web-based GIS to engage citizens in collecting wildlife information. *Ecology and Society* 11 (1) [online] URL: www.ecologyandsociety.org/vol11/iss1/art11/.

Lillesand, T.M. and Kiefer, R.W. (2000). *Remote Sensing and Image Interpretation*, (4th edn). New York: John Wiley & Sons.

Longley, P.A., Goodchild, M.F., Maguire, D.J., and Rhind, D.W. (2001). *Geographic Information Systems and Science*. New York: John Wiley & Sons.

Luoto, M., Rekolainen, S., Salt, C.A., and Hansen, H.S. (2001). Managing radioactively contaminated land: implications for habitat diversity. *Environmental Management* 27 (4), 595–608.

Lyman, P. and Varian, H.R. (2003). *How Much Information*. Retrieved on 30 May 2006 from: www.sims.berkeley.edu/how-much-info-2003.

Mace, R.D., Waller, J.S., Manley, T.L., Ake, K., and Wittinger, W.T. (1999). Landscape evaluation of Grizzly bear habitat in Western Montana. *Conservation Biology* 13, 367–377.

——, ——, ——, Lyon, L.J., and Zuuring, H. (1996). Relationships among grizzly bears, roads and habitat in the Swan Mountains, Montana. *Journal of Applied Ecology* 33, 1395–1404.

MacKinnon, J. (ed.) (1996). *Review of Biodiversity Conservation in the Indo-Malayan Realm*. Draft. Prepared by the Asian Bureau for Conservation in collaboration with the World Conservation Monitoring Centre. Washington, DC: The World Bank.

Margules, C.R. and Pressey, R.L. (2000). Systematic conservation planning. *Nature* 405, 243–253.

Mas, J.F. (1999). Monitoring land-cover changes: a comparison of change detection techniques. *International Journal of Remote Sensing* 20 (1), 139–152.

Mayaux, P. and Lambin, E.F. (1997). Tropical forest area measured from global land-cover classifications: inverse calibration models based on spatial textures. *Remote Sensing of Environment* 59 (1), 29–43.

Miller, A.B. (2001). Managing data to bridge boundaries. In D. Harmon (ed.), *Crossing Boundaries in Park Management: Proceedings of the 11th Conference on Research and Resource Management in Parks and on Public Lands*, pp. 316–320. Hancock, MI: The George Wright Society.

Mladenoff, D.J., Sickley, T.A., and Wydeven, A.P. (1999). Predicting gray wolves landscape recolonization: logistic regression models vs. new field data. *Ecological Applications* 9, 37–44.

——, ——, Haight, R.G., and Wydeven, A.P. (1995). A regional landscape analysis and prediction of favorable gray wolf habitat in the northern Great Lakes region. *Conservation Biology* 9 (2), 279–294.

Moilanen, A., Rungeb, M.C., Elith, J., Tyred, A., and Carmele, Y., et al. (2006). Planning for robust reserve networks using uncertainty analysis. *Ecological Modelling* 199, 115–124.

NATURE-GIS. (2001). *NATURE-GIS white paper*. Retrieved 20 September 2005 from www.gisig.it/nature%2Dgis/documents/Files/White_paper_nature-gis.doc.

Nix, H.A., et al. (2000). The BioRap Toolbox: national study of biodiversity assessment and planning for Papua New Guinea. Consultancy Report to the World Bank. Canberra: Center for Resource and Environmental Studies, Australian National University.

Noss, R.F. (1992). The Wildlands Project: land conservation strategy. *Wild Earth* (Special Issue), 10–25.

Oetting, J.B., Knight, A.L., and Knight, G.R. (2006). Systematic reserve design as a dynamic process: F-TRAC and the Florida Forever program. *Biological Conservation* 128, 37–46.

Pacas, C., Bernard, D., Marshall, N., and Green, J. (1996). *State of the Banff Bow Valley: a compendium of information.* Prepared for the Banff Bow Valley Study. Ottawa, ON: Department of Canadian Heritage.

Paquet, P.C., Wierzchowski, J., and Callaghan, C. (1996). Effects of human activity on gray wolves in the Bow River Valley, Banff National Park, Alberta. In J.C. Green, C. Pacas, L. Cornwell, and S. Bayley (eds), *Ecological Outlooks Project: a cumulative effects assessment and futures outlook of the Banff Bow Valley.* Prepared for the Banff Bow Valley Study. Ottawa, ON: Department of Canadian Heritage.

Pearce, J. and Ferrier, S. (2000). Evaluating the predictive performance of habitat models developed using logistic regression. *Ecological Modelling* 133, 225–245.

Pierce, S.M., Cowling, R.M., Knight, A.T., Lombard, A.T., Rouget, M., and Wolf, T. (2005). Systematic conservation planning products for land-use planning: interpretation for implementation. *Biological Conservation* 125, 441–458.

Possingham, H., Ball, I., and Andelman, S. (2000). Mathematical methods for identifying representative reserve networks. In S. Ferson and M. Burgman (eds), *Quantitative Methods for Conservation Biology*, pp. 291–306. New York: Springer.

Prato, T. (2005). Accounting for uncertainty in making species protection decisions. *Conservation Biology* 19 (3), 806–814.

Prendergast, J.R., Quinn, R.M., and Lawton, J.H. (1999). The gaps between theory and practice in selecting nature reserves. *Conservation Biology* 13, 484–492.

Pressey, R.L., Possingham, H.P., and Day, J.R. (1997). Effectiveness of alternative heuristic algorithms for identifying indicative minimum requirements for conservation reserves. *Biological Conservation* 80, 207–219.

——, Whish, G.L., Barrett, T.W., and Watts, M.E. (2002). Effectiveness of protected areas in north-eastern New South Wales: recent trends in six measures. *Biological Conservation* 106, 57–69.

——, Humphries, C.J., Margules, C.R., Vane-Wright, R.I., and Williams, P.H. (1993). Beyond opportunism: key principles for systematic reserve selection. *Trends in Ecology and Evolution* 8, 124–128.

Rodrigues, A.S., Cerdeira, J.O., and Gaston, K.J. (2000). Flexibility, efficiency, and accountability: adapting reserve selection algorithms to more complex conservation problems. *Ecography* 23, 565–575.

—— et al. (2004). Effectiveness of the global protected area network in representing species diversity. *Nature* 428 (8), 640–643.

Rotenberry, J.T., Knick, S.T., and Dunn, J.E. (2002). A minimalist approach to mapping species' habitat: Pearson's planes of closest fit. In J.M. Scott, P.J. Heglund, M.L. Morrison, J.B. Haufler, M.G. Raphael, W.A. Wall, and F.B. Samson (eds), *Predicting Species Occurrences: issues of accuracy and scale*, pp. 281–289. Washington, DC: Island Press.

Rothley, K.D. and Rae, C. (2005). Working backwards to move forwards: graph-based connectivity metrics for reserve network selection. *Environmental Modeling and Assessment* 10, 107–113.

Salomona, A.K., Ruesinka, J.L., and DeWreedeb, R.E. (2006). Population viability, ecological processes and biodiversity: valuing sites for reserve selection. *Biological Conservation* 128, 79–92.

Sample, V.A. (1994). *Remote Sensing and GIS in Ecosystem Management.* Washington, DC: Island Press.

Scott, J.M., Davis, F., Csuti, B., Noss, R., Butterfield, B., et al. (1993). Gap analysis: a geographic approach to protection of biological diversity. *Wildlife Monograph* 123, 1–41.

Smithsonian National Zoological Park (2006). *Monitoring and Assessment of Biodiversity.* Retrieved 22 March 2006, from http://nationalzoo.si.edu/ConservationAndScience/MAB/default.cfm.

Spellerberg, I.F. (2005). *Monitoring Ecological Change*, (2nd edn). Cambridge: Cambridge University Press.

Stafford, S.G. (1993). Data, data everywhere but not a byte to read: managing monitoring information. *Environmental Monitoring and Assessment* 26, 125–141.

Tischendorf, L. (2001). Can landscape indices predict ecological processes consistently? *Landscape Ecology* 16 (3), 235–254.

—— and Fahrig, L. (2000). How should we measure landscape connectivity? *Landscape Ecology* 15, 633–641.
Tobalske, C. (2002). Effects of spatial scale on the predictive ability of habitat models for the Green Woodpecker in Switzerland. In J.M. Scott, P.J. Heglund, M.L. Morrison, J.B. Haufler, M.G. Raphael, W.A. Wall, and F.B. Samson (eds), *Predicting Species Occurrences: issues of accuracy and scale*, pp. 197–204. Washington, DC: Island Press.
Trigg, S.N., Curran, L.M., and McDonald, A.K. (2006). Utility of Landsat 7 satellite data for continued monitoring of forest cover change in protected areas in South-east Asia. *Singapore Journal of Tropical Geography* 27 (1), 49–66.
Turner, W., Spector, S., Gardiner, N., Fladeland, M., Sterling, E., and Steininger, M. (2003). Remote sensing for biodiversity science and conservation. *Trends in Ecology and Evolution* 18 (6), 306–314.
United Nations (2005). *Progress towards the Millennium Development Goals, 1990–2005: GOAL 7 – Ensure environmental sustainability.* New York: UN Department of Economic and Social Affairs, Statistics Division.
Urban, D. and Keitt, T. (2001). Landscape connectivity: a graph-theoretic perspective. *Ecology* 82, 1205–1218.
Verbyla, D.L. and Litvaitis, J.A. (1989). Resampling methods for evaluating classification accuracy of wildlife habitat models. *Environmental Management* 13 (6), 783–787.
Vreugdenhil, D. and Mateus, M.D. (2002). *Ecosystems and Protected Areas Monitoring Database in MS Access, version 4.* Washington, DC: CCAD, Word Bank, WICE. Available at: www.birdlist.org/nature_management/monitoring/Mon_dbase_version_4_0_eng.zip.
——, Terborgh, J., Cleef, A.M., Sinitsyn, M., Boere, G.C., Archaga, V.L., and Prins, H.H.T. (2003). *Comprehensive Protected Areas System Composition and Monitoring.* Shepherdstown, USA: World Institute for Conservation and Environment.
Vujakovic, P. (1987). Monitoring extensive 'buffer zones' in Africa: an application of satellite imagery. *Biological Conservation* 39, 195–207.
Wade, P. (2000). Bayesian methods in conservation biology. *Conservation Biology* 14 (5), 1308–1316.
Wade, R. (2006). *Spirit of the Web : the age of information from telegraph to Internet.* Toronto: T. Allen.
Ward, S.A., Weaver, R.E., and Brown, R.W. (1989). Monitoring heather burning in the North York Moors National Park using multi-temporal Thematic Mapper Data. *International Journal of Remote Sensing* 10, 1151–1153.
Weber, R. (2005). The grim reaper: the curse of e-mail. *MIS Quarterly* 28 (3), i–vii.
Webster, F. (2005). Making sense of the information age: sociology and cultural studies. *Information, Communication & Society* 8 (4), 439–458.
Williams, P.H. (1998). Key sites for conservation: area methods for biodiversity. In E. Mace, G.M. Balmford, and J.R. Ginsberg (eds), *Conservation in a Changing World*, pp. 211–249. Cambridge: Cambridge University Press.
Wilson, E.O. (2003). Biodiversity in the information age. *Issues in Science and Technology* 19 (4), 45–47.
Woodley, S. and Forbes, G. 1995. Ecosystem management and protected areas: principles, problems and practices. In T.B. Herman, S. Bondrup-Nielsen, J.H.M. Willison, and N.W.P. Munro (eds), *Ecosystem Monitoring and Protected Areas. Proceedings of the Second International Conference on Science and the Management of Protected Areas, held at Dalhousie University, Halifax, Nova Scotia, Canada, 16–20 May 1994* (pp. 50–58). Wolfville, NS: Science and Management of Protected Areas Association.
Wright, A. and Fielding, A.H. 2002. Modeling wildlife distribution within urbanized environments: an example of the Eurasian Badger *Meles meles* L. in Britain. In J.M. Scott, P.J. Heglund, M.L. Morrison, J.B. Haufler, M.G. Raphael, W.A. Wall, and F.B. Samson (eds), *Predicting Species Occurrences: issues of accuracy and scale*, pp. 255–262. Washington, DC: Island Press.
Wurman, R.S. (1989). *Information Anxiety*. New York and Toronto: Doubleday.
Zadeh, L.A. (1965). Fuzzy sets. *Information and Control* 8, 338–353.

List of relevant websites

Biodiversity Conservation Information System: www.biodiversity.org
The Conservation Commons: http://conservationcommons.org
Conservation International Geographic Information Systems (CIGIS) Online: http://gis.conservation.org
The Convention on Biological Diversity Protected Area Site: www.biodiv.org/programmes/cross-cutting/protected/default.asp
Global Biodiversity Information Facility: www.gbif.org/
The Global Transboundary Protected Area Network: www.tbpa.net
IUCN Protected Areas Learning Network (PALnet): www.parksnet.org/index.php
NASA Protected Area Archive: http://asterweb.jpl.nasa.gov/PAA/
Society for Conservation GIS: www.scgis.org/
UNEP World Conservation Monitoring Centre: www.unep-wcmc.org
UNEP-WCMC World Database on Protected Areas website: http://sea.unep-wcmc.org/wdbpa/
World Commission on Protected Areas (WCPA): www.iucn.org/themes/wcpa/
World Database on Protected Areas (WDPA) 2005 CD-ROM Online: http://parksdata.conserveonline.org/website/gis_prod/data/IMS/WDPA_viewer/English/WDPA2005.html

Chapter 5
Anthropological contributions to protected area management

Melissa J. Remis and Rebecca Hardin

The challenges of protected area management are now widely acknowledged as being both ecological and social. Anthropology, a field with subfields that span natural and social science approaches and methods, has had a long history of engagement in the study of human–environment relationships. We will argue in this chapter that anthropology, and particularly collaborative, cross-subfield practices of anthropology, can be well suited to improving protected area management (see also Brosius, 2006; West and Brockington, 2006). It captures nuances and variations in formal and informal or illicit resource use patterns by both humans and animals over time, and helps trace the cultural and the ecological significance of such patterns. This, in turn, can enable more intelligent formulation of policy for complex multi-use sites. Fine-grained and long-term anthropological studies enhance our understanding of micro-regional shifts in density, ecology, and behaviour and how to put these insights into practice for conservation.

Increasingly, anthropologists trace the relevance of community institutions for understanding the larger domains of public institutions, the private sector, and the State. This generation of anthropologists is studying not 'primitives' but, rather, the sorts of tourists, experts, and managers who formulate and implement conservation policies. We can now look through an anthropological lens at the problem of integrating science into policy – to address the flow problems blocking information transfer between scientific and policy domains which, to date, have never been resolved. In deployment of available information to policy, anthropology is uniquely positioned to contribute to three essential aspects of conservation:

- understanding the historical and contemporary influences on the changing animal–human relationship;
- understanding cultural impediments to conservation policy; and
- understanding impediments to conservation policy in international policy agendas.

Today, a range of work falls within the broad rubric of environmental anthropology, and the field bears traces of many approaches within and across the older ecological anthropology, political ecology, and others (Orlove, 1980; Peet and Watts, 1996; Fairhead and Leach, 1996; Kottak, 1999; Brosius, 1999) while combining them in new ways, and moving beyond them into new practical and critical terrain. A growing corpus of long-term, fine-grained ethnographic and ecological fieldwork and cross-cultural and cross-species approaches are well suited for addressing issues of biodiversity and current debates on how to improve conservation approaches (Milton, 1993; Orlove and Brush, 1996; Sponsel et al., 1996; Fuentes and Wolfe, 2002; Paterson and Wallis, 2005).

Further, the practice of anthropological research complements the development of management concepts. Drawing on insights from our own ongoing research in the Dzanga-Sangha Dense Forest Reserve (RDS) in south-western Central African Republic (CAR) we will illustrate the relevance of this 'basic' research to management issues. We will suggest first how anthropological analysis advances our understandings of *integrated conservation and development*; second, how subtle change in animal and human forest use for monitoring and evaluation can work within *adaptive frameworks*; third, we will consider how its field-based partnerships can serve within complex local communities to improve *conservation capacity*; and, finally, we will explore how anthropology's attention to intersections of symbolic and material economies provides a broad framework for contemplating particular species in transnational as well as local contexts as *cross-cultural and ecological keystones*.

Figure 5.1 *Central African Republic and the Dzanga-Sangha Reserve within the context of Central Africa. It is flanked by two other protected areas that together make up the Tri-National Sangha River Protected Area Complex*

The Dzanga-Sangha Special Reserve, Central African Republic

The RDS is one of the most recent protected areas established in CAR, and one of only two sizeable forest reserves in that country. Initial estimates of forest animals' population density indicated that elephants and gorillas are more abundant per square kilometre than are humans in this part of the world (Carroll, 1988; Fay, 1989). Abundant large clearings in the forest enable wildlife viewing (and hunting), but also access to trees and other resources. The area, thus, now supports several rival economies including local community hunting and gathering, alongside international timber extraction, nature tourism, scientific research, and safari hunting (Osaki and Abbaté, 1995).

Dzanga-Sangha's principal tourist attraction is one of the large saline clearings in the region, called Dzanga. It is 'maintained' by the activity of the elephants that fell trees and visit in large numbers to use the marshy clearing and its streams as a source of minerals, for mud baths, and socializing. Initial estimates of forest animals' population density in the area indicated that elephants (*Loxodonta africana cyclotis*) and gorillas (*Gorilla gorilla gorilla*) were more abundant per square kilometre than are humans (Carroll 1988; Fay 1989). Since these early censuses, ongoing ecological monitoring has been initiated in the interest of assuring preservation of the wildlife (Blom *et al.*, 2004b).

The current Reserve is in an area that was previously logged by several generations of transnational logging interests.[1] Since 1990, the study region has been designated a forest reserve with multiple sub-sectors. Initially intending to replace a logging economy in the area, conservation professionals have, instead, had to contend with remarkable rivalry from loggers over the past three decades. Recent logging activities appear to be escalating forest fragmentation and more minor edge effects that had resulted from the initial road building efforts in the 1980s, and create a perpetual backdrop of change that makes forest conservation efforts all the more urgent (Hardin and Remis, 2006).

Reserve regulations

The RDS National Park and Reserve were gazetted in 1990, but conservation and anti-poaching activities began in the area in 1987, through training of guards and the initiation of regular patrols for the enforcement of national hunting codes. At present, the Central African government, in conjunction with the WWF and the GTZ,[2] manages the protected area through a series of fairly complex interior regulations intended to govern human activity differently in several separate sectors (see Figure 5.2). RDS policy is intended to favour the continued existence of hunting and gathering lifestyles for forest residents, whose use of the forest has, in fact, been constrained by the advent of conservation measures. As such, revenues generated by forest use by outsiders in growing local industries of ecotourism and trophy hunting are distributed among residents. For example, under the Reserve's interior regulations (approved in 1991), about 90 per cent of tourism revenues remain within the Reserve, although, to date, tourism has not generated as much income as hoped (Blom, 2000). While these revenues provide new options for some residents, many have mixed feelings about the fact that the national park area is currently off limits for all activities save research, tourism, and education. These issues pose the most difficult challenge to sustainable use in the region, and are intimately bound up with the history of logging. Within the multiple-use managed hunting sector, however, gathering and traditional spear, crossbow, and net hunts, as well as regulated shotgun hunting of less vulnerable species, are all permitted.

Legal access to core areas, and even to hunting within buffer areas, has become easiest for foreigners and for certain elites. Codes pertaining to firearms favour hunting by those, usually not long-term, residents who have the financial, educational, and political resources to seek and purchase legal permits and hunting quotas. In addition, annual quotas are currently allotted by the national government to foreign safari hunting tour operators, which

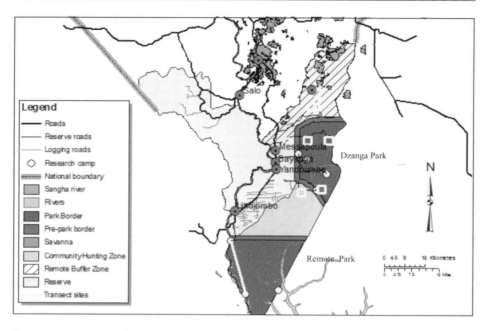

Figure 5.2 Dzanga-Sangha Reserve Study Sectors, Central African Republic. Towns appear as circles. Research camps, study sites, and blocks appear in yellow

operate in the area only about five to six months out of each year, leaving 'their' sectors somewhat open to part-time multiple uses by locals. Activity has proliferated in the area, luring further immigrants and altering the frequency with which, and conditions under which, humans and animals interact.

As of this writing, conservation activities have likely prevented the sorts of local extinctions of wildlife experienced in towns further north and south of CAR. The use of wire cable snares and the killing of elephants, gorillas, chimpanzees (*Pan troglodytes troglodytes*), leopards (*Panthera pardus*), and other vulnerable species are outlawed throughout the country (Carroll, 1998). But such legal restrictions are not always effectively enforced, due to demand for leisure hunting by national and international elites, or for wild game meat to feed growing communities in the area (Wilkie and Carpenter, 1999).

Overall, protection of wildlife is still far from equitably and efficiently accomplished in the RDS area. Management practices intended to safeguard subsistence hunting and gathering against the growth of commercial markets are imperfectly applied. Infractions persist, requiring constant fine-tuning of the system. In addition, as we have suggested, certain inequitable relations of access are embedded in the very basic assumptions of the rules and codes. Some Reserve residents are prevented from using the most efficient and powerful technologies to hunt and gather, while others are enabled to extract more efficiently for profit.

Reserve residents

In the Dzanga-Sangha Reserve area, the BaAka (Pygmy) foragers and their non-BaAka neighbours, locally known as 'Bilos', have linked histories of migration, subsistence, and labour in colonial regimes, despite speaking different languages. The terms 'villagers' and 'farmers' no longer provide meaningful distinctions, as many BaAka also live in villages, at least part of the time, and plant fields of their own (Kretsinger and Hardin, 2003). The first residents of the Reserve's largest town, Bayanga, appear to have been groups of horticulturalists-fishers-traders who settled on the banks of the Sangha River during the upheaval of colonial exploration and initial trades in ivory and other animal products at the turn of the twentieth century (Hardin, 2000). Each of these commodity cycles transformed migration patterns and landscapes; each had experienced booms, then busts (see also Barnes, 2002). Like these colonial companies, today's logging operations have been suspended and then resumed in recent years. Nevertheless, the demand for tropical wood on world markets has made logging the major industry in the area for the past three decades. Thus, Bilos and BaAka who have − or historically had − hereditary trade relations now experience professionalization and commodification of work in the forest.

As diamond towns in south-western CAR such as Nola and Berberati have grown, independent communities such as that of Babongo, just north of Bayanga within the Reserve, have settled defiantly on the outskirts of such booming economic activity, continuing to live a largely autonomous lifestyle based on combining horticulture, trapping, fishing, and other uses of the abundant natural resources, with occasional wage labour in regional industries. This accounts, in part, for their fierce opposition to strict park management which, to them, seems to curtail their activities.

The RDS Reserve was established as a protected area that would contain and contribute to the livelihoods of its inhabitants in an area that contains many temporary and permanent human settlements. The largest is the town of Bayanga, whose population was estimated at 5,000 in 2001 but had fluctuated between approximately 1,000 (1988) and as many as 10,000 (2004) in association with changes in the local economy (Loudiyi, 1995; Noss, 2000; Carroll, pers. comm., RDS project records). Bayanga's first and most recent influx of immigrants came to work at the sawmill. More recent immigrants have been attracted by the promise of rich natural resources or employment in the protected area. Babango, ten kilometres north of Bayanga, is one of the longer established and least ethnically diverse villages in the region. In recent years some of the strongest oppositions to conservation and highest numbers of elephant poachers have come from this village. Flanked by both a logging company and a nature tourism and Reserve administration complex, the neighbourhoods of Bayanga are filled with a variety of local people who work in one or both of these major local industries.

A tendency for BaAka to now remain roadside is only reinforced by the construction of schools, wells, and other development initiatives carried out by the RDS administration in roadside 'villages' such as Yandoumbé and Messapoula. Yet, BaAka continue to use the forest as both a natural and a cultural resource. The majority of households in Bayanga were involved in hunting activities in 2002, according to a rapid survey commissioned by World Wildlife Fund at RDS in 2002 (Ghiurghi and Lakara, 2002). When in the forest, they remain dependent on the abundance of wild game and plant resources not only for nutrition, but also for certain senses of identity and independence from wage labour. It is in this sense that the decline in game meat species can be considered a crisis, not only for animal populations, but among those humans who depend upon them.

Forest wildlife: fluctuating responses to human activities

Gorillas are traditionally hunted in this region, and play an increasingly large role in the bushmeat trade (Bowen-Jones, 1999; Auzel and Wilkie, 2000; Peterson and Ammann, 2004). This commercial trade increases in logging zones in the region (Robinson *et al.*, 1999), and is also becoming more visible at RDS, largely linked to the role of in-migrant hunters and traders (Daspit and Remis, unpublished). Gorilla abundance has been found to be inversely correlated with human disturbance at other sites, and even during some of our earlier census years at RDS (Remis, 2000). Nevertheless, our 2005 data show this relationship has broken down and yielded to spatial diversification of human and gorilla activities across sectors at RDS. Now both humans and gorillas concentrate their activities and navigate along the grid of recently opened logging roads. While many other primates and other mammals decline with logging (Struhsaker, 1997; Chapman et al., 2000), we have been both surprised and dismayed at the high amount of gorilla feeding signs and nests in recently logged zones that are herb rich but also bustling with human activity and snares (also Laurance et al., 2006). Gorillas appear to be sometimes, temporarily, able to co-exist with intense human activity. Unfortunately, our longitudinal data suggest that gorillas' proximity to hunters along roadsides proves to be risky for them once other more preferred prey decline (Blom, 2005; Remis et al., 2006). This vulnerability, even within protected areas, exacerbates consequences of looming threats of Ebola and other disease outbreaks for gorilla populations as a whole (Wallis and Rick, 1999; Walsh et al., 2003).

During initial surveys of the RDS area (using methods articulated in Tutin and Fernandez, 1984; White and Edwards, 1999; Remis, 2000), gorillas were found to be much more numerous than chimpanzees, a pattern found elsewhere in other selectively logged forests in the region (Carroll, 1988). In 1985, the highest density of gorilla nests in the Dzanga region was along abandoned logging roads (Carroll, 1988). But herbs decline with forest succession, and four to six years later gorillas had shifted their activities away from the roads (Remis, 1997). Thus, any logging-related increases in herbaceous resources for gorillas are temporary; these foods decline before gorillas, with their slow reproductive rates, respond. Unlike gorillas, chimpanzees decline after logging at most sites across Africa. At Lope chimpanzees attempted to move out of areas being actively logged; but those fleeing a logging zone are likely to be exterminated by a neighbouring chimpanzee community (White and Tutin, 2001). Long-term multidisciplinary field studies are essential in order to understand the full effects of logging on apes, and to make predictions about long-term population stability.

Recent studies of elephant mobility in this region indicate that they migrate across international borders and over remarkably long distances to visit the Dzanga saline (Turkalo and Fay, 1996; Blake and Hedges 2004). After the conception of the Reserve, and local enforcement of anti-poaching laws, there was a community perception of increase in elephant activity in Bayanga and its surrounding agricultural fields and forests (Kamiss and Turkalo, 1999). It is interesting to wonder whether, in those initial years, elephants were concentrating their activities within Dzanga, relative to more heavily hunted zones outside of the core protected area; but any such benefits of the Reserve may have been short-lived. In recent years there has been a marked decline in elephant abundance and activity within Bayanga and its surrounds (Remis, 2002); though elephants continue to use the Dzanga clearing many of their major paths in the neighbouring community hunting sector have now grown over (Remis et al., 2006).

Like the elephants and gorillas, other herbivores including forest buffalo (*Caffir nanus*), sitatunga *(Tragelaphus spekei)*, several species of duiker (*Cephalophus*), giant forest hogs (*Hylocherus meinertzhageni*), and bush pigs (*Potamochoerus porcus*) may also use the streams, sunbathe, and graze on tender re-growth vegetation in the clearings around Dzanga and on recently opened logging roads in the region. Ironically these 'benefits' of light levels of selective logging likely increase the vulnerability of all of these species to the human hunters who use the grid of logging roads as arteries from which to enter the forest and launch their hunting activities (Blom, 2005). It is more efficient to move quickly along these open roads and hunters more easily locate fresh animal trails here than in dense low-visibility forest where travel is cumbersome. Likewise, those tracking wildlife for ecotourism and research also benefit from the road grid and clearings that provide both wildlife viewing and hunting opportunities. The abundance of elephants and other wildlife were negatively correlated with proximity to secondary logging roads at RDS and in Southern Gabon (Laurance, 1995; Blom, 2005; Laurance et al., 2006).

Collaborative anthropological research at RDS

We have studied changing human perceptions and uses of the forest within this protected area, focusing here primarily on life in, and work within, hunting camps and spatial patterning of wildlife abundance in the same sectors. We have documented in more detail elsewhere some of the changes that the aforementioned policies and political economic developments have made in the abilities of families to meet their needs in the forest (Hardin and Remis, 2006). These changes include, most notably, spatial shifts in, and intensification of, hunting and gathering as people are integrated into larger commercial economies. Heterogeneity in hunting practices in different zones and change over time have worked to reduce wildlife populations across multiple sectors, especially duikers, which are the preferred prey in the region (Eves and Ruggiero, 2000; Remis *et al.*, 2006).

With respect to wildlife, we focus here on shifts in behaviour and abundance of two species that are of ecological and symbolic significance to all stakeholders and even visitors in the area: elephants and gorillas. We have noted elsewhere that research and conservation can have positive effects for animals and humans, valorizing expert tracking, and creating interesting cross-cultural rapport among people as they interact with wildlife (Hardin, 2000). Our present work shows animals and humans responding to different ecological and economic pressures differently in different sectors over time, such as between buffer and core but also North/South, logging/tourism and overlap areas where multiple economic activities are booming at one time. We will briefly summarize those results here, in four different sectors of the protected area, providing very condensed accounts of the individuals, groups, and animals with which we worked in each sector.

Core Park Sites for research, tourism, and protection

The Core Dzanga Sector Park area is generally richer in wildlife than the other sectors studied. Elephant sign, gorilla nests and duiker sign were all very common in our initial 1997 census, as well as earlier censuses (Remis, 2000). Currently most visible human activities on transects around the primary research sites in this sector appear to be related to research and tourism, in accordance with Park and Reserve regulations, but hunters do disregard interdictions fairly widely and are regularly encountered here by researchers and guards. Transects at sites less centrally located within the core park reveal heavy poaching pressure. By 2005, gorilla, elephant, and duiker abundances were reduced there relative to the better protected Core Park Sites (Figure 5.3).

The Dzanga clearing serves as a major tourist attraction and research site at Dzanga-Sangha (Figure 5.2). Its elevated platforms afford safe, reliable wildlife viewing, though visitors do need to walk two kilometres along elephant paths to get there. The long-term success of much of the other research and tourism here, as in other parks, depends on successful habituation of gorillas to the presence of humans. This process takes many years (Blom *et al.*, 2004a), and even once achieved does not guarantee the safety of the animals or their continued availability for viewing (Butinski and Kalina, 1998). For example, gorilla research and ecotourism have been disrupted after silverbacks from habituated groups have died following aggressive encounters with other males. Chimpanzees and gorillas routinely endure close contact with researchers or tourists; in one of the most unfortunate cases this has led to the death of a young gorilla at RDS who fell while fleeing from a party of well-intentioned tourists. Nevertheless, the sometimes negative impacts of conservation and research-related incidents cannot compare with the wholesale slaughter of apes that occurs in 'hotspots' of bushmeat extraction that most often accompany the extension of logging into remote forest areas (Rose *et al.*, 2003).

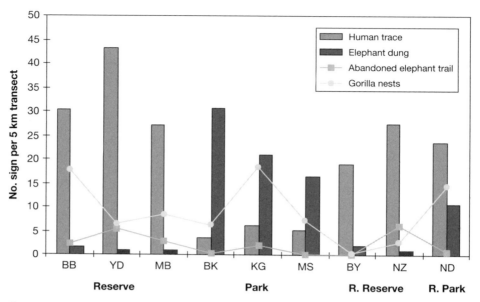

Figure 5.3 Gorilla and elephant sign in relationship to human activity by site at RDS, Central African Republic, 2005. The nine sites along the x-axis increase in distance from the main town of Bayanga. Babango (BB), Youndombe (YD), and Mabongo (MB) are Core Reserve Sites. Bakombe (BK), Kongana (KG), and Messapoula (MS) are Core Park Sites. Beyobe (BY) and Nzoboko (NZ) are Remote Reserve Sites, and Ndoki (ND) is the Remote Park Site

Buffer Zone: dynamics of diverse subsistence and scarcity

The BaAka hunting camp we visited in the community hunting zone in 1997 was small and intimate, though surrounded by a dense network of paths of other human users of the forest. Demand by local, regional, and national elites, as well as foreign tourist desires, are the latest forces contributing to the fabulous variety of hunting techniques among the BaAka in this sector. Even in 1997, the camp was a remarkably complete living 'museum' of hunting and gathering practices adopted and adapted from a range of interpenetrating influences over time, a stark contrast with the more limited repertoires of neighbouring regions. Despite declining wildlife, BaAka have since varied and intensified their harvesting strategies to take advantage of commercial opportunities provided by tourists or regional game markets (see Hardin, 2000; Remis, 2000; Noss and Hewlett, 2001). Net-hunts have more participants and can travel by truck to less-disturbed sectors when travelling with tourists, likely increasing their yields. By 2002 and 2005 the heaviest sector of human activity in the Buffer Zone had shifted several kilometres to the east, logged in 2002, and the site of a 32-hut BaAka camp that was geared for commercial hunting and trade.

The community hunting zone transect sites had the lowest elephant abundance across censuses, likely due to high human presence (Figure 5.3). Lack of elephant activity in the area has actually been a boon to BaAka who have cultivated fields for the first time since moving to Youndombe. Agriculture and logging impact habitat quality for wildlife (Laurance, 1995; Struhsaker, 1997). Logging temporarily increases forest resource availability for gorillas (Remis, 1997; Malcolm and Ray, 2000). We have twice observed the pattern of simultaneous increases in human hunting and gorilla nesting and feeding in herb-rich recently logged sites followed by a decline once more preferred game species are reduced in these areas.

BaAka elders recognize the influence of others on their community, and the ever-harder work of hunting due to scarcity of overexploited game (Hardin and Remis, 2006). Our longitudinal study suggests that initial adaptive responses to such changes may permit certain animal populations, including gorillas and elephants, to maintain themselves, at least initially. But it also suggests that if such diverse regimes of forest use persist, animal populations encounter a threshold beyond which they plummet, in effect 'stranding' those humans who depend upon them as economic and cultural keystone resources.

In 1997, we visited another, more remote, Buffer Zone site, on the Park border. Kongana is permanently staffed by conservation guards but their presence has not prevented others from seeking perceived opportunities provided by protection in this sector. During our stay, camp life was intense: 38 individuals worked together, smoking up to 20 animal carcasses at one time and staying awake almost through the night to finish baskets for the transport of meat to market. Despite the mixed strategies of exploitation and protection co-occuring at this park border site (KG), human sign on transects is relatively rare. Wildlife populations remain more intact here than other study sectors closer to town (BB, MB, YD), but even here gorilla nest numbers and group sizes were reduced in 2002 and 2005 relative to previous years (Figure 5.3; Remis, 2000; Remis *et al.*, 2006).

Remote Park Sector at RDS

To the south of Bayanga is a fascinating stretch of the Sangha River, dotted with logging towns that boom, then bust, leaving behind the skeletal remains of hangars, mills, and villas. They also leave behind a small but fierce breed of African merchants who run small commercial stores, as well as running diamonds, ivory, and other resources from the inner forests out through their merchant networks to larger towns such as Douala. This area of the Sangha, which flows from southern CAR into northern Congo Brazzaville, is ethnographically and ecologically under-documented, but is becoming crucial from a conservation perspective. Because this area has long been 'the end of the road' where northern and southern trade networks do not quite meet, it has long been a haven for animals and for particular groups of people. For instance, groups of 'Sangha Sangha' fishermen and horticulturalists, who seek refuge from government and other controls on their activities in order to live by their own views of subsistence, call the area their own. Other inhabitants include former combatants in the Congolese opposition, who seek to remain under the radar of the current government.

Elephant abundance had been low in the southern Ndoki sector during a 1986 census, largely attributed at that time to high numbers of poachers in the area (Fay, 1989, 1991). Elephant protection has since increased in the Remote Park area of RDS since the original survey, with the ivory ban and creation of the Tri-National protected areas in this border region of Cameroon, Congo, and CAR (Blake, 2005). Now, elephants are more numerous in this southern sector, while their numbers have declined dramatically in community hunting zones close and far from Bayanga (Figure 5.3). Nevertheless, during our census work in this sector in 2005 we encountered numerous hunting camps and abandoned elephant carcasses whose tusks had been removed, suggesting the rebound in their numbers will be short-lived (Remis *et al.*, 2006). Gorillas are also apparently targeted by hunters in this area. Fay (1989) found the number of gorilla nests per kilometre in Ndoki sector in 1986 to be 3.1. Average group size at nest sites was 3.9. In 2005, in this sector we found only 2.8 gorilla nests per kilometre, with a reduced average group size of only 1.4.

Remote Buffer Zone: intersection of wildlife, diamonds, and safari hunting

Much about Dzanga-Sangha's real economy remains impossible to quantify, or even to accurately estimate. An ethnographic perspective can be extremely useful for discerning networks and commodity flows about which people are unlikely to respond in a survey format. For instance, let us take the widow of a former sawmill worker who settled in the region several years ago and was involved in a national trial in 1995. She had been discovered hoarding nearly 80 elephant tusks in the latrine behind her house. She claims rights to contraband local forest resources due to her 'precarious' economic condition, her gender, her Central African nationality, and her long residence in the region. She avidly manipulates non-local kin networks for transport or commercial aid via government and logging company connections, and is a widely acknowledged master at network management.

Africans from further north (Chad, Mauritania, Senegal) also get along well with the widow in question. They are, as she is, part of a group of African 'expatriates' in the region. They participate in public transport networks that span the country of CAR, thereby also enabling the lucrative flows of illegal animal products, diamonds, and arms and ammunition across international borders (CITES report, 2005). Their trade presents both opportunities and threats to the fabric of Bayanga's social life, dependent as it is on a rich and varied resource base that these African outsiders, like so many Europeans, are using for private profit.

The diamond trade that has flourished to the north of Bayanga represents intricate interpenetrations of regionally rooted communities. As with tourism, it would not have been able to flourish as easily without the road networks and light plane traffic that the base economy of logging has provided. The settlement of many BaAka on mission stations immediately to the north of the Reserve area has also facilitated labour and a ready supply of agricultural produce and game meat for alluvial diamond fields.

During our census work in this Remote Buffer Zone sector we surveyed areas adjacent to the safari hunting concession at the Libwe clearing and the location of a diamond hunting village which predates the creation of the Reserve but has now grown to contain perhaps more than 500 people. During our 2005 census, abandoned elephant trails, low elephant dung abundance, low gorilla nest abundance, and small gorilla group sizes all confirm the very high levels of extraction here (BY and NZ sites, Figure 5.3), and low levels of anti-poaching patrols or sanctions, despite its status within the Reserve borders. Further confirming a long-term pattern of extreme game scarcity distinct from other RDS sectors, we found evidence that hunters here have diversified their prey species; there were rat traps on transects, and hornbill and bat carcasses on smoking racks. Hunters who tried to hold our research team hostage (mistaking them for Park guards) also bemoaned the scarcity of large mammalian prey.

Toward an anthropology of protected area management

Over the past two decades RDS has been managed as an integrated conservation and development project (ICDP) with multiple-use zones. Yet our results suggest that conservation and development are not, in fact, integrated, but actually compete here within separate zones within the reserve, as though parallel logging and conservation concessions. This experiment in co-existence yields interesting results that can provide insight for protected area management in Africa and elsewhere.

Management challenges at RDS include growing in-migration due to ongoing logging activities and a growing conservation and research sector; and flourishing, largely illicit commercial trades that span the Reserve area, connecting it to broader national and regional economic circuits for the exchange of meat, arms and ammunition, and diamonds (TRAFFIC, 2005). To better understand and respond to these major management challenges, conservation initiatives must integrate more comprehensive monitoring and assessment of complex human and animal populations. Anthropological approaches can focus such monitoring on changes in behaviour by animals and people within protected areas, but also on cultural, political, economic, and ecological trends or shifts that can structure their behaviour.

A multi-factored approach to ecological, social, and demographic influences on wildlife behaviour illustrates how these micro-level factors articulate with management forms over time. This type of fine-grained analysis reminds us to be mindful of the importance of the continued viability of wildlife populations, even as we pursue management experiments. Although our 1997 wildlife surveys suggested gorilla populations were initially holding their own in areas of high human activity in the RDS Reserve sectors, more recent wildlife surveys there suggest declines in gorillas and other species. Elephants, in particular, have declined dramatically at RDS, from an average of 8.15 dung per kilometre along pre-1990 transects (Carroll, 1988; Fay, 1989) to a mean of only 1.39 dung per kilometre in 2005 (see Remis and Hardin, in prep.). These and other studies further emphasize indications of hunting sign and vulnerability for most wildlife even within the Core Park Sectors at RDS (Blom, 2005; Blake, 2005; Hardin and Remis, 2006).

Implications for specific wildlife species

Another clear contradiction stands out: that between local perceptions of the protected area as a reserve of prey for difficult economic times, and the perceptions of non-local managers and leisure users of the protected area as a space for conservation of global patrimony toward education and experiential enrichment of many (Kellert, 1995; also Sicotte and Uwengeli, 2002). These larger perceptual issues can be linked analytically to behavioural and density issues, especially with respect to elephants. When evaluating the effectiveness of conservation programmes, increases in elephant vulnerability to hunting within the core of protected areas are particularly troubling. Recent MIKE (Monitoring the Illegal Killing of Elephants) surveys detected numerous poached elephant carcasses and poachers camps in the heart of the Park sector around the Dzanga saline where elephant abundance is the highest (Blake, 2005). Further, reports of increased nervousness when humans are scented, and aggressiveness of elephants in the Core Park Sector at RDS, also suggest that elephant vulnerability has increased. The Elephant Trade Information System documents that after an initial decline international trade in ivory has increased from 1995 onwards and signalled Cameroon and DRC as two of the most important players in the internal illicit trade in ivory. Southern CAR is implicated as a major source country for Cameroon (Blake, 2005; TRAFFIC no. 23 February 2005, accessed at www.traffic.org/dispatches/DispNo23.pdf).

Even researchers with long experience in perceiving and responding to elephants have been vulnerable to persistent pursuit or attack at RDS this past year. In the worst of a series of recent elephant charges, a student gorilla researcher was seriously injured at the Bai Hokou tourist site (Carroll, pers. comm.). This also heightens concern for tourists in the region, who must walk two kilometres along elephant paths through 'elephant infested' forest (as Delia Akeley once said) to reach the safety of the Dzanga saline tourist viewing platform. In order for ecotourism to be wisely or successfully implemented, basic safety must be assured for wildlife, park staff, and tourists alike through effective conservation measures.

At sites in the Buffer Zone, wire snare and firearm hunting co-exists with longer-term net hunting, and plant or honey gathering activities, albeit in varying proportions. Chimpanzee nests and nest-sites are rare, possibly related to habitat structure, hunting or logging, and road infrastructure-related loss of large canopy trees in the area (Remis, 2000). Our long-term data indicate that the accuracy of line-transect censuses and resulting gorilla densities in this and other studies are influenced by within- and between-site variation in gorilla nest construction patterns (Remis, 1993). Initially, in 1997 gorillas did not appear to be primary targets of hunters within the study sectors, though they were reported to be targets in more remote areas inside and beyond the Reserve borders. Indeed, our average numbers of gorilla nests per kilometre for the RDS Reserve were markedly similar to those recorded by Carroll in a 1985 survey (1.8 gorilla nests per kilometre, Carroll, 1986).

Thus, the 1997 census data provide a snapshot of potential ability of gorillas to subsist, at least temporarily, in a multiple-use zone, alongside light levels of logging, which temporarily increases the availability of herbaceous foraging material. Perhaps gorillas are especially likely to co-exist with humans where research and tourism efforts focus awareness on, and increase the value of, this species (see Weber and Vedder, 2001; Sicotte and Uwengale, 2002). Nevertheless, the subsequent 2002 and 2005 census data show that gorillas' presence and mean nest group size in most sectors appear dramatically reduced relative to the earlier surveys. The abundance of gorilla nests per kilometre of transect fell in all sectors (Core Park Sites: pre-1992 = 2.5, 2005 = 0.7; Remote Reserve Sectors: pre-1992 = 1.6, 2005 = 0.2; Remote Ndoki Park Sector: pre-1992 = 3.1, 2005 = 1.4). Further, in the mid-1980s, gorilla nest size averaged more than four adult-sized nests but had fallen to just over one in the Remote Reserve Sectors in 2005 (Carroll, 1986, 1988; Fay, 1989; Fay and Agnana, 1991; Remis et al., 2006; Remis and Hardin, in prep.).

The recent data show that human activity has increased in the remote sectors, and is no longer concentrated close to villages. Reduced game densities apparently require hunters to travel longer distances in search of more productive forest zones. BaAka hunters in 2002 and 2005 noted that they were hunting within the Park borders because of higher animal abundance in these less exploited zones. The Park sites we surveyed revealed different patterns of human use. The Kongana Park Sector was regularly patrolled by Park guards in 2002. Nevertheless, heavy hunting pressure by commercial Bilo hunters at other sites (at the Messapoula headwaters Park site, and in the southern remote Ndoki Park sector) were apparently already impacting animal densities there (especially the main prey of duikers), despite protection status and relative inaccessibility of these sectors. The Ndoki site appeared to be, once again, a commercial centre for meat and ivory hunting, much of which is likely exported through Cameroon (TRAFFIC, 2005). These cross-sector insights about key species, as they are interacting with a range of human uses of the area, also prompt us to reflect on several management issues or approaches.

Integrated conservation and development projects

There is often a divide between the conservation community with a natural science ecosystem perspective and social science researchers championing human and property rights. Current conservation debates point to the need for an alternative to the guards with guns approach to preservationist conservation efforts . . . at the same time, the integrated conservation and development approach (ICDP) has faltered without providing adequate levels of success on either conservation or development fronts (see research reviewed in Hughes and Flintan, 2001; Salafsky and Margoluis, 2004). The ICDP model's baseline assumptions about links between development and conservation objectives remain unproved, largely due to imperfect monitoring and evaluation efforts (Barrett and Arcese, 1995).[3] Current conservation policy emphasizes the continued need to jointly consider human needs and conservation, but to come up with new methods for more successful implementation and realistic expectations (Wells and McShane, 2004). Although some projects have added ecological or sociological monitoring to the lists of things that they do (Kreman et al., 1994), at RDS there seems to be little analysis of those data or direct flow of information between assessments and decisions about practice.

Conservationists (within RDS and beyond) throw up their arms about how difficult it is to run an effective ICDP conservation project in Bayanga because of the nature of the complex socio-economic–political context. Indeed, political instability and remoteness of the Reserve have made it difficult to increase ecotourism revenues (Blom, 2000). Further challenges include a large proportion of immigrants in the growing population, unrealistic expectations about compensations and benefits that should be dispersed to stakeholders and interest groups, general unwillingness to comply with forest regulations, and ease of access to forest resources (loggers, trucks, and a grid of logging roads maintained by both logging and conservation).

The ICDP model has been criticized as ineffective for conservation, and as embodying an incompatibility between development and conservation goals (Oates, 1999). Proponents of the approach, however, note that it has never been evaluated on a more long-term basis. Rather, due to the political economy of project funding, much debate is centred around superficial reviews of ICDPs across contexts or with fairly shallow time-depth. At the same time, the politics of more stringently protected protected areas, with their resultant displacement of human populations, are growing more and more volatile despite their long histories (Anderson and Grove, 1987; Schmidt-Soltau, 2003). If it is possible to render the ICDP model more effective, then the very terms on which ICDPs are formulated as 'projects' must, we argue, be adjusted to reflect their status as complex, intercultural, and cross-species experiments. The lens of anthropology with its combined biological and cultural approaches provides a unique standpoint from which to launch this effort.

Our results show clearly that the integration of conservation activities into the Bayanga area has caused different distributions of constraint and opportunity among even the sub-set of Reserve residents known as BaAka. While new sources of logging or conservation revenue provide new options for some residents, many have mixed feelings about the fact that the national park area is currently off limits for all activities save research, tourism, and education, although hunting is permitted in other sectors. Because of its reliance on distinct zones, the ICDP approach has weakened the BaAka community from Messapoula in terms of their capacity to spend time in the forest in ways described for residents of the Mabongo sector. And yet, that relative cultural and economic impoverishment does seem to coincide with greater faunal diversity, and the cultivation of expertise about that fauna, among a select few in the Messapoula community. We shall further consider the problem of zones below.

Zoning: distinctions between buffer and core; Reserve and Park

One central irony did emerge from our reflection and analysis. Tourism or research in a place such as Bai Hokou could not be carried out had a logging company not created the road that leads there from town. Given the legacy of roads from logging, intensive tourism and research could be the only alternatives to the Buffer Zone scenario of apparently increasing pressure and wildlife scarcity, even in the fairly distant camps where pedestrian traffic for forest use is limited.

Research camps in the core Dzanga Park sector are unique examples of the ways in which new activities affect power relations and human–animal interaction. As distant from town as other Core Park Sites, but more integrally protected, the harvesting of plants and animals at Bai Hokou is minimal, but not non-existent. In fact, illegal activity appears to be on the rise in the Park sector as Reserve residents recognize its high game densities, as punitive measures taken against Park encroachers weaken, and as human population density in the surrounding area increases due to upswings in economic activity.

The comparison of the different sectors at RDS provides insights into the difficulty of safeguarding both cultural and biological biodiversity in this area, and calls for reconsideration of the literature on hunting and tropical forest conservation. Obviously, as our findings confirm, multiple-use buffer zones are locations for a vast array of human activity, both legal and illegal, whose intensity can increase in proximity to towns. But it is important to note the longer-term trajectory noted here whereby the radius of exploitation reaches beyond the buffer zone into the core. Research that seeks patterns for regional-level planning purposes, asserting proximity to roads and/or population pressure as the fundamental factors determining hunting pressure, may miss the mosaic and longitudinal effects we are describing here for both humans and animals – gorillas and elephants in particular

(Fitzgibbon et al., 1995; also Laurance et al., 2006). Consideration of various legal, political, infrastructural, and ecological factors at the micro-regional level might help explain significant differences between sectors within a single protected area, and should not be overlooked in analysis of change or attempts at management.

As is true for people, impacts, reactions, and outcomes of wildlife can vary widely within the protected area. Where hunting is most intense, small ungulates (duikers) are extremely scarce or absent and monkeys are present but cryptic. Hunting and other human activities impact elephant ranging patterns and abundance at RDS despite the protected status of elephants (Barnes et al., 1991). Gorillas are likewise beginning to show declines at RDS, with group sizes and numbers of nests per kilometre now lower than before the inception of the protected area. In fact, by 2005, distance from town had broken down as a major factor explaining spatial patterning of wildlife at RDS, alongside a diversification of hunting locations but decrease in the variety of resource use and ritual practices by particular human communities. When more commercial hunting combines with intensive subsistence-level practices, the resultant pressure seems likely to render the long-term practice of either type of hunting unsustainable within Reserve sectors.

In these diverse efforts to make ends meet, all come up against newly created (if weakly enforced) borders between 'sectors' of the protected area, as well as against newly reinforced and patrolled national borders. This broader study demonstrates that if the census data had been analysed independently of (1) long-term data collection on climate, fruiting patterns, or gorilla socio-ecology in the region, and (2) cultural anthropological data on family histories, gender roles, and property relations, interpretations would have run the risk of being both inaccurate and overly simplistic (Hardin and Remis, 2006). We might, for instance, have failed to recognize the impacts of variations in microhabitat use by large mammals, or local variation in gorilla nesting patterns in interpreting gorilla census data. We might have simply considered BaAka forest foragers and Bilo farmers/hunters as 'indigenous people', rather than examining various sub-groups as they relate to market shifts, technological innovations, formal policies, and micro-regional ecological variations.

At the planning stages, such complex thresholds and adaptations were not considered. Rather, the reasons why the 'community hunting' area of the Reserve zone was located in the southern, rather than the northern part of the Park was dictated because of the location of valuable ecological resources for tourism at the Dzanga clearing. This has created one BaAka community more involved with logging and commercial hunting; the other with conservation and research. And now we begin to see mobility in more permanent ways across these communities.

In terms of reducing negative impacts of logging on wildlife, comparative fine-grained data provide some suggestions for more ecologically minded extraction practices. For example, chimpanzees fared better at Budongo, Uganda where logging occurred in small blocks than at Lope or RDS (Plumptre and Reynolds, 1994). Examination of fine-grained data on ape abundance in areas under different logging regimes suggest that careful planning of extraction could significantly reduce ecological damage caused by logging and other industries (also Tabarelli and Gascon, 2005).

Multidisciplinary impact studies, research-driven conservation policies for more sustainable use, and governmental monitoring of logging operations in protected areas, in combination, could all significantly further conservation of biodiversity in the Buffer Zones surrounding national parks including those at RDS. This sort of approach is captured by the turn toward adaptive management in much recent policy and practice.

Adaptive management

In the case of RDS, much of the necessary data already exist, but are often ignored or underutilized. As researchers and collaborators to our field-based colleagues, we have often encountered frustrations about continued and repeated claims from project staff that scientific reports were never filed. Others complain that copies no longer exist in project archives or libraries when national or expatriate researchers come to look for results of previous studies. We have personally encountered resistance to the dissemination and sharing of our research results at RDS despite our extensive efforts to file reports, present results, and engage with local conservation managers at the completion of each of our projects. During our 2005 visit, project staff charged with the daily nuts and bolts of protected area management asked us to again present the 2002 results, and only at that time began to engage with these now outdated results, no longer as useful for modifying conservation practice, given the escalating and continued deterioration of wildlife abundance in the Reserve.

At stake is neither the compliance, nor the conscience, of either researchers or project personnel here. Rather, it is the value accorded to particular kinds of knowledge in the challenging arenas of project management (which, of course, is often closely linked to information management). In this case, the low value of information impeded its circulation, and made for a clumsy and ineffectual interface between researchers and managers. In other cases, it is the high value of information that can impede its circulation and render such interfaces ineffectual. ICDPs were created with very little role for research, either scientific or social-science oriented. Many managers subsequently realized the need to integrate particular scientific methods into their monitoring practices, towards solving problems of implementations and unrealistic expectations.

Internationally, anthropology is increasingly being used to accomplish big business agendas to alter products, and to fine tune marketing strategies in accordance with knowledge about consumers. In the conservation world, where access to financial resources has always been a challenge, the possibility that basic research might trumpet flaws in an imperfect, ambitious, and rapidly changing process must be weighed against the possibility

---

Box 5.1 Specific ways in which the role of anthropological research might better inform policy in the RDS

- Planning where to target anti-poaching patrols should be systematically informed by continuous assessment of animal abundance and human activities in different sectors of the Reserve. This would reduce the influences of social pressures on decisions about where to patrol.
- Sustainable use guidelines should be modified in conjunction with local demographic and socio-economic information on patterns of use as they change over time.
- Setting quotas for safari hunting offtake should be informed by existing data on animal behaviour, populations, and use to avoid over-exploitation.
- Adapting wildlife regulations and permitted uses should be informed by information about community hunting, marketing, and trade patterns over time and space.
- Adapting regulations for extractive industry should be done in close association with independent monitoring and research on local forestry practices.
- Setting guidelines for community access to forest resources or compensations for limitations to access should be based on ethnographic data describing patterns of use, levels of dependence, and local understandings that could be integrated into guidelines for developing feasible and culturally relevant alternatives to overuse.

that such insights might provide valuable information for the adjustment of policies and planning processes.

In the case of the RDS, a 'Comité de la Recherche Scientifique' was formed during the early 1990s, in order to enable project staff and active researchers to review and accept or reject incoming proposals for scientific research. This process was intended to address serious concerns on behalf of several broader communities: project staff who simply could not provide logistical support for an infinite number of researchers; residents of the Reserve who had experienced intrusive or culturally inappropriate research methods as objectionable; and, finally, researchers themselves, who might seek input from those familiar with the field setting. The committee, however, had fairly high turnover rates, and occasionally encountered conflicts of interest between researchers or professionals currently in place and those who proposed to come. There was little or no protocol for such issues to be addressed and, ultimately, the unwieldy process was unable to assert itself as a definitive mechanism for management and integration of research and conservation.

Capacity building

The practice of science as power is intricately bound up with the colonial and postcolonial histories of many protected areas, and cannot be ignored (Coquery-Vidrovitch, 1998). Certain inequitable relations of access are embedded in the very basic assumptions of formal research, as well as in the rules and codes that serve to govern many protected areas. Such structural inequalities must be considered in relation to capacity building and the politics of knowledge in the area. Conflicts between conservationists and local communities have been increasing in recent years (see Giles-Vernick, 1999). During our field research in the mid-1990s some mornings in the town of Bayanga saw groups of angry women marching to the front door of conservation offices with the crushed remnants of their corn or manioc plants, complaining about the nightly ravages of elephants in their fields. As we have discussed elsewhere, hunter/gatherers and women are two groups who are likely to suffer most from the alienation of forest resources for tourism and research.

And yet, in part due to the exigencies of field research, our experience in field camps with the cultural/health/social benefits of family participation in research camp life introduced us to a different kind of model for forest management and research than the male-dominated, military-style research camps common throughout the region (see Hardin and Remis, 2006). While a set of clearly articulated feminist reflections on conservation practice has not yet emerged (and is most urgently needed), we can safely note that the combination of cultural and biological anthropology offers remarkable possibilities for the identification of alternatives to dominant and increasingly standardized modes of monitoring, surveillance, and discipline within protected areas, enhancing some of the more socially progressive elements of conservation practice.

The anthropology of indigenous knowledge has, itself, become more reflective of the transnational and inter-group dynamics of knowledge construction in these complex and natural-resource rich sites (Agrawal, 1995). For conservationists to effectively recognize the contributions of local experts – both those who are formally trained and those who are not – remains a challenge. In some cases, recognizing those contributions can also craft changes in them. The project has been a force for social change, and its hiring practices reflect that. Perhaps as an expansion of their traditional roles as large-game hunters, a male-centred professionalism is emerging around conservation-related encounters with wildlife in this part of the core protected area at RDS. The BaAka research trackers impart information to novice expatriates, Bilo researchers, and younger BaAka during camp life, also expanding their own knowledge bases and skills. This contrasts markedly with their roles as porters and trackers for patrolling Bilo Park guards, about which many BaAka complain bitterly due to the lack of respect they are accorded in that context. Researchers, now, often hire BaAka on a two-weeks-on and two-weeks-off schedule, in an effort to accommodate their other subsistence activities. In addition, the Reserve administration has recently hired its first two female guards.

Terrain Dimali, BaAka tracker who studied literacy and guide skills with Hardin's research team in 1992, 1995, and 1997, has become the first from his social group of hunters and gatherers to be a formally employed Reserve guide within the RDS. Sylvain Dangholo, a child of regional ethnic groups known for their wandering fishing practices, completed a Masters degree in Geography at the University of Bangui, collaborating with Hardin on her dissertation research, and is now employed in the Rural Development branch of the RDS, implementing policies where his Master's thesis critically engaged them. On a more ecological front, Bruno Bokoto DeSombeli, a former member of the Remis research team during his university student days, is now responsible for ecological monitoring at RDS. In all these ways, and more, we have seen anthropological training work hand in hand with more straightforward conservation monitoring and marketing practices, building local participants' abilities and experience in long-term qualitative and quantitative research, and thereby strengthening the capacity of local actors and institutions to respond to changing conditions by blending old and new skills and perspectives on the forests they inhabit.[4]

And yet there is a long way to travel with respect to these issues, and to the conflicts currently coming to a head around them. Conservation management practice results in alienation of people from their resource base with only mixed success at preserving that resource base. As a result, women turn to agriculture and other economic activities which pose new management challenges and have negative consequences for conservation. They extend their territory into areas where new economies promise profitable livelihoods but endanger the sustainability of the animal populations upon which they depend.

The anthropology of protected area management is just now experiencing its first flush of full-length monographs (see, for example, Walley, 2004; West, 2006). Other social science approaches that emerge from geography, history, or political science are in conversation with anthropology about the complexities of current alienation and reconfiguration in relationships of resource access in rural worlds (see, for example, Giles-Vernick, 2002; Agrawal, 2005; Moore, 2005). An informed anthropological approach to perceptions of alienation from resources recognizes human dependence on those animal populations being conserved and a need for a middle-ground approach. An environmental anthropological approach to conservation problems facing the RDS in the ICDP case study we describe leads us to: (1) consider the ecosystem consequences of human activities and how to better approach the interdependence of humans and animal populations in a protected area; and (2) integrate and enable women's participation in conservation-related employment and benefits.

Cross-cultural keystones

Garibaldi and Turner (2004a) introduce the concept of cultural keystones as those species whose removal alters the structure of a community, not only ecologically but economically, or in other ways. Not only are the subsistence foundations of many human groups directly tied to the continued availability and use of particular plant or animal species (Cristancho and Vining, 2004), but examining the linkages between the role of particular species in cultural and ecological systems emphasizes the interdependence of animal and human communities (Berkes and Folke, 1998; Dove, 2001; Berkes *et al.*, 2003; also Wolch and Emel, 1998). Power *et al.* (1996) advocate that recognition of cultural keystones might help strengthen social systems, leading to improved efforts to help maintain their ecological integrity.

We are also interested in the role that particular species play in the cultural politics and practices of identity, not only at the level of clan or tribe, but also at the level of projects, corporations, and the curiosity of travellers or investors. We have seen our informants and collaborators manipulate and embrace such cross-culturally significant symbols of their forest homes, such as gorillas or elephants.

To our mind, then, keystones exemplify a management concept (Simberloff, 1998) currently in vogue that can be invigorated and improved through engagement with anthropology, making its use less evocative of the need for some people to be managed by others. Anthropology is uniquely positioned to infuse such considerations into the current and

Figure 5.4 Photograph of tracker Moyekoli, who becomes a gorilla in dance and song. Not fully visible in this photo is Ellemo who has become a gorilla researcher, pretending to juggle stopwatch, binoculars, notebook, and removal of sweat bees from under her contact lenses (photo: Melissa Remis)

effervescent debates around issues of cultural keystones. Gorillas are prime candidates for consideration not only as ecological and cultural keystones *within* the context of the western Congo basin (Lewis, 2002) but beyond with the remaking of King Kong, a metaphor for cross-cultural contact, for human difference, fear of the other, and villager/pygmy divide (stories about gorillas abducting villagers and taking them back to the forest) (Haraway, 1989; Browne, 2006).

Gorillas are thus enormously important in the tourist imaginary about RDS (Hardin, 2000). At the same time, traditional BaAka stories, songs, and dance about gorillas and chimpanzees are important forms of transmission of forest lore and knowledge to younger generations. They continue to be central even as BaAka transform their perception of, and relationships to, these animals, and their folktales and coming-of-age stories focus on animal viewing rather than hunting experiences. But now as gorillas and other key wildlife decline at RDS, at forest research camps gorilla folktales are often replaced by heated disgruntled discussions of conservation policies and wildlife scarcity in communal hunting zones. In nearby Equatorial Guinea, those displaced from direct contact with gorillas in protected areas attributed their increased hunting of gorillas to their displacement outside the protected forest area, which reduced their concerns about nightly spiritual retribution by killed gorillas (Schmidt-Soltau, 2003).

We argue that investigating the possibilities of capitalizing on the cultural importance of gorillas holds great potential, for mobilizing it in ways that would be useful in the management of protected areas, and getting locals on board with conservation goals. Anthropology is uniquely qualified to analyse the human–primate relationship in light of both changing economic and ecological relationships and the complex cultural meanings of gorillas in colonial and postcolonial contexts.

Conclusion

The conditions under which conservation regimes confer tenuous and temporary protection to particularly prominent (or mediatized) animals raise an important set of methodological questions for further research on complex human/wildlife interactions. We have written this chapter within the framework of protected area management despite the fact that much recent attention, particularly to primate populations, suggests that protected areas may not be the solution to their survival (e.g. Walsh *et al*., 2003).

Pressures on wildlife and other natural resources in the CAR have escalated in the past ten years during a period of particular economic and political instability, including lack of wages paid to civil servants. At odds with Western notions of the recreational and aesthetic values and uses of protected areas (Kellert, 1995), most community members interviewed at RDS have regarded allowing wildlife stocks to multiply to ensure future use as the primary incentive for creation and continued protection of RDS (Carroll, 1997; Remis field notes; also Sicotte and Uwengeli, 2002). For many, with few alternatives in the country as a whole, that day has arrived. The challenge of how to ensure more sustainable use and long-term preservation of these wildlife stocks remains unmet.

Rapid assessments can complement, but must not replace finer-grained monitoring over time. Our work speaks to the current debates about how to move conservation science forward to provide the foundations needed to simultaneously conserve biological and cultural diversity and implement conservation programmes that benefit traditional people (Sanderson and Redford, 2003; Robinson, 2006; Wilkie *et al*., 2006).[5] Our combined analyses highlight the complex ways resources are used over time, on micro-regional levels. Because what initially looks like species abundance could be seasonal or otherwise temporary, we need to have a long-term window into behavioural changes and impacts on the dynamics of animal populations. Our work points to the unique contribution of a reflexive understanding of the complexity of the animal/human dynamic. We emphasize the need to think beyond management categories often projected onto the landscape, instead focusing on how particular areas are truly utilized by people and animals. Our work provides snapshots that look beyond either animals or humans as individual actors, and attempts to discern broader patterns that reflect ecological and social change, and reveal differences between stated attitudes or preferences, and practice.

Notes

1 For a more detailed historical presentation of logging and conservation-related activities in the region, see Hardin and Remis, 2006.
2 WWF is the World Wildlife Fund; GTZ is a German development agency, Gemeinschaft für technische Zusammenarbeit.
3 Some researchers have called attention to the ICDP's lack of incorporation of adequate scientific research and tendency to rely on 'rapid' appraisals (Newmark and Hough, 2000). Further, few ICDPs have attempted to link biodiversity with socio-economic monitoring during project implementation, preventing analysis of, for example, whether alternative income generation has resulted in more sustainable management of biodiversity resources (Hughes and Flintan, 2001).
4 See Hardin and Remis, 1997. Available at www.yale.edu/sangha on the 'conferences' page (last accessed 1 June 2006).
5 For further resources on these issues, see the Anthropology and Environment section of the AAA at www.eanth.org/onlineresources2.php; the Society for Applied Anthropology also has a page at www.sfaa.net/eap/abouteap.html; and the University of Georgia is building an excellent programme of study and research on these precise issues (see www.uga.edu/eea). All sites accessed 1 June 2006.

Literature cited

Agrawal, Arun (1995). Dismantling the divide between indigenous and western knowledge. *Development and Change* 26 (3), 413–439.
—— (2005). *Environmentality: technologies of government and the making of subjects*. Durham, NC: Duke University Press.
Alpert, Peter (1996). Integrated Conservation and Development Projects: examples from Africa. *BioScience*, 46 (11), 845–855.
Anderson, David and Grove, Richard (1987). *Conservation in Africa: people, policies and practice*. Cambridge: Cambridge University Press.
Auzel, Philippe and Wilkie, David (2000). Wildlife use in Northern Congo: hunting in a commercial logging concession. In John G. Robinson and Elizabeth Bennett (eds), *Hunting for Sustainability in Tropical Forests*, pp. 413–426. New York: Columbia University Press.
Barnes, Richard F.W. (2002). The bushmeat boom and bust in West and Central Africa. *Oryx* 36, 236–242.
——, Barnes, K.L., Alers, M., and Blom, A. (1991). Man determines the distribution of elephants in the rain forests of northeastern Gabon. *African Journal of Ecology* 29, 54–63.
Barrett, C.B. and Arcese, P. (1995). Are Integrated Conservation-Development Projects (ICDPs) sustainable? On the conservation of large mammals in sub-Saharan Africa. *World Development* 23 (7), 1073–1084.
Berkes, F. and Folke, C. (eds) (1998). *Linking Social and Ecological Systems: management practices and social mechanisms for building resilience*. New York: Cambridge University Press.
——, Colding, J., and Folke, C. (2000). Rediscovery of traditional ecological knowledge as adaptive management. *Ecological Applications* 10, 1251–1262.
——, ——, and —— (eds) (2003). *Navigating Social-ecological Systems: building resilience for complexity and change*. Cambridge: Cambridge University Press.
Blake, S. (2005). *Long term system for monitoring the illegal killing of elephants (MIKE). Central African Forests*. Final Report on Population Surveys 2003–2004. A report by the Wildlife Conservation Society, New York. Last accessed 9 June 2006 at www.cites.org/common/prog/mike/survey/central_africa_survey03-04.pdf.
—— and Hedges, S. (2004, October). Sinking the flagship: the case of forest elephants in Asia and Africa. *Conservation Biology* 18 (5), 1191–1202.

Blom, Allard (2000). The monetary impact of tourism on protected area management and the local economy in Dzanga-Sangha (Central African Republic). *Journal of Sustainable Tourism* 8 (3), 175–189.

—— (2005). Factors influencing the distribution of large mammals within a protected central African forest. *Oryx* 39 (4), 381–388.

——, Cipolletta, Chloe, Brunsting, Arend M.H., and Prins, Herbert H.T. (2004a). Behavioral responses of gorillas to habituation in the Dzanga-Ndoki National Park, Central African Republic. *International Journal of Primatology* 25 (1), 179–196.

——, van Zalinge, Robert, Mbea, Eugene, Heitkonig, Ignas M.A., and Prins, Herbert H.T. (2004b). Human impact on wildlife populations within a protected Central African forest. *African Journal of Ecology* 42, 23–31.

Boserup, Ester (1989). Population, the status of women, and rural development. 'Rural Development and Population: Institutions and Policy', issue supplement. *Population and Development Review* 15, 45–60.

Bowen-Jones, E.S.P. (1999). The threat to primates and other mammals from the bushmeat trade in Africa, and how this threat could be diminished. *Oryx* 33, 233–246.

Brosius, J. Peter (1999). Analyses and interventions: anthropological engagements with environmentalism. *Current Anthropology* 40 (3), 277–309.

—— (2006). Common ground between anthropology and conservation biology. *Conservation Biology* 28 (3), 683–685.

Browne, Janet (2006). Science in culture: a bigger picture of apes. *Nature* 439, 142 (12 January 2006) doi: 10.1038/439142a.

Butinski, Thomas M. and Kalina, Jan (1998). Gorilla tourism: a critical look. In E. J. Milner-Gulland and R. Mace (eds), *Conservation of Biological Resources*, pp. 294–313. Oxford: Blackwell Science.

Carroll, Richard W. (1986). *The status, distribution and density of the lowland gorilla (Gorilla gorilla gorilla), forest elephant (Loxodonta africana cyclotis) and associated dense forest fauna in southwestern Central African Republic: research towards the establishment of a reserve for their protection*. Unpublished MS, Yale School of Forestry and Environmental Studies.

—— (1988). Elephants of the Dzanga-Sangha dense forest of south-western Central African Republic. *Pachyderm* 10, 12–15.

—— (1997). Feeding ecology of lowland gorillas in the Dzanga-Sangha Dense Forest Reserve of the Central African Republic. Ph.D. Dissertation. School of Forestry, Yale University.

Chapman, Coline, Balcolmb, S.R., Gillespie, T.R. et al. (2000). Long-term effects of logging on African primate communities: a 28 year comparison from Kibale National Park, Uganda. *Conservation Biology* 14, 207–217.

CITES (2005). Taking stock: experiences of elephant ivory and rhino horn stockpile management in Africa. *CITES World* 16, 13–15. Available online at www.cites.org/eng/news/world/16.pdf. Accessed 2 July 2007.

Coquery-Vidrovitch, Catherine (1998). The Upper Sangha in the time of concession companies. In H.E. Eves, R. Hardin, and S. Rupp (eds), *Resource Use in the Trinational Sangha River Region, Equatorial Africa (Cameroon, Central African Republic, Congo): histories, knowledge systems, institutions*. Forestry and Environmental Studies Bulletin Series, 102. New Haven, CT: Yale University.

Cristancho, Sergio and Vining, Joanne (2004). Culturally defined keystone species. *Human Ecology Review* 11 (2), 153–164.

Daspit, Leslie and Remis, Melissa J. (2004). *Primate Conservation and the Intersection of the Bushmeat Trade*. Unpublished report to World Wildlife Fund-US and The Department of Sociology and Anthropology, Purdue University.

Dove, Michael R. (2001). Interdisciplinary borrowing in environmental anthropology and the critique of modern science. In C.L. Crumley (ed.), *New Directions in Anthropology and Environment: intersections*, pp. 90–110. Walnut Creek, CA: Altamira Press.

Eves, Heather E. and Ruggiero, Richard G. (2000). Socio-economics and the sustainability of hunting in the forests of Northern Congo (Brazzaville). In J.G. Robinson and E. Bennett (eds), *Hunting for Sustainability in Tropical Forests*, pp. 427–454. New York: Columbia University Press.

Fairhead, J. and Leach, M. (1996). *Misreading the African Landscape: society and ecology in a forest-savannah mosaic*. Cambridge: Cambridge University Press.

Fay, J. Michael (1989). Partial completion of a census of *Gorilla gorilla gorilla* in Southwestern Central African Republic. *Mammalia* 53 (2), 203–214.

—— (1991). An elephant (*Loxodonta africana*) survey using dung counts in the forests of the Central African Republic. *Journal of Tropical Ecology* 7, 25–36.

—— and Agnana, M. (1991). A population survey of forest elephants (*Loxodonta africana cyclotis*) in northern Congo. *African Journal of Ecology.* Nairobi 29 (3), 177–187.

Fitzgibbon, C.D., Mogaka, H., and Fanshawe, J.H. (1995). Subsistence hunting in Arabuko-Sokoke Forest, Kenya and its effects on mammal populations. *Conservation Biology* 9, 1116–1126.

Fuentes, Agustin and Wolfe, Linda D. (2002). *Primates Face to Face: conservation implications of human and nonhuman primate interconnections*. New York: Cambridge University Press.

Garibaldi, A. and Turner, N. (2004a). Cultural keystone species: implications for ecological conservation and restoration. *Ecology and Society* 9 (3), 1 [online] URL: www.ecologyandsociety.org/vol9/iss3/art1.

—— and —— (2004b). The nature of culture and keystones. *Ecology and Society* 9 (3), r2. [online] URL: www.ecologyandsociety.org/vol9/iss3/resp2/.

Ghiurghi and Lakara (2002) Wild game hunting in the Dzanga Sangha Reserve. Unpublished report to GTZ/WWF.

Giles-Vernick, Tamara (1999). Leaving a person behind: history, personhood, and struggles over forest resources in the Sangha Basin of Equatorial Africa. *International Journal of African Historical Studies* 32, 311–338.

—— (2002). *Cutting the Vines of the Past: environmental histories of the Central African rain forest*. Charlottesville, VA: University of Virginia Press.

Haraway, Donna (1989). *Primate Visions: gender, race, and nature in the world of modern science*. New York: Routledge.

Hardin, Rebecca (2000). *Translating the Forest: tourism, trophy hunting, and the transformation of forest use in Southwestern Central African Republic (CAR)*. Ph.D. Dissertation, Department of Anthropology, Yale University.

—— and Remis, Melissa (1997). Research and rural development work sessions: Bayanga, RCA, 31 July–2 August 1997. *Report to Worldwide Fund for Nature-US and Central Africa Regional Program for the Environment (USAID)*. Electronic document, www.umich.edu/~infosrn/confer/CNF_SET.html (click 1997 in page frame); accessed 17 January 2005.

—— and —— (2006). Biological and cultural anthropology of a changing tropical forest: a fruitful collaboration across subfields. *American Anthropologist* 108 (2), 273–285.

Hughes, Ross and Flintan, Fional (2001). *Integrating Conservation and Development Experience: a review and bibliography of the ICDP literature*. London: International Institute for Environment and Development.

Kamiss, A. and Turkalo, A. (1999). *Summary: elephant crop raiding in the Dzanga-Sangha Reserve*. URL: www.iucn.org/afesg/hec/pdfs/hecdzren.pdf.

Kellert, Stephen R. (1995). *The Value of Life: biological diversity and human society*. Washington, DC: Island Press

Kottak, Conrad (1999). The new ecological anthropology. *American Anthropologist* 101 (1), 23–35.

Kreman, C., Merenlender, A.M., and Murphy, D. (1994). Ecological monitoring: a vital need for integrated conservation and development projects in the tropics. *Conservation Biology* 8 (2), 388–397.

Kretsinger, Anna and Hardin, R. (2003). Watersheds, weddings and workforces: migration, sentarization and social change among the BaAka of Southwestern Central African Republic. *African Studies Monographs* 28, 123–141.

Laurance, William (1995). Effects of logging on wildlife in the tropics. *Conservation Biology* 11, 311–312.

——, Croes, Barbara M., Tchignoumba, L., Lahm, Sally A., Alonso, Alfonso, et al. (2006). Impacts of roads and hunting on Central African rainforest mammals. *Conservation Biology* June online early. DOI 10.1111/j.523-1739.2006.00420.x. Accessed 14 June 2006.

Lewis, Jerome (2002). Chimpanzees and gorillas: ethnic stereotyping in the Ndoki Forest, Northern Congo-Brazzaville. 9th International Conference on Hunting and Gathering Societies, September 2002. Electronic document, www.abdn.ac.uk/chags9/1lewisJ.htm, accessed 17 September 2005.

Loudiyi, Dounia (1995). *Census of the Population of the Dzanga Sangha Dense Forest Reserve, Southwestern Central African Republic*. Washington, DC: World Wildlife Fund.
Malcolm, Jay and Ray, Justina C. (2000). Influence of timber extraction routes on Central African small mammal communities, forest structure and tree diversity. *Conservation Biology* 14, 1623–1638.
McShane, T.O. and Wells, M.P. (eds) (2004). *Getting Biodiversity Projects to Work: towards more effective conservation and development*. NY: Colombia University.
Mills, L.S., Soulé, M.E., and Doak, D.F. (1993). The keystone species concept in ecology and conservation. *BioScience* 43, 219–224.
Milton, K. (1993). Environmentalism and anthropology. In K. Milton (ed.), *Environmentalism: the view from anthropology*, pp. 1–17. New York: Routledge.
Moore, D. (2005). *Suffering for Territory: race, place and power in Zimbabwe*. Durham, NC: Duke University Press.
Newmark, W.D. and Hough, John L. (2000). Conserving wildlife in Africa: integrated conservation and development projects and beyond. *Bioscience* 50 (7), 585–592.
Noss, Andrew (2000). Aka of the Central African Republic. In W. Weber, L.J.T. White, A. Vedder, and L. Naughton-Treves (eds), *African Rain Forest Ecology and Conservation*, pp. 313–333. New Haven, CT: Yale University Press.
—— and Hewlett, Barry (2001). Contexts of female hunting in Central Africa. *American Anthropologist* 103 (4), 1024–1040.
Oates, John (1999). *Myth and Reality in the Rain Forest: how conservation strategies are failing in West Africa*. Berkeley, CA: University of California Press.
Orlove, Ben (1980). Ecological anthropology. *Annual Review of Anthropology* 9, 235–273.
—— and Brush, Stephen B. (1996). Anthropology and the conservation of biodiversity. *Annual Review of Anthropology* 25, 329–352.
Osaki, John and Abbaté, Mike (1995). *Tourism in Dzanga Sangha: an interpretation and visitor services plan*. Washington, DC: World Wildlife Fund.
Paterson, J.D. and Wallis, Jannette (eds) (2005). *Commensalism and Conflict: the human primate interface*. Special Topics in Primatology 4, American Society of Primatologists, 483 pages.
Peet, Richard and Watts, Michael (1996). *Liberation Ecologies: environment, development, social movements*. London: Routledge.
Peterson, Dale and Ammann, Karl (2004). *Eating Apes*. Berkeley, CA: University of California Press.
Peterson, J.T. (1978). *The Ecology of Social Boundaries: Agta foragers of the Philippines*. Urbana, IL: University of Illinois Press.
—— (1993). *An Anthropological Critique of Development*. New York: Routledge.
Pickering, Andrew (ed.) (1992). *Science as Practice and Culture*. Chicago, IL: University of Chicago Press.
Plumptre, A.J. and Reynolds, V. (1994). The effect of selective logging on the primate populations in the Budongo Forest Reserve, Uganda. *Journal of Applied Ecology* 31 (4), 631–641.
Power, M.E., Tilma, D., Estes, J.A., Menge, B.A., Bond, W.J., et al. (1996). Challenges in the quest for keystones. *BioScience* 46, 609–620.
Remis, Melissa J. (1993). Nesting behavior of lowland gorillas in the Dzanga-Sangha Reserve, Central African Republic: implications for population estimates and understandings of group dynamics. *Tropics* 2 (4), 245–256.
—— (1997). Western lowland gorillas as seasonal frugivores: use of variable resources. *American Journal of Primatology* 43, 87–109.
—— (2000). Preliminary assessment of the impacts of human activities on gorillas: *Gorilla gorilla gorilla* and other wildlife at Dzanga-Sangha Reserve, Central African Republic. *Oryx* 34 (1), 56–65.
—— (2002). *Preliminary analyses of phase II of study on the impacts of human disturbance on mammals at Dzanga-Sangha Reserve, Central African Republic*, Jan–May 2002. Report to World Wildlife Fund-US, Central African Ministry of Scientific Research 2002.
——, Kpanou, J.B., and Otto, K. (2006). *Impacts of Human Activities in the Dzanga-Sangha Reserve*. Report to the Central African Government.
Robinson, J. (2006). Conservation biology and real world conservation. *Conservation Biology* 20 (3), 658–669.
——, Redford, Kent H., and Bennett, Elizabeth L. (1999). Wildlife harvest in logged tropical forests. *Science* 284 (5414), 595–596.

Rose, Anthony, Mittermeier, Russell A., Langrand, Olivier, Ampadu-Agyei, Okyeame, and Butynski, Thomas M. (2003). *Consuming Nature, A Photo Essay on African Rain Forest Exploitation*. Palos Verdes Peninsula, CA: Altisima Press.

Salafsky, N. and Margoluis, R. (2004). Using adaptive management to improve ICDPs. In Thomas O. McShane and Michael P. Wells (eds), *Getting Biodiversity Projects to Work: towards more effective conservation and development*, pp. 372–394. New York: Columbia University Press.

Sanderson, S.E. and Redford, K.H. (2003). Contested relationships between biodiversity conservation and poverty alleviation. *Oryx* 37 (4), 389–390.

Schmidt-Soltau, Kai (2003). Conservation-related resettlement in Central Africa: environmental and social risks. *Development and Change* 34 (3), 525–551.

Sicotte, Pascale and Uwengeli, Prosper (2002). Reflections on the concept of nature and gorillas in Rwanda: implications for conservation. In A. Fuentes and L.D. Wolfe (eds), *Primates Face to Face: the conservation implications of human-nonhuman primate interconnections*. Cambridge and New York: Cambridge University Press.

Simberloff, D. (1998). Flagships, umbrellas, and keystones: is single-species management passé in the landscape era? *Biological Conservation* 83 (3), 247–257.

Sponsel, Leslie E., Bailey, Robert C., and Headland, Thomas N. (1996). Anthropological perspectives on the causes, consequences, and solutions of deforestation. In L.E. Sponsel, T.N. Headland, and R.C. Bailey (eds), *Tropical Deforestation: the human dimension*, pp. 3–52. New York: Columbia University Press.

Struhsaker, Thomas T. (1997). *Ecology of an African Rain Forest: logging in Kibale and the conflict between conservation and exploitation*. Gainesville, FL: University Press of Florida.

Tabareilli, M. and Gascon, C. (2005). Lessons from fragmentation research: improving management and policy guidelines for biodiversity conservation. *Conservation Biology* 19 (3), 734–739.

TRAFFIC (2005). Available at www.traffic.org/dispatches/DispNo23.pdf.

Turkalo, A. and Fay, J.M. (1996). Studying forest elephants by direct observation: preliminary results from the Dzanga clearing, Central African Republic. *Pachyderm* 21, 45–54.

Tutin, Carolyn E.G. and Fernandez, Michele (1984). Nationwide census of gorilla (*G. g. gorilla*) and chimpanzee (*Pan t. troglodytes*) populations in Gabon. *American Journal of Primatology* 6, 313–336.

——, Parnell, R.J., White, L.J.T., and Fernandez, M. (1995). Nest building by lowland gorillas in the Lope Reserve, Gabon – environmental influences and implications for censusing. *International Journal of Primatology* 16, 53–76.

Walley, C.J. (2004). *Rough Waters: nature and development in an East African marine park*. Princeton, NJ: Princeton University Press.

Wallis, Janette and Rick, Lee D. (1999). Primate conservation: the prevention of disease transmission. *International Journal of Primatology* 20 (6), 803–826.

Walsh, Peter, Abernathy, Kate A., Bermejo, Magdalena, Beyers, R., DeWachter, P., et al. (2003). Catastrophic ape decline in Western Equatorial Africa. *Nature* 422, 611–614. 10 April. doi: 10.1038/nature01566. Accessed 20 January 2006.

Weber, W. and Vedder, A. (2001). Trends and challenges in applied research and conservation. In W. Weber, L.J.T. White, A. Vedder, and L. Naughton-Treves (eds), *African Rain Forest Ecology and Conservation*, pp. 547–556. New Haven, CT: Yale University.

Wells, M. and McShane, T.O. (2004). Integrating protected area management with local needs and aspirations. *Ambio* 33 (8), 513–519.

West, P. (2006) *Conservation is our Government Now: the politics of ecology in Papua New Guinea*. Durham, NC: Duke University Press.

—— and Brockington, D. (2006). An anthropological perspective on some unexpected consequences of protected areas. *Conservation Biology* 20 (5), 609–616.

White, Lee G.T., with Edwards, Ann (1999). *Conservation Research in African Rain Forests: a technical handbook*. New York: Wildlife Conservation Society.

—— and Tutin, Caroline E.G. (2001). Why chimpanzees and gorillas respond differently to logging: a cautionary tale from Gabon. In William Weber, Lee J.T. White, Amy Vedder, and Lisa Naughton-Treves (eds), *African Rain Forest Ecology and Conservation*, pp. 449–462. New Haven, CT: Yale University Press.

Wilkie, David with Auzel, Philippe (2001). Defaunation not deforestation: commercial logging and market hunting in Northern Congo. In A. Grajal, J.G. Robinson, and A. Vedder (eds), *The Impact of Commercial Logging on Wildlife in Tropical Forests*, pp. 375–400. New Haven, CT: Yale University Press.

—— and Carpenter, J.F. (1999). Bushmeat hunting in the Congo basin: an assessment of impacts and options for mitigation. *Biodiversity Conservation* 8, 927–955.

——, Morelli, G., Demmer, J.L., Starkey, M., Tefler, P., and Steil, M. (2006). Parks and people: assessing the human welfare effects of establishing protected areas for biodiversity conservation. *Conservation Biology* 20 (1), 247–249.

Wolch, Jennifer R. and Emel, Jody (1998). *Animal Geographies: place, politics, and identity in the nature-culture borderlands*. New York: Verso.

Chapter 6
Steering governance through regime formation at the landscape scale:
evaluating experiences in Canadian biosphere reserves

Rebecca M. Pollock, Maureen G. Reed, and Graham S. Whitelaw

Introduction

Advocates of an ecosystem approach to establishing and managing protected areas recognize the complex dynamics between natural and social systems. This complexity includes the need for people to help restore and maintain ecological integrity and biological diversity while preserving a sustainable livelihood for themselves and for their communities (Slocombe, 2003; Dorcey, 2003; Ellsworth and Jones-Walters, 2006). This understanding is accompanied by a call to increase democratic processes for making decisions about the management of those areas, in particular to include local people in decisions that affect them directly (Cortner and Moote, 1999; Bagbey and Kusel, 2003). Community participation could range from education and stewardship projects to negotiated co-management agreements for governing natural resources, such as fisheries or forests. Francis (this volume) provides a more global overview of governance and systems perspectives that influence or impact upon protected areas. We portray some of the ways these larger-scale factors are exemplified more immediately within protected areas situated in regional landscapes.

From a social and political perspective, parks and protected areas are not places in nature that stand apart from human use. Rather, they represent institutional arrangements that are created through the interaction of government agencies, management authorities, environmental movement organizations (EMOs), industry, local interests, external pressures, and a variety of other relationships. These multi-level governance arrangements illustrate how regional systems are open to external influences (from both local and global pressures) and how the institutional layers for managing them might be fragmented among separate, and sometimes competing, organizations. As Draper (2004: 229) explains, addressing the challenge of open systems and institutional fragmentation 'requires learning how to strengthen existing relationships, forge new partnerships, incorporate different kinds of knowledge, and institute new co-management (governance) processes. [It] also entails understanding and managing complex relationships among ecosystems and people.'

Establishing these relationships requires the involvement of communities in creating and maintaining governance regimes. Yet, it is not clear how communities should act to address both internal and external pressures on protected areas, nor is it clear how they should establish effective links among local, regional, and external governing organizations. We define communities as inclusive of local residents, those people and organizations within and outside an immediate locality or landscape (e.g. residents of a city that become involved in the protection of an outlying area), and the widest group of stakeholders that partake in governance activities. We consider under what conditions these communities can help to shape the governing regimes for protected areas and some of the factors that allow environmental organizations, in particular, to steer governance directly. We also assess local community capacity to address the two challenges of open systems and institutional fragmentation. We describe the process of regime formation in three protected areas in Canada to address these questions.

The focus is on UNESCO biosphere reserves because they contain protected areas at their core, yet must incorporate adjacent areas and the inhabited surrounding 'working landscapes' to integrate conservation with sustainable development. Canadian biosphere reserves provide a unique opportunity to look at the processes of integration, protection, and community participation, and to understand how landscapes become a focal point for multi-stakeholder collaboration. By evaluating Canadian examples of biosphere reserve formation with different institutional arrangements, we can assess the extent to which local communities can effectively participate in the governance of regional landscapes. The discussion first reviews ideas about governance and regime formation, that leads us to focus on the fluid boundaries of *open systems* and institutional *fragmentation* as key challenges and opportunities for communities to participate in steering governance of protected areas. We illustrate these dynamics and how they were addressed with three Canadian cases at Niagara Escarpment and Oak Ridges Moraine (Ontario) and Redberry Lake (Saskatchewan).

Governance for sustainability

A multi-stakeholder approach to management represents a fundamental shift in relationships where government agencies are not viewed as the sole institutions of governance. Instead, governance now refers to institutional arrangements that extend beyond government to include private-sector and other non-governmental organizations, as well as the rule systems under which these different actors operate (Francis, 2003). A shorthand for this understanding is Knight's phrase: 'collective decision-taking and action in which government is one stakeholder among others' (Knight *et al.*, 2002: 131, cited in Dorcey, 2003: 535). The style is more like a set of working relations, or mode of interactive behaviour, than a fixed system or formal institutional framework. The new context of governance demands citizen participation for legitimacy, non-regulatory institutions and policies, and relies heavily on social norms for compliance.

Indeed, for government to be effective it must partner with citizens and other sectors and share decision-making. 'The relevance of federal governments is determined by both how well they govern and how well they partner with citizens and other sectors in bringing about good governance' (Ellsworth and Jones-Walters, 2006: 5). If decisions about protected areas are shared, then more organizations have the opportunity to participate and influence the outcomes of specific decisions and the overall trajectory of management and/or development. Yet, it is not solely the quality of decisions for governance that is important, but also who participates in the process. Increasingly,

> the quality of governance is determined by the design of institutional arrangements (such as treaties, laws, and organizations) and by the way in which decisions are made. Who makes those decisions is also a major factor shaping the quality of governance.
>
> (Kreutzwiser and de Loë, 2004: 189)

In this way, non-governmental partners may influence decisions and broker new norms in order to steer governance in a particular direction.

Steering as a metaphor for governance captures the complex processes of social interaction (Kooiman, 2003) as opposed to state intervention and control. This perspective underscores how the relationships between actor systems actually form the basis of governance. As the number of actors and institutions involved multiplies, effective governance requires the navigation of many different types of institutional layers. As Jessop (1975: 575) explains, governance is best understood as 'the complex art of steering multiple agencies, institutions and systems which are both operationally autonomous from one another and structurally coupled through various forms of reciprocal interdependence'. Some communities are beginning to understand that governance is an evolution to which they can contribute by steering particular decision-making priorities and objectives (e.g. habitat protection, sustainable resource use). Activities such as building networks, collaborating on integrative projects, and producing accessible and transparent science, lead to information and understanding that can change values and influence policy and decision-making (Pollock and Whitelaw, 2005).

According to Ellsworth and Jones-Walters (2006: 5):

> communities are at the heart of this governance transition. As places, they experience issues as a web of interrelated problems. As people, they live with direct effects, indirect effects, side effects and cumulative effects of policies. As relationships, they are the product of rewarding interactions.

This turn to the community has been a powerful trend, with academics and practitioners promoting and implementing concepts such as self-organization and self-governance (Kooiman, 2003); community capacity (Kusel, 2001; Mendis, 2004); and collaborative planning (Healey, 1997, 2003).

Despite the prospect of these related approaches to governance, both community capacity and institutional capacity of other actors, such as governments, are crucial. There are concerns that promoting capacity at one level reduces capacity at others. For example, some observe that promoting active citizenship at the community level inadvertently encourages governments to abrogate their responsibility for economic, social, and environmental well-being (Reed, 1997; Rice and Prince, 2000), reducing the overall capacity available for the governance system. Swift (1999: 9) observes that: 'In Canada, where a fashionable neo-liberal ideology has meant a return to laissez-faire, government often promotes the idea that "the community" should take upon itself the tasks of providing services once delivered by the welfare state.' The rhetoric of shared governance may act to *dis*empower communities if resources are reduced while expectations mount (see also Graham and Phillips, 1998; Smith, 2005). Furthermore, the turn to the local may not necessarily be more democratic as power relations within communities might stifle debate and/or foster a form of paternalism that is locally generated (Reed, 1995).

The landscape-scale governance challenge

Landscapes have been chosen as the unit of analysis because they are powerful cultural constructs that reflect human–environment relations and thus take on diverse organizational forms to reflect social values, human history, and sense of place. Campbell (2005: 202) notes how:

> People and nature continually respond to and redefine one another. We need to be able to distinguish where humans have imposed on the environment and where they have adapted to it, and recognize that a landscape is a product of both dynamics.

The concept of 'working landscapes' integrates the protection of biological diversity and ecosystem functions with resource use to support human livelihoods (UNESCO, 2002) as demonstrated by biosphere reserves. These sites explicitly recognize that landscapes sustain societies by providing ecosystem goods and services (e.g. clean water, air, soil, fuel, minerals, etc.), including intangible aesthetic, spiritual, and recreational values (Millennium Ecosystem Assessment, 2005). For landscapes to be sustained in turn, they often require the use of specific governance regimes or institutional arrangements for regulating resource use and development.

In his book, *Planning at the Landscape Scale*, Paul Selman (2006: 69) says: 'the main attraction of the landscape scale as a framework . . . is its holistic nature, and its capacity to integrate human and environmental systems with identifiable and distinct places. However, this also makes for great, perhaps overwhelming, complexity.' We maintain that biosphere reserves provide a focal point for 'getting our heads around . . . whole landscape units' (Selman, 2006: 69). Indeed, the growing number of biosphere reserves in Canada attests to an explicit attempt to advance sustainability in certain communities within their larger and more complex ecosystems. People involved in environmental or development projects often see the need for recognizing interrelationships at multiple geographic scales and apply their efforts at the local and landscape scale (Pollock, 2004).

The literature on landscape planning and management recognizes at least two other important challenges for governance of protected areas. First, our existing political jurisdictions do not reflect the organizing principles of landscape ecology and conservation biology, but are *fragmented* both politically and physically. Francis (1995: 149) explains that 'boundaries pose a major challenge to governance: the jurisdictional, administrative and proprietal boundaries rarely make ecological sense, and environmental problems are frequently pervasive'. Certain landforms or watersheds might be nested regionally, between local and provincial jurisdiction, but then wildlife corridors or coastlines transcend those same boundaries at still larger scales. Indeed, for the governance of transboundary issues, such as air pollution or habitat fragmentation, an appreciation of landscape science and ecology is crucial.

Second, these landscapes are vulnerable to both local and global environmental and economic trends (as outlined by Francis in this volume) due to their fluid and porous boundaries. 'With respect to spatial scales, all ecosystems are "*open*" systems, and thus receive impacts from neighbouring systems. Effective management, therefore, necessitates the involvement of levels of authority from the local to the global' (Rapport, 2003: 50, emphasis added). To manage these challenges, a single landscape might be carved up into countless institutional layers of governmental agencies, research and monitoring bodies, conservation organizations, and citizens' groups. Each organization addresses its own important piece of a particularly complex puzzle for protected areas. Organizations also make use of policy instruments to guide human behaviour toward their desired ends. These complex layers of organizations may be working collaboratively or they may be working at cross-purposes, or both simultaneously. As a result, the capacity of communities to participate in governance at landscape scales will be affected by institutional arrangements and influences from open systems.

However, the extent to which citizens can help to steer governance for protected areas is still not well understood. It is not clear, for example, whether the large number of actors, who operate across the porous boundaries of landscape, help communities to consolidate or act to further fragment multi-level governance arrangements. While these organizational layers are sometimes appropriate for particular functions at particular scales, such as monitoring, regulation, or enforcement, there is still a need for research to assess whether landscape-scale institutions are able to create and maintain a broader perspective on socio-ecological change toward sustainability goals.

Indeed, landscape-scale organizations are seen as umbrellas that 'do not comfortably fit into the established framework of local, state and federal governments' (McKinney et al., 2002). One hypothesis in this study is that landscape institutions, such as biosphere reserves, might be able to integrate *and* transcend existing fragmented political jurisdictions in order to provide this 'big picture' perspective to help govern sustainability. Jessop (2002) refers to this capacity as meta-governance or the overall institutional system of rules that govern the distribution of power, authority, and responsibilities within society. It 'involves managing the complexity, plurality, and tangled hierarchies found in prevailing modes of coordination' (Jessop, 2002: 6). In this way, landscape-scale institutions help communities to keep pace with the developments happening at both the smaller and larger scales of complex multi-level governance systems. A collection of regionally scaled organizations that overlay the landscape and provide linkages between local and global issues might also be able to create new regimes for sustainability at the landscape scale.

What is meant by a governance regime? Regimes are 'the system of rules and norms . . . that govern institutional behaviour' (Francis, 1988: 110). As social constructs they can evolve and change. According to Reed (2006), regimes can take on diverse forms and change character depending on ecological conditions, shifting interests and alliances, available logistical resources and activities, and management efforts at local, provincial, and federal levels. Regimes are used to guide regulations, international agreements, and collaborative management plans. The result is a complex web of regulatory codes, laws and customs, treaties and accords, and multi-stakeholder management frameworks. Regimes such as international agreements try to create common ground from which countries can then work together more closely (Young, 1997). Regimes can exemplify generally accepted rule systems either with or without the organizational capacities to foster compliance, for example, international treaties and accords.

Governance systems for protected areas can be steered by regimes that are both formal and informal. Certain institutional arrangements may be formally established by regional or state authorities, or through international agreements, which typically include laws, regulations, and enforcement mechanisms. At the same time, a suite of informal social practices help to govern parks and protected areas. These might include traditions, habits, sanctions, stories, customs, or consensus – all considered types of shared understandings or norms. Both formal and informal social institutions will work together to establish governance regimes that promote or undermine sustainability. This understanding is consistent with Göran Hydén's definition of governance that reflects 'the conscious management of regime structures' (Hydén, 1992: 7) through formal and informal institutional arrangements that are made legitimate by the favourable exercise of authority, reciprocity, trust, and accountability (Hydén, 1999). Thus, formal and informal institutional arrangements create regimes that help to steer the governance of protected areas.

It is clear that there is a role for communities in creating and maintaining appropriate regimes for landscape governance. The ability to establish links among local, regional, and extra-regional governing organizations is a significant challenge. The examples that follow illustrate three different types of institutional arrangements and levels of institutional capacity and help us to understand the factors that contribute to, and hinder, effective governance of protected areas and the role of communities.

Steering governance: comparing three Canadian cases

Biosphere reserves are geographic areas designated because of the expressed desire of local communities to work toward sustainability. Residents seeking biosphere reserve status for their region must have it nominated at the local level, endorsed by provincial and national governments, and, finally, recognized by UNESCO (see Box 6.1). While biosphere reserves are intended to be community-based and locally driven (UNESCO, 2000), individual reserves

Box 6.1 Biosphere reserves

UNESCO world biosphere reserves contain three zones: (1) a *core* that must be protected by legislation; (2) a *buffer* where research and recreation uses compatible with ecological protection are allowed; and (3) a *transition* zone where sustainable resource use is practised. In Canada, the outer zone is also referred to as an 'area of cooperation'. Biosphere reserves are created to demonstrate three functions: environmental protection; logistical provisioning for scientific research; and sustainable resource use (UNESCO, 2000). The buffer zone(s) surrounding the core area

> demonstrates the same ecosystem organized to meet human needs, particularly by traditional means. It is meant to be a place of reconciliation, a model of a human community in harmony with the natural world. . . . Beyond all of these zones is the more customary multiple-use area, where human communities are encouraged to cooperate and be open to some of the lessons learned in the inner zones. Boundaries are often indefinite and fluctuate over time, depending on the scope and character of human activity. There are no fees or hours of entry, for the park is meant to overlay land and landscape.
>
> (Wilson, 1991: 239)

typically include several municipalities and interests that extend beyond the boundaries of local jurisdiction. Thus, in Canada, biosphere reserves effectively become regional in scope and are nested within provincial and federal areas of jurisdiction.

Biosphere reserves are typically established on the basis of watersheds or other landscape-level features that extend beyond the boundaries of local human communities. They may also reflect a strong sense of place, recognizing the cultural heritage and current 'working landscapes' that sustain traditional and contemporary livelihoods. As a recent UNESCO (2005: 2) publication explained:

> Biosphere reserves constitute innovative approaches to governance at multiple levels. Locally, biosphere reserves are a potent tool for social empowerment and planning; nationally, they serve as hubs of learning for replication elsewhere in the country; internationally they provide a means of cooperation with other countries. They also provide a concrete means of addressing international obligations such as Agenda 21, the Convention on Biological Diversity, the Millennium Development Goals . . .

Since all communities, not just those situated in biosphere reserves, are subject to the pressures of both local and global forces, a research focus on the regional or landscape scale helps to account for multi-directional dynamics and multi-level governance responses. Specifically, researchers are in a position to study complex drivers of change and stressors on ecological and social systems; likewise strategies for adaptation and resilience can be observed – concepts central to the Millennium Ecosystem Assessment and the related discourse of sustainable development.

> In governance terms, a biosphere reserve . . . is structured by rules, consisting of formal ones such as property rights, aboriginal rights, jurisdictions, and admini-

Table 6.1 Characteristics of the three case studies

	Niagara Escarpment	Oak Ridges Moraine	Redberry Lake
Population	120,000	250,000	1,000
UNESCO designation	1990	N/A	2000
Approximate size (ha)	190,270 (725 km corridor)	198,000 (160 km corridor)	112,200
Terrestrial ecozone	Mixed wood plains	Mixed wood plains	Prairies
Major activities	Wine production, fruit farming, cattle farming, tourism	Mixed agriculture, aggregate mining, 'greenbelt' management	Agriculture, livestock raising, wildlife protection

Sources: UNESCO, 2006; STORM 2006.

strative authorities, and informal ones that guide local 'politics' for cooperation, decisions, and dispute resolution (except when recourse to formal rules becomes necessary). This overlay of rules constitutes the governance or 'management regime' for the area of the biosphere reserve.

(Francis, 2004: 15)

Three examples of regime formation will highlight the role of communities in steering governance for protected areas in two UNESCO biosphere reserves in Canada, Niagara Escarpment and Redberry Lake, and a third protected area, called the Oak Ridges Moraine. The Oak Ridges Moraine is in the process of having a nomination prepared for international biosphere reserve designation. Although the three cases highlight divergent social, economic, and ecological contexts (Table 6.1), we explore how communities are involved in addressing pressures from their open systems and overcoming institutional fragmentation to steer governance at the landscape scale.

In particular, the three cases introduce the role of environmental movement organizations (EMOs) that create and maintain governance regimes for their respective protected areas. Both Niagara Escarpment and the Oak Ridges Moraine are in the southern portion of the province of Ontario (Figure 6.1). They have highly complex institutional arrangements for protection and are under significant pressures from urbanization. By contrast, Redberry Lake, in the prairie province of Saskatchewan, reflects an agricultural landscape with rural communities facing enormous pressures to maintain livelihoods and economic viability.

Niagara Escarpment case[1]

The Escarpment is a 725-kilometre-long landscape stretching from Lake Ontario (near Niagara Falls) to the tip of the Bruce Peninsula (between Lake Huron and Georgian Bay) and has significant elevation changes either associated with exposed cliff face and talus slopes, or rolling and hummocky terrain (Tovell, 1992). The Escarpment has extensive natural areas with high biodiversity. These natural areas contrast with mainly agricultural areas in much of southern Ontario. Open landscape and scenic values are associated with the remaining agricultural lands and views up to, and down from, the Escarpment. These characteristics have led to extensive recreational use of the Escarpment (Whitelaw et al., 2005).

As with all other areas, the Niagara Escarpment is subject to influences at various scales from the global to the local. Global influences include immigration to southern Ontario resulting in residential development pressures and aggregate companies seeking resources to meet needs in and around the City of Toronto area and the southern portion of the Escarpment. Although only 120,000 people live in the area, including 1,000 First Nations, that number rises to approximately 1.2 million when including the populations of surrounding urbanized areas. Land use conflicts occur with regard to tender fruit agriculture and

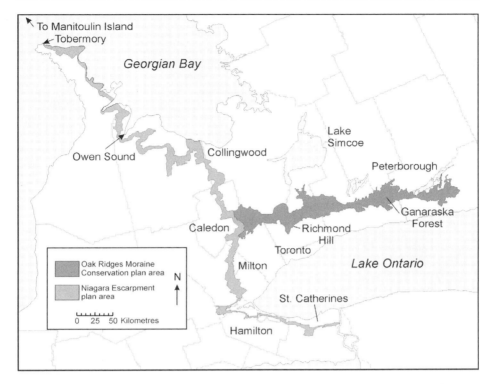

Figure 6.1 Location map of Oak Ridges Moraine Conservation Plan and Niagara Escarpment Plan areas (source: Whitelaw, 2006)

development in the south; urban expansion in the regions of Niagara, Halton, and Peel (e.g. the cities of Burlington and Milton); agricultural viability of rural areas; and recreational and tourism pressure in the northern parts of the Escarpment.

In 1988, Professor George Francis, then Chair of Canada's Man and the Biosphere (MAB) Working Group on Biosphere Reserves, informally raised the idea of a biosphere reserve designation for the Niagara Escarpment with the Chair of the Niagara Escarpment Commission (NEC). This led to consultations with Parks Canada, Bruce Peninsula National Park, and Fathom Five National Marine Park. Favourable comments were received and the Niagara Escarpment was designated a biosphere reserve by UNESCO in February 1990 (Francis and Whitelaw, 2002). What formed the basis for the designation was the regime that had evolved based on a 20-year planning process along one of Ontario's most significant landforms. Major events had included: a comprehensive planning study (Niagara Escarpment Study Group, 1968); passage of the Niagara Escarpment Planning and Development Act in 1973 (Government of Ontario, 1973); creation of the NEC; and plan development, associated hearings, and final passage of the Niagara Escarpment Plan (NEP) in 1985 (Niagara Escarpment Commission, 1985). The Plan has been formally updated twice since then, in 1994 and 2005, and its protective features improved.

Management regimes

The NEC has responsibility for the Niagra Escarpment Biosphere Reserve (NEBR) designation. The Niagara Escarpment is representative of the Lake Erie Lowland ecoregion in the south and the Manitoulin-Lake Simcoe ecoregion in the north. The core areas of the NEBR include the most environmentally sensitive lands: those identified as Niagara Escarpment Parks and Open Space (e.g. Bruce Peninsula National Park, Provincial Parks, Conservation Areas) and lands designated Escarpment Natural in the NEP. Some of the lands designated Escarpment Natural lie within 131 Niagara Escarpment Parks and Open Spaces, administered by various management agencies (Ontario Parks, Ministry of Natural Resources, conservation authorities, and municipalities). The buffer areas include all lands designated Escarpment Protection in the NEP (lands considered less environmentally significant than Escarpment Natural but significant enough to preclude residential development or aggregate extraction). Areas of transition include lands designated Escarpment Rural in the NEP. These are mainly agricultural lands. Urban areas are also within the transition zone. With the exception of lands in the Parks and Open Space System, most lands within the NEP, including a majority of Escarpment Natural lands, are privately owned (Francis and Whitelaw, 2002).

The land use planning regime of the Niagara Escarpment is unique. The Niagara Escarpment Planning and Development Act created the NEC, which then assumed the majority of municipal land use planning control from Niagara Escarpment municipalities. There have been repeated unsuccessful attempts to delegate planning control back to the local municipalities. Jurisdictional fragmentation issues are typical and extensive. Conservation authorities along the Escarpment provide input and review to the NEC as they do with municipalities outside the NEP area in southern Ontario. The Ontario Ministries of Natural Resources, Municipal Affairs, Environment, Transportation, Culture, and others have legislation that affects the Niagara Escarpment (e.g. Aggregate Resources Act, Parks Act, Water Resources Act, Planning Act, Municipal Act). The federal government also has various Acts that influence Niagara Escarpment activities including the National Parks Act and the Canadian Environmental Assessment Act. Even though land use planning has been standardized across the Escarpment, other legislation, policy, and programmes influence Niagara Escarpment activities, contributing to a fragmented jurisdictional environment.

Steering governance

Community groups, and EMOs in particular, have played significant roles in the evolution of the Niagara Escarpment regime. EMOs created landscape value for the Escarpment and this led to subsequent actions to protect the region by the government. This was not a traditional Not-In-My-Back-Yard (NIMBY) process involving local citizens mobilizing to stop a particular development in their area. Rather, it was driven by a group of naturalists who launched an effort to build landscape value for what they recognized as an important regional feature. In fact, there was no recognition of the Niagara Escarpment as a distinct landscape feature, geographic space, or political domain prior to the 1960s by most planners, managers, or the general public. Landscape value was primarily achieved through Bruce Trail development activities. In 1960, the Hamilton Field Naturalists formed a citizen's committee to investigate the creation of a footpath to run along the Escarpment from Queenston in the south to Tobermory in the north. The view was that by providing people the opportunity to hike the Niagara Escarpment, 'they would gain an appreciation of it and thereby want to protect it' (Plaunt, 1978: 11). This citizen's committee embarked on a campaign to set an agenda with the Government of Ontario to establish the Bruce Trail. In March 1963, the Bruce Trail Association (BTA) was formed and today we have the internationally renowned footpath known as the Bruce Trail. General awareness about the Niagara Escarpment also increased due to development threats to the Escarpment from aggregate extraction and subdivision development, and government initiatives in regional planning (Government of Ontario, 1962; Plaunt, 1978; Bruce Trail Association, 2006).

This growing concern for the Escarpment placed increased pressure on the government to address Niagara Escarpment protection. In 1967, the provincial government launched the

Niagara Escarpment Conservation and Recreation Study (Niagara Escarpment Study Group, 1968). The purpose of the study was to recommend a new planning regime for the Niagara Escarpment (Niagara Escarpment Study Group, 1968). The mapping and information collection carried out for this work provided extensive information on the Niagara Escarpment for the first time. The government passed the Niagara Escarpment Protection Development Act in 1973 based on the recommendations of the Escarpment Study and this legislation is still the basis for the regime that exists today (Whitelaw and Hamilton, 2003).

Currently, Niagara Escarpment EMOs are mainly involved in Niagara Escarpment activities in two broad areas: lobbying government and coordinating biosphere reserve logistical activities. The Coalition on the Niagara Escarpment (CONE) is the main suite of Escarpment EMOs and includes numerous provincial and national EMOs including Ontario Nature (formerly the Federation of Ontario Naturalists) and the Nature Conservancy of Canada. CONE's lobbying activities include extensive work through meetings with politicians, participation at Niagara Escarpment Board hearings, and legal activities. For example, one current issue involves a proposed major expansion to the Dufferin Aggregates quarry near Milton, Ontario. CONE lobbied the NEC to turn the application down, participated in the most expensive hearing process in the history of Ontario (Murzin, pers. comm., 2006), and has petitioned the Ontario Cabinet to reject the application.

Biosphere reserve logistical activities carried out by Niagara Escarpment EMOs include stewardship, monitoring, and education. Stewardship and educational activities are similar to other EMO activities carried out elsewhere including land acquisition, restoration, and celebration events. The monitoring activities are notable. CONE carries out Plan monitoring that includes tracking, evaluating, and reporting planning decisions made by the NEC through the press. This type of policy monitoring plays a critical role in regime maintenance. The former Assistant Director of the NEC at CONE's 25th anniversary celebration described their role as follows: 'CONE has blazed a new approach to EMO interaction through monitoring, being a regulatory watch dog with a big bite, constantly surveying the legislation and Plan implementation, something that is unique, no other organization does this' (Louis, 2003).

The policy monitoring carried out by CONE provides the information used to ensure the government enforces the Act and implements the Plan (Calderisi, 2003). Major budget and staff reductions at the provincial level after 1995, brought in by the Conservative government limited the ability of the NEC to implement biosphere reserve activities. Out of necessity, the NEC focused on its core mandate of managing: the permit system for new development the plan amendment process, and five-year plan reviews. This left little time for biosphere reserve activities (Francis and Whitelaw, 2002).

Governance analysis

In the Niagara Escarpment, environmental and citizens' groups have filled the biosphere reserve management void left by government agencies due to these cutbacks. Although CONE and the other Niagara Escarpment EMOs had little to do with the process of biosphere reserve designation, EMOs subsequently embraced the concept and used it as a vehicle to implement activities along the Escarpment. CONE is active establishing biosphere reserve regimes and norms, through signage, education, and reporting on policy monitoring (Pim, 2003). Some examples of their activities include the publication called *Protecting the Niagara Escarpment: a citizen's guide* (Coalition on the Niagara Escarpment, 2005), newsletters, and monitoring of the NEP (Murzin, 2003; Coalition on the Niagara Escarpment, 2006). Citizens wanting to undertake biosphere reserve activities formed an exemplary group called the Bruce Peninsula Biosphere Association (BPBA). The group contributes in the northern part of the biosphere reserve through education, stewardship, monitoring, and greater ecosystem park management (Francis and Whitelaw, 2002; BPBA, 2005). The ten-year review of the NEBR was recently carried out by the not-for-profit Canadian Biosphere

Reserves Association on behalf of Canada MAB. The main conclusion of the reviewers was that the Niagara Escarpment merits continued membership in the world network of biosphere reserves (Francis and Whitelaw, 2002). Much of this merit can be attributed to community groups and EMOs of the Niagara Escarpment.

In summary, Niagara Escarpment EMOs have contributed to steering governance of the Niagara Escarpment in a number of ways. The system is open to influences from the global to the local and has required EMOs to take on strong regime maintenance roles (e.g. resisting development applications to the NEP driven by global flows of people into Southern Ontario). Mainly, EMOs carry out this regime maintenance function by upholding the vision of the Escarpment as a continuous natural area, monitoring new policies and the overall Plan, and marketing the Escarpment to government, the private sector, and the public. The coalition of EMOs also forces more integration of the fragmented jurisdictions along the Escarpment, as individual organizations have come to realize the risks of not taking the bigger landscape picture and value of the Escarpment into consideration.

Oak Ridges Moraine case[2]

The Oak Ridges Moraine (ORM) stretches 160 kilometres from east to west and runs just north of the City of Toronto. The ORM is representative of the Manitoulin-Lake Simcoe ecoregion. It is a landscape with significant elevation changes associated with glacial moraine features. The Moraine has extensive natural areas with high biodiversity. These natural areas, similar to the Niagara Escarpment, contrast with mainly agricultural areas in much of southern Ontario. The ORM also has open landscape and scenic values associated with the agricultural lands (Figure 6.2).

Again similar to the Escarpment, the ORM is subject to influences at various scales from the global to the local. Global influences include significant immigration to the Toronto area resulting in intense residential development pressures north of the city and need for

Figure 6.2 Oak Ridges Moraine (photo: Rick Harris)

sand and gravel from the western, central, and eastern portions of the Moraine. In the western portion of the ORM regional issues emanate from intense development pressures. In the east, agricultural viability, aggregate mining, and economic development dominate where development pressure is low. The area has yet to be designated a biosphere reserve, although there are currently activities underway to explore the potential of the ORM as a biosphere reserve and to prepare background information in support of a designation request from UNESCO (McCarthy, 2006).

Management regimes

The main legislation affecting the ORM is the ORM Conservation Act (ORMCA) and Conservation Plan (ORMCP). The vision for the ORM is 'a continuous band of green rolling hills that provides form and structure to south central Ontario, while protecting the ecological and hydrological features and functions that support the health and well-being of the region's residents and ecosystems' (Government of Ontario, 2002). The objectives of the ORMCA and ORMCP deal mainly with protecting and restoring the ecological and hydrological integrity and function of the Oak Ridges Moraine Plan Area through land uses that improve, or are compatible with, the vision. The Plan has four land use designations: Natural Core, Natural Linkage, Countryside, and Settlement Areas. The Natural Core includes areas with a high concentration of natural heritage features and hydrologically sensitive features or landform conservation areas. The designation is restrictive with no new housing subdivisions or aggregate extraction activities permitted. The Natural Linkage designation forms part of a central corridor system that supports, or has the potential to support, movement of wildlife. This designation also has restrictive development policies although new aggregate mining is permitted under specific policies. The Countryside designation includes rural land uses, recognizes existing hamlets or similar existing small communities, but does not allow for new subdivision development. Residential development is mainly limited to the Settlement designation (eight per cent of the Plan area) where urban development is focused, including a range of residential, commercial, industrial, and institutional uses (Government of Ontario, 2002). The vast majority of lands across the ORM are in private ownership.

For the proposed biosphere reserve, potential core areas might include all lands designated ORM Core and Linkage. These are the most environmentally significant areas and were delineated based on principles of conservation biology. Buffer areas might include all lands designated Countryside lands, mainly consisting of agricultural land uses. Transition areas might include all urban settlement areas, Greenbelt Plan lands (Ontario Ministry of Municipal Affairs, 2006) adjacent to the Oak Ridges Moraine, and the river valleys flowing from the ORM north and south administered by conservation authorities (Francis, pers. comm., 2006).

Other legislation, plans, policies, and programmes that influence the ORM are similar to the Niagara Escarpment due to the proximity of the two landforms and their locations in Ontario. A major difference is the implementation model for land use planning between the Niagara Escarpment and the ORM. The ORMCP is implemented through municipal official plans and municipal council decision-making, rather than an arm's length provincial commission as is the case with the Niagara Escarpment. The differences in scale and style of decision-making hold major implications for standardization of land use policies, protection of natural heritage, and tracking ORMCP implementation across the 32 municipalities.

Steering governance

EMO and community activities have had significant influence over the evolution of the ORM regime. Similar to the Escarpment, the creation of landscape value for the ORM played a critical role in the evolution of ORM governance. The process started in the late 1980s when numerous local grassroots EMOs emerged to fight local battles, based initially on the

NIMBY response. Reactionary protest and the emergence of small grassroots EMOs was the first step in the creation of a broad-based network called Save the Oak Ridges Moraine (STORM) Coalition, which came to understand the ORM as an important landscape worth protecting. The key role played by the STORM Coalition in the early days of the campaign was agenda setting; specifically they created a vision for the Moraine and communicated the need for its protection. Already established provincial and national EMOs also became involved in the ORM campaign over the course of the effort, including the Federation of Ontario Naturalists (now Ontario Nature), Nature Conservancy of Canada, and Earthroots.

The STORM Coalition steered government toward recognition of the ORM by recommending its protection in three government studies. They were: *The Adequacy of the Existing Environmental Planning and Approvals Process for the Ganaraska Watershed* (Environmental Assessment Advisory Committee, 1989); *Spaces for All: an option for a greater Toronto area greenlands strategy* (Kanter, 1990); and the *Royal Commission on the Future of the Toronto Waterfront* led by David Crombie (Royal Commission on the Future of the Toronto Waterfront, 1992). The efforts of STORM led to government action. In 1990, the Liberal government declared provincial interest in the ORM and issued Implementation Guidelines (1991). The subsequent New Democratic Party (NDP) government launched an extensive study of the ORM and developed a protection strategy through a collaborative process (Oak Ridges Moraine Technical Working Committee, 1994). With the election of the Conservative government in 1995, the STORM Coalition withdrew from agenda setting and collaborative processes to focus mainly on educational activities. This strategic decision was made in response to the Conservative government's dismantling of environmental programmes and the STORM Coalition's hope that not bringing attention to the ORM issue would leave the existing ORM guidelines in place (Government of Ontario, 1991). Re-engagement of the environmental movement occurred during the lead-up to the 1999 provincial election. STORM activities included media engagement, polling, political action, use of internet and e-mail, direct marketing through mail-outs, use of the scientific community, and participation in the ORM Richmond Hill Ontario Municipal Board Hearing that was addressing large housing subdivision development that threatened to sever the ecological continuity of the ORM at Yonge Street. These activities led to the government suspending the hearings and establishing a collaborative process through the ORM Advisory Panel to develop a solution for long-term protection. The panel presented recommendations to the government on how to protect the ORM which led to the ORM Conservation Act and Conservation Plan (Advisory Panel for Consultations on the ORM, 2001).

Governance analysis

EMOs have played, and continue to play, important roles across the ORM. The STORM Coalition is currently involved in governance on two main fronts. The first involves a watchdog function. The second involves collaboration with government and the private sector on development process issues (e.g. working with the Regional Municipality of York) to achieve more informed infrastructure planning. Other EMOs are involved in a wide variety of stewardship activities including restoration and land acquisition (ORM Foundation, 2006). Similar to the Niagara Escarpment, EMOs in the ORM are playing a regime maintenance role. The STORM Coalition is involved with other EMOs in a Monitoring the Moraine project designed to track whether the ORM is achieving its goals of ecological integrity and whether the Plan is being implemented as intended by the legislation (Monitoring the Moraine, 2006). Monitoring political accountability provides a unique opportunity for citizens' groups to participate in tracking landscape-scale change and steering government toward a desired response.

The STORM Coalition is playing an active role in exploring the biosphere reserve designation and the majority of biosphere reserve logistical activities are already being carried

Regime formation at the landscape scale

out by EMOs supported by the ORM Foundation (ORM Foundation, 2006) and others. The activities of a wide variety of actors and organizations are all keeping attention focused on the ORM. This attention and promotion of the Moraine's landscape value helps address the issue of jurisdictional fragmentation. The collaborative activities with York Region are an example. These efforts use strategic environmental assessment principles to better integrate environmental assessment processes with land use planning as outlined in official conservation plans (Regional Municipality of York and STORM, 2006).

Redberry Lake case[3]

Located in the central Canadian prairie province of Saskatchewan, Redberry Lake was designated a biosphere reserve by UNESCO in 2000. Much of the biosphere reserve's ecological significance rests with the aquatic environment that supports waterfowl and shorebird populations that are considered globally and nationally significant (Schmutz, 1999). As part of the Aspen Parkland region, many habitats and wildlife that would have been present at Redberry were extirpated at the time of the fur traders and settlement. Thus, by the late nineteenth and early twentieth centuries, the region was already a highly modified landscape (McGillivray, 1998). Continuous regional modification is a result of settlement, agricultural production, and industrialization (Figure 6.3).

According to hydrographic data, the saline lake, for which Redberry is named, has dropped 4.15 metres in the past 37 years (Saskatchewan Watershed Authority, 2003), reducing the habitat for shorebirds who use the lake for nesting or staging. Interestingly, the most significant drop in lake level occurred during the latest period of drought, from 2000 to 2003. The buffer zone of the biosphere offers little protection from the changes in the lake itself or from the agricultural inputs from nearby farms. Few members of the public know about the status of the lake and few federal officers come to enforce the legislation. Beginning in 1989, the Canadian Wildlife Service (CWS), a federal agency, contracted the

Figure 6.3 Canola fields (photo: Redberry Lake Biosphere Reserve)

Redberry Pelican Project to monitor the effects of visitors on nest disturbance, undertake regular patrols of the bird sanctuary, and report violations to authorities. The project also provided public awareness and education to local residents through an interpretive centre built by the lake in 1992. During the 1990s, funding from the CWS declined and by 2001, funding and activities associated with lake monitoring and patrols were terminated (Anonymous, 2002).

Management regimes

The core of the Redberry Lake Biosphere Reserve (RLBR) is the lake itself, protected by the federal Migratory Birds Convention Act, with islands protected as provincial wildlife reserves. The core area composes only five per cent of the biosphere reserve. The buffer zone of the Redberry Lake biosphere reserve is relatively small, composing only 6,300 hectares immediately surrounding the lake, or six per cent of the total reserve area. Of this, 920 hectares are protected under the provincial Wildlife Habitat Protection Act. A regional park was created for the remainder to ensure that land developments at the lakefront would not jeopardize habitat provided for waterfowl and shorebirds. The transition zone, where the town of Hafford and the rural municipality of Redberry are located, composes 89 per cent of the land area of the biosphere reserve. The dominant resource activity is agriculture and, therefore, most of the land in the transition zone of the biosphere reserve is privately owned.

At Redberry Lake, there are very few public instruments for environmental conservation and relatively few EMOs present in the region. Most land is held privately and legislation regulates individual users rather than regional priorities. Acts and regulations established for agriculture in the province of Saskatchewan tend to focus on operational issues of individual properties (e.g. Environmental Farm Plans) with issues such as nutrient management and integrated pest management identified as key components of shifting toward more sustainable agriculture. Other initiatives, such as the Canada-Saskatchewan Farm Stewardship Plan, encourage landscape-scale planning within a larger geographic area defined by physical boundaries (e.g. a watershed) to address local agri-environmental issues. Upon completion of the group plan, producer members of the group are eligible to apply for financial incentives to assist them in implementing best management practices for the issues that they have chosen.

Despite these types of government programmes, the initial lack of government interest in the Redberry region was striking, given the necessity of provincial and federal endorsement of the UNESCO nomination. Although civil servants working out of Regina (the provincial capital) received funding requests by RLBR members, neither the provincial nor the federal level of government has been directly involved with the biosphere reserve since it was created. Additionally, the Redberry Lake Biosphere Reserve committee had not yet been successful in securing any Aboriginal involvement through their requests for participation from the First Nation, with headquarters located in the nearby city of Saskatoon.

Steering governance

Until 2006, nine regular volunteers, of whom six are board members, served the Redberry Lake Biosphere Reserve committee. Since its inception, the committee made annual requests to the provincial government to provide core funding to maintain research, education, and public information activities, and to keep its interpretive centre in operation. Until 2005, these requests met with very modest levels of funding, provided to the biosphere reserve on an annual basis. In 2005, multi-year funding was specified in the provincial budget, providing an opportunity for the committee to meet some of the programme objectives it set out in the *Redberry Lake Community Sustainability Plan* (Sian, 2001).

As a voluntary organization, the focus of the RLBR committee was to continue lobbying government for more operating funds. For the first six years of its existence, the committee retained the same membership of mostly farmers or local business people, representing the various communities and regional municipalities within the biosphere reserve. While this strategy may be beneficial for continuity, it risks stagnation, particularly as many members of the committee were also committed to other community organizations in the area. However, the committee restructured its board in 2006 and invited government and university scientists to help promote the biosphere reserve and link local interests with broader research priorities at the provincial university and (federal and provincial) government research agencies.

Two national environmental groups with conservation mandates have become active in the Redberry region: Ducks Unlimited Canada (DUC) and the Nature Conservancy of Canada (NCC). These organizations undertake important conservation work within and outside biosphere reserves by working with landowners to protect habitat. Joint ownership, land donations, stewardship agreements, and conservation easements are all part of these EMOs' strategy for placing land under a protective management regime (Ducks Unlimited Canada, 2004a; Nature Conservancy of Canada, n.d.). Both the NCC and DUC establish their own national conservation priorities prior to negotiating with individual landowners or acquiring land themselves for stewardship values rather than traditional extractive practices.[4] Both organizations place a high value on the 'best available conservation science' (Nature Conservancy of Canada, 2001; see also Ducks Unlimited Canada, 2004b) to identify priority landscapes for their activities.

In Redberry Lake, these organizations worked on projects together to protect riparian habitat at Oscar Creek within the core area. However, some members of the community expressed concern that national EMOs negotiate land donations, stewardship agreements, or conservation easements with individual landowners without the knowledge of the adjacent owners. Farming communities might perceive land acquisition as competitive with traditional uses and feel threatened if they are excluded from the governance (i.e. the planning and decision-making phases) of conservation.

Neither the NCC nor DUC were able to connect with the Redberry Lake Biosphere Reserve Committee, despite their shared interests in principle. Proposals on the part of these EMOs were greeted with a sense of distrust by members of the committee who were reluctant to establish partnerships with external organizations that would not provide any financial benefits to build local capacity. According to one board member who is appointed by the regional municipality, some committee members had a fear of opening the door to national partnerships for fear of losing local control. Another explanation for the lack of partnerships may lie in the philosophies, goals, and management approaches of these two national EMOs which appear to be in marked contrast to interviewees' interpretations of the broader prairie cultural tradition that focuses on the importance of the small producer and *collective* goodwill. Consequently, the external EMOs decided to undertake their own projects without further reference to the committee, and the biosphere reserve more broadly, as they found it was more effective to work on their own. Thus, through mutual distrust and claims of exclusion, partnerships for more effective governance of the biosphere reserve as a whole failed to become established.

Governance analysis

The lack of collaboration between the RLBR committee and national EMOs, reinforced by private property regimes, combined with neglect from provincial government and other provincial EMOs (e.g. Saskatchewan Environmental Society; Nature Saskatchewan) reinforced a form of environmental management based on privately driven institutional arrangements. In the particular case of Redberry Lake, both the tools that have been used (property instruments) and the ways of wielding them (in confidential negotiations and closed planning processes) are more private than public in character. The result is that critical elements of environmental management practice (including assessing ecological significance of particular properties and regions; determining the required level and type of intervention; composing and implementing enforcement and monitoring strategies) are being undertaken by EMOs in the absence of broader community participation in steering the regional regime. Thus, protected area governance has been driven by a few organizations without the benefit of government interest or local community capacity to engage more fully.

In 2005, persistent lobbying of the provincial government coincided with the government's policy of supporting a 'green economy'. The biosphere reserve succeeded in obtaining core funding for the next five years. With this funding, the reserve renovated its original interpretive centre and re-opened it in 2006 as a research and education centre. New connections with the local school were established and the school was granted a national award for its work in environmental education and activism. The committee publicly stated that it sought new partnerships with organizations working in the region, including NCC and DUC. Public recognition of shared interests is an important first step, but must be accompanied by explicit invitations to develop shared work to address the gap that has grown up between organizations.

To address this gap, the committee secured funding for an initiative to promote ecologically sensitive farming practices through subsidies that would provide farmers with financial incentives to protect their 'piece' of the watershed. This project involves partnerships with six rural municipalities, the Saskatchewan Watershed Authority, DUC, and the Prairie Farm Rehabilitation Administration (PFRA) under the federal Department of Agriculture. The committee also restructured its board, inviting government and university scientists to join to help promote the biosphere reserve and link the locality with research priorities at the provincial university and (federal and provincial) government research agencies. It appears that the financial recognition brought greater confidence and capacity for engagement on a number of levels, helping biosphere reserve members to embrace their biosphere reserve as a system open to outside players and external institutional arrangements. Most importantly, the local committee is newly poised to address institutional fragmentation by beginning to serve as brokers for several landscape-scale initiatives under more complex organizational partnerships.

Conclusions

As traditional government-led management of protected areas gives way to diverse forms of collaborative governance, communities face the dual challenges of navigating complex institutional and organizational arrangements while managing pressures on their open systems from many different scales. For communities to help steer governance in any meaningful way, they need to take an active role in creating and maintaining landscape regimes. Despite the many differences among the cases explored above, each case presented the conditions in which communities can work toward creating governing regimes for protected areas, particularly when private properties are involved. Several themes help to illustrate the factors that allow communities and EMOs to steer governance more directly. These themes include: (1) overcoming fragmentation of institutional arrangements; (2) building capacity through collaboration; and (3) realizing the potential for communities in open systems to address sustainability concerns beyond their core protected areas.

The complexity of institutional arrangements for a given landscape will affect the ability of a community to create formal and informal regimes. Some community groups struggle to make sense of the fragmentation of institutional organizations, instruments, and regimes within their particular landscape. To participate effectively in shaping governance regimes for protected areas, communities must first navigate the existing governance system

(Francis, pers. comm., 2006) and then form or maintain a regime that will help to steer the governance system toward their desired goals. Niagara Escarpment and the ORM share a similar institutional context in terms of having complex overlays of organizations, agencies, and EMO coalitions working below, above, and at the landscape scale. Each of these protected areas is governed by a combination of regimes, such as legislation, regulation, and planning guidelines. Yet, the presence of informal institutions, such as EMOs (and their education campaigns, lobbying, media attention, and political leverage), ensure that the governance regimes of protected areas continue to evolve in order to keep pace with external threats that flow into the system (e.g. resource extraction and urban growth pressure).

Conversely, in the Redberry Lake case, where formal and informal institutional mechanisms are somewhat lacking, individual EMOs have sought to create conservation regimes consistent with their own mandates, and often at the expense of deeper community engagement. It appears that the absence of effective institutions and organizations to fulfil management functions creates a unique type of fragmentation due to *gaps* in civic and governmental participation. If there is a vacuum in regional governance regimes, then organizations (either private or public) with sufficient capacity might begin to steer governance in a particular direction without significant community participation.

Ontario's experience can also inform other regions interested in effective landscape governance. The cases reveal that as protected areas, such as biosphere reserves, move from designation to implementation, they need to critically assess the potential for strategic combinations of private/public/civic governance arrangements with organizations and agencies within and outside their locality. For communities to get involved, biosphere reserve committees can provide opportunities for greater public scrutiny and debate of management alternatives, such as the work of CONE to oppose aggregate extraction and development proposals. Such efforts require that management committees and other EMOs reach out to one another to negotiate the terms of local environmental regimes.

Public debate about the future of the Niagara Escarpment as a whole landscape unit was partially inspired by the collaboration of community groups steering the development of the Bruce Trail, both a formal and physical presence on the land and a symbolic one. Where local governments previously had traditional jurisdiction over planning and management regimes in each municipality, the creation of the NEC and its associated legislation created a much broader governance regime for the whole landscape unit. EMOs in the ORM built even more broadly based community networks to share information, set agendas, and negotiate protection regimes. Using the media to leverage public support gave their interactions more power and influence. The collaborative nature of the STORM Coalition gave it credibility as an institution to actually steer government toward conservation planning and create strong governance regimes to address urban sprawl.

A second and related lesson that the cases suggest is that community capacity to address institutional fragmentation can begin to be addressed by multi-stakeholder collaboration. One major role for communities lies in creating and maintaining appropriate regimes for landscape governance. As Francis (2003: 238) explains, it is critical that protected area managers and affected communities 'seek organizational arrangements that are sufficient to carry out sustainable development'. For some communities, the ability to establish links among local, regional, and external organizations is accomplished through the work of EMOs with a common vision and purpose. For others, bridging their local concerns and conservation priorities with those of other actors (e.g. government, science, EMOs) in a way that simultaneously builds reciprocity, trust, and accountability, as Hydén (1999) observed above, remains a significant challenge. Currently, these characteristics appear to be strongest within the STORM Coalition and weaker for the Redberry Lake community. However, with continued levels of modest funding for local efforts, the capacity to create new governance regimes could be enhanced, resulting in more fruitful partnerships with organizations operating in the biosphere reserve as a whole. By including other stakeholders, such as government agencies and university researchers, capacity can also be enhanced to address the broader challenges of an open system, such as global environmental change and the effects of neoliberal economic restructuring on rural livelihoods.

The potential for science to communicate the challenges of open systems was seen in the cases of Niagara Escarpment and ORM. Both showed how the capacity for new environmental regimes could be created when compelling science is presented to an interested public and attentive decision-makers. Having a common monitoring regime, such as *Watershed Report Cards* or *State of the Environment* reports, can establish a credible source for information about landscape-scale change. At the local level, for example, water- and air-quality monitoring can become institutionalized when it is community-driven by cultural values that connect ecological and human health (Pollock and Whitelaw, 2005). In both Ontario cases, science was used to legitimate protection proposals and helped to instil landscape values for the public. Yet, this was not the case at Redberry Lake because much of the scientific work was not easily accessible and the Biosphere Reserve committee lacked the capacity to coordinate their interests with other organizations. Consequently, there emerged an imbalance in efficacy among organizations and a rift between external EMOs and the local Biosphere Reserve committee.

Collaboration between civic organizations and public agencies contributes to both private and public property regimes, builds institutional capacity, and helps to steer governance. The interplay between EMOs and government is key to understanding the efficacy of the environmental management regime at Redberry Lake. Like Niagara Escarpment and the ORM, the strong presence of private property regimes offered a pragmatic opportunity for private land conservation by EMOs to undertake conservation projects. Nevertheless, acting on these opportunities without previous local discussion hindered collaboration at Redberry Lake.

To generate improved trust and collaboration at the landscape scale, biosphere reserves might consider becoming repositories of scientific research. Such action would require local committees to become more actively involved in identifying and advertising research needs, linking scientists undertaking research locally, and ensuring data collected is retained publicly.[5] This is a difficult challenge that requires sustained funding, committed staff *and* volunteers, as well as efforts to find common ground and to create incentives to work together on conservation initiatives. However, such a role would give local communities greater capacity to make links between individual EMOs and ultimately to take part in steering regional governance activities themselves.

Communities are beginning to realize the benefits of tracking sustainability at the landscape scale. Indeed, for the agricultural communities of Redberry Lake, and the Countryside areas of the ORM, the viability of traditional farming (including the incremental shift to some organic farming) is a key component of governance for sustainability not addressed in these cases. The *Monitoring the Moraine* project creates a new regime to measure not only biodiversity and ecological integrity on the ground, but also the consistency between political objectives (as stated in official documents) and political decisions. In the case of Redberry Lake, the *Community Sustainability Plan* could help local communities to define formal and informal regimes that reflect their concerns for rural sustainability. As UNESCO (2006) has noted about Redberry: 'There exists a strong potential to undertake the development of new, sustainable agriculture, livestock, and silviculture products that could be marketed under the "brand" of the biosphere reserve, such as "model" farms and natural prairie grass cultivation for seedstocks.' In response to the fragmented and open nature of these landscape systems, communities are beginning to think about what types of institutional arrangements give them greater opportunities for participation in governance, in order to secure fundamentally important livelihoods as part of a broader vision for landscape protection.

Governance regimes for protected areas seem to go far beyond traditional management of parks and natural resources. Some communities seem to recognize the interconnections between protected areas and sustainable livelihoods, between environmental and human health, and between local pressures and global change. These emergent forms of regime creation rely on community engagement and, indeed, leadership, in order to steer governance in new ways. Jamison (2001: 152) urges us to see that:

> sustainable development is not merely about environmental problems: it is also about local governance, about making democracy work . . . Community activism today is about synthesizing local knowledge and experiences with global challenges, and it is important that we understand the difficulties involved.

Understanding the process of regime formation at the landscape scale will help communities, researchers, and practitioners to identify how best to steer governance arrangements. Much more research is required. Our experience suggests that the complex institutional arrangements in any governance system need to be first understood, then tested for their democratic practice, and, finally, engaged by communities through collaborative institutional arrangements. Regimes that address institutional fragmentation and channel the pressures from open systems have the potential to steer the governance of protected areas at a landscape scale. In this way, governance can become a pathway for the conservation of protected areas within the much broader context of achieving sustainable development.

Acknowledgements

The authors would like to thank those who participated in the original research for these case studies. We especially appreciate the comments from the community reviewers Richard Murzin (Niagara Escarpment), Debbe Crandall (Oak Ridges Moraine), and Andrew Hawrysh (Redberry Lake) who helped bring the information up to date. We would also like to acknowledge the anonymous reviewers who helped to improve the original manuscript with their thoughtful suggestions. And, finally, we wish to thank Professor George Francis for his insights and support.

Notes

1. The Niagara Escarpment case study is based on dissertation research carried out by Graham Whitelaw during 2004 and 2005. The research included review of policy documents, legislation, and secondary research (Plaunt, 1978). Personal involvement by Graham in Niagara Escarpment planning activities between 1990 and 2006 also contributed to the case study.
2. The Oak Ridges Moraine case study is based on dissertation research carried out by Graham Whitelaw during 2004 and 2005. The research included review of policy documents and semi-structured interviews. Personal involvement by Graham in Oak Ridges Moraine planning activities between 1991 and 2006 also contributed to the case study.
3. The Redberry Lake case study is based on field observations conducted between 2002 and 2004 by Maureen Reed and Sharmalene Mendis. The analysis involved participation in Biosphere Reserve Committee meetings, review of committee documents, the official nomination papers submitted to UNESCO, and related government documents. A total of 44 semi-formal interviews and 13 additional interviews were conducted with community, government, and environmental movement organizations.
4. The NCC has chapters across Canada that work with local land trusts and they have been active in several Canadian biosphere reserves. The NCC has worked with cattle ranchers in south-western Alberta's Waterton biosphere reserve to secure land under threat of urban development. With provincial environment staff in Ontario, they have supported Ecological Surveys of the Niagara Escarpment (Riley et al., 1996) and the Eastern Georgian Bay (Jalava et al., 2005). In June 2006, the

NCC provided the financial instrument to double the size of the St Lawrence Islands National Park, which makes up the core area of the Thousand Islands-Frontenac Arch biosphere reserve in Ontario (Birtch, pers. comm., 2006).
5 The Canadian Biosphere Research Network encourages communities, researchers, and institutions to exchange knowledge, information, and experience in working in, and with, national biosphere reserves (CBRN, 2006).

Literature cited

Advisory Panel for Consultations on the Oak Ridges Moraine (2001). *Workshop Guides*. Toronto, Ontario: Ogilvie and Company.

Anonymous (2002). *15 Years: 15 years towards sustainable development at the Redberry Lake Biosphere Reserve, 1998–2002*. (Copy available from the Redberry Lake Biosphere Reserve, PO Box 221, Hafford, SK, Canada S0J 1A0).

Bagby, K. and Kusel, J. (2003). *Civic Science Partnerships in Community Forestry: building capacity for participation among underserved communities*. Taylorsville, CA: Pacific West Community Forestry Center.

Berkes, F. (2004). Rethinking community-based conservation. *Conservation Biology* 18, 621–630.

Birtch, J. (2006). Personal communication. 8–11 June 2006.

Bruce Peninsula Biosphere Association (2006). Niagara Escarpment Commission. Website viewed on 19 June 2006. www.escarpment.org/pdfandword_files/BPBA_Brochure.pdf.

Bruce Trail Association (2006). Website viewed on 19 June 2006. www.brucetrail.org.

Calderisi, D. (2003). *Speech to the Niagara Escarpment 25th Anniversary Reception*, 15 October 2003, Faculty Club, University of Toronto, Toronto: Ontario Nature.

Campbell, C. (2005). *Shaped by the Westwind: nature and history in Georgian Bay*. Vancouver: UBC Press.

Coalition on the Niagara Escarpment (2006). Website viewed on 19 June 2006. www.niagara.escarpment.org.

CBRN (2006). The Canadian Biosphere Reserve Research Network. Website viewed on 19 June 2006. www.biosphere-research.ca.

Cortner, H. and Moote, M. (1999). *The Politics of Ecosystem Management*. Washington, DC: Island Press.

Dorcey, A.H.J. (2004). Sustainability governance: surfing the waves of transformation. In B. Mitchell (ed.), *Resource and Environmental Management in Canada: addressing conflict and uncertainty* (3rd edn), pp. 528–554. Don Mills, Ontario: Oxford University Press.

Draper, D. (2004). Marine and freshwater fisheries. In B. Mitchell (ed.), *Resource and Environmental Management in Canada: addressing conflict and uncertainty* (3rd edn), pp. 200–232. Don Mills, Ontario: Oxford University Press.

Ducks Unlimited Canada (2004a). *Our Wetland and Wildlife Conservation Progress*. Website viewed on 19 June 2006. www.ducks.ca/aboutduc/progress/index.html.

—— (2004b). *Research Guiding Conservation*. Website viewed on 19 June 2006. www.ducks.ca/aboutduc/how/research.html.

Ellsworth, J.P. and Jones-Walters, L. (2006). *Journeys in Governance: the role of federal governments in addressing tough community issues and their underlying causes*, Working Paper, March 2006.

Environmental Assessment Advisory Committee (1989). *The Adequacy of the Existing Environmental Planning Approvals Process for the Ganaraska Watershed*, Report 38 to the Minister of the Environment. Toronto: Environmental Assessment Advisory Committee (members: P. Byer, R.B. Gibson, and C. Lucyk).

Francis, G. (1988). Institutions and ecosystem redevelopment in Great Lakes America with reference to Baltic Europe. *Ambio* 17 (2), 106–111.

—— (1995). Ecosystems. In L. Quesnel (ed.), *Social Sciences and the Environment*, pp. 145–171. Ottawa: University of Ottawa Press.

—— (2003). Governance for conservation. In F.R. Westley and P.S. Miller (eds), *Experiments in Consilience: integrating social and scientific responses to save endangered species*, pp. 223–379. Washington, DC: Island Press.

—— (2004). Biosphere Reserves in Canada: ideals and some experience. *Environments* 32 (3), 3–26.
—— (2006). Personal communication. Numerous discussions between 2005 and 2006.
—— and Whitelaw, G. (2002). *Niagara Escarpment Biosphere Reserve Periodic Review*. Canadian Biosphere Reserves Association, Reviewers on Behalf of the Canadian Commission for UNESCO and Canada/MAB.
Government of Ontario (1962). *Metro Toronto and Region Transportation Study*, Toronto: Ontario Ministry of Municipal Affairs.
—— (1985). *The Niagara Escarpment Plan*. Georgetown, Ontario: Niagara Escarpment Commission. Website viewed on 19 June 2006. www.escarpment.org.
—— (1991). *Oak Ridges Moraine Implementation Guidelines*. Toronto: Ontario.
—— (2001). *Oak Ridges Moraine Conservation Act*. Toronto: Ontario Ministry of Municipal Affairs. Website viewed on 19 June 2006. www.e-laws.gov.on.ca/DBLaws/Statutes/English/01o31_e.htm.
—— (2002). *Oak Ridges Moraine Conservation Plan*. Toronto: Ontario Ministry of Municipal Affairs. Website viewed on 19 June 2006. www.mah.gov.on.ca/userfiles/HTML/nts_1_6846_1.html.
—— (2006). *Greenbelt Plan, Toronto*. Ontario Ministry of Municipal Affairs. Website viewed on 19 June 2006. www.mah.gov.on.ca/userfiles/HTML/nts_1_22087_1.html#greenbelt.
Graham, K. and Phillips, S. (eds) (1998). *Citizen Engagement: lessons in participation from local government*. Toronto: Institute of Public Administration of Canada.
Healey, P. (1997). *Collaborative Planning: shaping places in fragmented societies*, Vancouver: UBC Press.
—— (1998). Building institutional capacity through collaborative approaches to urban planning, *Environment and Planning A* 30, 1531–1546.
—— (2003). Collaborative planning in perspective, *Planning Theory* 2 (2), 101–123.
Hurley, P.T. and Walker, P.A. (2004). Whose vision? Conspiracy theory and land-use planning in Nevada County, California. *Environment and Planning A* 36, 1529–1547.
Hydén, G. (1992). Governance and the study of politics. In M. Bratton and G. Hydén (eds), *Governance and Politics in Africa*, pp. 1–26. Boulder, CO: Lynne Rienner.
—— (1999). Governance and the reconstitution of political order. In R. Joseph (ed.), *State, Conflict and Democracy in Africa*, pp. 179–196. Boulder, CO: Lynne Rienner.
Jalava, J.V., Cooper, W.L., and Riley, J.L. (2005). *Ecological Survey of Eastern Georgian Bay*. Peterborough, Ontario: Nature Conservancy of Canada and the Ontario Ministry of Natural Resources.
Jamison, A. (2001). *The Making of Green Knowledge: environmental politics and cultural transformation*. Cambridge, UK: Cambridge University Press.
Jessop, B. (1997). Capitalism and its future: remarks in regulation, government and governance. *Review of International Political Economy* 4 (3), 561–581.
—— (2002). *Governance and Metagovernance: on reflexivity, requisite variety, and requisite irony*. Department of Sociology, Lancaster University, UK. Website viewed on 19 June 2006. www.lancs.ac.uk/fss/sociology/papers/jessop-governance-and-metagovernance.pdf.
Kanter, R. (1990). *Overview: options for a Greater Toronto Area Greenlands Strategy*. Toronto: Ontario Ministry of Natural Resources.
Knight, B., Chigudu, H., and Tandon, R. (2002). *Reviving Democracy: citizens at the heart of governance*. London: Earthscan.
Kooiman, J. (2003). *Governing as Governance*. London: Sage Publications.
Kreutzwiser, R. and Loë, R. de (2004). Water security: from exports to contamination of local water supplies. In B. Mitchell (ed.), *Resource and Environmental Management in Canada: addressing conflict and uncertainty* (3rd edn), pp. 166–194. Don Mills, Ontario: Oxford University Press.
Kusel, J. (2001). Assessing well-being in forest dependent communities. In G.J. Gray, M.J. Enzer, and J. Kusel (eds), *Understanding Community-Based Forest Ecosystem Management*, pp. 359–382. New York, London, and Oxford: The Haworth Press.
Louis, C. (2003). *Speech to the Niagara Escarpment 25th Anniversary Reception*. 15 October 2003, Faculty Club, University of Toronto.
McCarthy, D. (2006). *An Exploration for a UNESCO World Biosphere Reserve Designation: Oak Ridges Moraine/Greenbelt Areas*. Proposal to the Oak Ridges Moraine Foundation. Team: Daniel McCarthy, Robert Gibson, Susan Wismer, Debbe Crandall, George Francis.

McGillivray, W.B. (1998). The Aspen Parkland: a biological perspective. In D.J. Goa and D. Ridley (eds), *Aspenland, 1998: local knowledge and sense of place*, pp. 95–103. Central Alberta Museums Network, Red Deer Alberta, Canada.

McKinney, C., Fitch, C., and Harmon, W. (2002). Regionalism in the West: an inventory and assessment. Prepared for the *Public Land and Resources Law Review*, pp. 1–20.

Mendis, S. (2004). Assessing community capacity for ecosystem management: Clayoquot Sound and Redberry Lake Biosphere Reserves. Unpublished Master's Thesis. Saskatoon: University of Saskatchewan.

Millennium Ecosystem Assessment (2005). Website viewed on 16 November 2006. www.maweb.org/.

Monitoring the Moraine (2006). Website viewed on 19 June 2006. www.monitoringthemoraine.ca/.

Murzin, R. (2003). *Speech to the Niagara Escarpment 25th Anniversary Reception*. 15 October 2003, Faculty Club, University of Toronto, Toronto: Ontario Nature.

—— (2006). Personal communication. 27 November 2006.

Nature Conservancy of Canada (2001). *Conservation Science: A science-based conservation program*. Website viewed on 19 June 2006. www.natureconservancy.ca/files/frame.asp?lang=e_and region=1andsec=science.

—— (2003). *Prairie Conservation Planning Update: highlighting ecoregional and site conservation planning across the prairies by the Nature Conservancy of Canada and its partners*. Regina: Nature Conservancy of Canada.

—— (no date) *A Conservation Toolchest for Landowners: investing in the land you love*. Website viewed on 19 June 2006. www.natureconservancy.ca.

Niagara Escarpment Study Group (1968). *Niagara Escarpment Study Conservation and Recreation Report*. Prepared for the Treasury Department of Ontario.

Oak Ridges Moraine Foundation (2006). Website viewed on 19 June 2006. www.ormf.com.

Oak Ridges Moraine Technical Working Committee (1994). *The Oak Ridges Moraine Strategy for the Greater Toronto Area: an ecosystem approach for long term protection and management*. Toronto, Ontario: Ontario Ministry of Natural Resources by the Oak Ridges Moraine Technical Working Committee.

Pim, L. (2003). *Speech to the Niagara Escarpment 25th Anniversary Reception*. 15 October 2003, Faculty Club, University of Toronto, Toronto: Ontario Nature.

Plaunt, M. (1978). *The Decision Making Process that Led Up to the Passing of the Niagara Escarpment Planning and Development Act*. Unpublished paper prepared for Public Policy Course 610, York University, Toronto, Ontario.

Pollock, R.M. (2004). Identifying principles for place-based governance in biosphere reserves. *Environments* 32 (3), 27–41.

—— and Whitelaw, G.S. (2005). Community-based monitoring in support of local sustainability. *Local Environment* 10 (3), 211–228.

Rapport, D.J. (2004). Ecosystem health and ecological integrity: foundations for sustainable futures. In B. Mitchell (ed.), *Resource and Environmental Management in Canada: addressing conflict and uncertainty* (3rd edn), pp. 24–53. Don Mills, Ontario: Oxford University Press.

Reed, M.G. (1995). Co-operative management of environmental resources: a case study from Northern Ontario, Canada. *Economic Geography* 71 (2), 132–149.

—— (1997). The provision of environmental goods and services by local non-governmental organizations: an illustration from the Squamish Forest District, Canada. *Journal of Rural Studies* 13 (2), 177–196.

—— (2003). *Taking Stands: gender and the sustainability of rural communities*. UBC Press: Vancouver.

—— (forthcoming). Uneven environmental management in Canada. Submitted to the *Journal of Environmental Management* January 2006.

Regional Municipality of York and Save the Oak Ridges Moraine Coalition (2006). *Exploring the (Potential) Application of Strategic Environmental Assessment in the Regional Municipality of York, Ontario, Canada, an Environmentally Significant Urban Periphery Area that Includes Portions of the Oak Ridges Moraine*. A proposal to the Canadian Environmental Assessment Agency's Research and Development Programme, Newmarket, Ontario, Canada.

Rice, J.J. and Prince, M.J. (2000). *Changing Politics of Canadian Social Policy*. Toronto: University of Toronto Press.

Riley, J.L., Jalava, J.V., and Varga, S. (1996). *Ecological Survey of the Niagara Escarpment Biosphere Reserve*. Nature Conservancy of Canada and the Ontario Ministry of Natural Resources.

Royal Commission on the Future of the Toronto Waterfront (1992). *Regeneration*. Final Report, Toronto, Ontario: Royal Commission on the Future of the Toronto Waterfront.

Saskatchewan Watershed Authority (2003). *Hydrographic Data for Redberry Lake near Krydor Levels 1966–2003*. Station: 05GD003.

Schmutz, J. (1999). *Community Conservation Plan for the Redberry Lake Important Bird Area*. Saskatoon: Centre for Studies in Agriculture, Law and the Environment, University of Saskatchewan.

Selman, P. (2006). *Planning at the Landscape Scale*. London: Routledge.

Sian, S. (2001). *Redberry Lake Biosphere Reserve: a community's plan for sustainability*. (Copy available from the Redberry Lake Biosphere Reserve, PO Box 221, Hafford, SK, Canada S0J 1A0).

Slocombe, D.S. (2004). Applying an ecosystem approach. In B. Mitchell (ed.), *Resource and Environmental Management in Canada: addressing conflict and uncertainty* (3rd edn), pp. 420–441. Don Mills, Ontario: Oxford University Press.

Smith, M. (2005). *A Civil Society? Collective Actors in Canadian Political Life*. Peterborough: Broadview Press.

STORM (2006). *Save the Oak Ridges Moraine*. Website viewed on 16 November 2006. www.stormcoalition.org/.

Swift, J. (1999). *Civil Society in Question*. Toronto: Between the Lines.

Tovell, Walter M. (1992). *Guide to the Geology of the Niagara Escarpment, with Field Trips*. Edited by L. Brown. Georgetown, Ontario: Niagara Escarpment Commission and the Ontario Heritage Foundation.

UNESCO (2002). *Biosphere Reserves: special places for people and nature*. UNESCO: Paris.

—— (2006). *World Network of Biosphere Reserves: North America: Canada*. Website viewed on 16 November 2006. www.unesco.org/mab/br/brdir/directory.

Whitelaw, G. (2006). *The Role of Environmental Movement Organizations in Land Use Planning: case studies of the Niagara Escarpment and Oak Ridges Moraine processes*. Unpublished doctoral dissertation, University of Waterloo, Waterloo, Ontario, Canada.

—— and Hamilton, J. (2003). *Evolution of Niagara Escarpment Governance*, Conference Proceedings of the 5th International Science and Management of Protected Areas Association, 11–16 May 2003, Victoria, BC, Halifax: SAMPAA.

——, Hamilton, J., and Milne, R. (2005). Historical recreation patterns along the Niagara Escarpment in Ontario and the challenges to heritage-based recreation and tourism. Proceedings *Heritage-based Recreation Along the Great Arc*. Edited by G.J. Nelson and J. Peter. Waterloo, Ontario, An Environments Publication, University of Waterloo.

Wilson, A. (1991). *The Culture of Nature: North American landscape from Disney to the Exxon Valdez*. Toronto: Between the Lines.

Young, O.R. (1997). Arctic governance: bringing the high latitudes in from the cold. *International Environmental Affairs* 9 (1), 54–68.

Part II
Critical perspectives

Chapter 7
Conflict and protected areas establishment:
British Columbia's political parks

Kevin S. Hanna, Roderick W. Negrave, Brian Kutas, and Dushan Jojkic

It would be an understatement to characterize natural resource policy on Canada's west coast as being quarrelsome. Canada's westernmost province, British Columbia (BC), is a jurisdiction where conflict about public land use plays a significant part in public policy dialogue. The use of public lands in Canada, and, indeed, across North America, has focused on deriving social and economic benefits from resource extraction (Perry, 1998). Despite the rhetoric of ecosystem management, integration, and multiple use that flows from managing agencies and industry, in practice the preservation of ecosystems is given considerably less attention (Bean, 1997; Beaty, 2000; Margules and Pressey, 2000). The industrial use of public lands also reflects the values of many citizens, especially in communities where well-being is tied to resource industries (Grumbine, 1997; Song and M'Gonigle, 2001). Conflict has emerged as alternative views of the values of public lands have gained prominence, particularly with respect to the designation of protected areas, which have, in turn, become integral conservation policy instruments (Noss and Scott, 1997). Alternative views also have a dualistic relationship with conflict – they flow in part from a more conflictual public policy setting, but they also intensify it. We can increasingly observe contexts where protected areas decisions are dictated not so much by ecology or science as by emotive conflict over place.

This chapter considers the role of conflict in the creation of protected areas. Four examples from BC are provided. These were chosen because they provide particularly compelling illustrations of conflict about land use, and in each a protected area designation was the result. The BC case will undoubtedly resonate with practitioners and scholars working in other jurisdictions. British Columbia is interesting because of the volatility of the parks and protected areas setting, and because series of local events have had a cumulative effect that has resulted not just in the designation of new protected areas, but also in larger changes to macro resource-management land use policies. Conflict in BC has also been focused on forested landscapes, an experience shared by many other global settings.

The chapter begins with a brief discussion of the ideas that guide our perspective on events, and then focuses on case studies from BC. Events in BC are then woven into a discussion of larger subsequent policy changes represented in several large-scale consultation and land use planning initiatives (BC's Commission on Resources and Economy, Land Resource Management Plans, and the Protected Areas Strategy). We conclude by outlining a conceptual process framework for characterizing conflict stages in parks and protected areas.

The place of conflict

Creating protected areas out of public lands is a political process – one that requires not only public support, but also the acquiescence of industry and a range of allied interests (Lippke and Bishop, 1999). Protected area designation is a social choice and conservation policies are not always an easy political sell. Protected areas can be contentious, especially when they conflict with uses that have established economic and social values (Priddle, 1982; Lanskail, 1984; Soulé, 1991; Beaty, 2000; Margules and Pressey, 2000; Dearden and Rollins, 2002). Protected areas policy has been guided by a certain place-sensitivity, imbued with a pragmatism that seeks to avoid industrially important areas in designation processes (Norton, 1999). This is not to say that North America's protected areas are devoid of ecological value, but the ecological benefits of many are variable (Shafer, 1995; Soulé and Sanjayan, 1998).

Protest movements and the fundamental differences in values among the diverse interests they often represent have emerged as common motivating factors for governments to withdraw land from extractive use (Wilson, 2001b). But from the perspective of protecting ecosystems or diversity, conflict-based decisions might not necessarily be the best ones. There has also been a tendency in green rhetoric to describe places in superlatives to convince audiences of their value. Protected areas designation not uncommonly comes about from conflict generated by aesthetics, the perception that an area may have a particular ecosystem value (which may or may not be supported by science), or a 'last of its kind' syndrome where arguments focus on the rationale that a place must be preserved because it is the last of its kind within a given region, nation, or even globally.

Conflict can serve tandem objectives. Environmental movements seek not only specific place preservation, but they often also seek larger policy change. For public agencies acquiescing to the preservation of a place of contention may become easier than fundamentally changing the political economic/ecologic foundations of public resource use. In BC it has been argued that despite protecting areas such as those described here, each of which was designated as a result of public protest, surrounding landscapes have continued to be transformed by human activities. Conflict may also inject hesitancy into the policy process, where agencies design policy around conflict mitigation, rather than from a detailed knowledge of the ecological qualities of places.

British Columbia examples

Context

In BC natural resources are for the most part publicly owned, while the means of production are privately controlled. The BC government owns about 95 per cent of productive forest land; private interests hold tenures to extract timber and other natural resources. Until the 1980s differences over forestry practices had been sporadic and localized, hardly posing a significant challenge to government policies. But starting in the 1980s and lasting into recent years plans to log areas such as the Stein Valley, areas on Vancouver Island, and, more recently, on BC's central and north coasts served as the catalysts for a large protest movement that sought not only to preserve these areas, but to change the fundamental structure of the province's forest economy and forest management practices. The catalyst for this movement came from intrinsic and evocative views of each of the above areas as unique places; centred on their aesthetics (as places of beauty), specific features (such as big trees), or the belief that they represented the last untouched examples of large watersheds or forests, giving each a *power of place*.

The stage for recent land use conflicts was set early by BC's process of forest tenure allocation. During the 1950s it was Tree Farm Licenses (TFL) issued by the provincial government to logging companies for harvesting rights over a relatively large area. The TFL system may have helped insure the stability of the timber supply for a while, but it also created a context of corporate concentration. The province had long been regulating and monitoring logging operations, adjusting the Annual Allowable Cut, and enforcing a range of regulations, but these were very much about supporting timber production and for many years this policy was well supported by British Columbians.

But public tastes changed. Logged landscapes were no longer seen as symbolic of prosperity and work, and, in the 1980s, the pact between communities, industry, and government weakened with declining forest employment. The ecological integrity of forest ecosystems became a cause célèbre in the Pacific Northwest just as many timber firms were expanding their logging operations into areas with significant aesthetic assets. This was to set the stage for years of protest (Cashore et al., 2000; Wilson, 2001a). Four places stand out in the BC conflict story: the Stein Valley, Clayoquot Sound, the Carmanah and Walbran valleys, and the 'Great Bear Rainforest' (see Figure 7.1).

Figure 7.1 British Columbia, with locations discussed in this chapter

Stein Valley Nlaka'pamux Heritage Park

The Stein Valley Nlaka'pamux Heritage Park covers an area of 107,101 hectares and is located along the south-eastern portion of the Coast Mountains in south-west mainland British Columbia (BC Parks Division, 2000). This area has great ecological significance within BC. It encompasses transition zones where plant life changes from the wet coast to the dry interior, while also changing from high mountains to lowland. The Stein has three outstanding features: its watershed; its physical diversity; and the physical nature of its cultural heritage (BC Parks Division, 2000). After protected area designation, the resulting Stein Nlaka'pamux Heritage Park encompassed the complete Stein River watershed, including tributary streams and associated glacier systems, from headwaters to its outlet into the Fraser River. This is essentially the last undisturbed watershed of any size within a day's drive of the large urban centres of the Vancouver region. The variety of ecosystems and environments contained within the park is rare. A number of variants from six of the 14 Biogeoclimatic Zones that occur in BC are found within the park (BC Parks Division, 2000). These zones include: Engelmann Spruce Subalpine Fir; Mountain Hemlock; Montane Spruce; Interior Douglas Fir; Ponderosa Pine; and Alpine Tundra. Environments range from alpine vegetation, including heath and grassland types, to closed forests of Engelmann spruce, subalpine fir, mountain hemlock, and lodgepole pine; to Douglas-fir forests; to open ponderosa pine–grassland complexes; to alluvial systems with floodplains dominated by black cottonwood. Only a few rivers in BC flow through as wide a range of landform, climate, and vegetation over such a short distance as does the Stein. The Stein Valley also provides linkages to other protected areas, such as nearby Mehatl Provincial Park.

The Stein Valley provides habitat for large 'heroic megafauna' such as grizzly bear, cougar, and wolverine, all living within relatively close proximity to Vancouver. The Stein River supports significant fisheries value, including resident sport species and anadromous species of commercial value, including Coho, Chinook, and Pink Salmon; Steelhead and Rainbow Trout; Rocky Mountain Whitefish; and Dolly Varden Char (BC Parks Division, 2000). The lower Stein Valley has the highest concentration of archaeological sites on the Interior Plateau of BC and one of the highest in the province.

Interest in protecting resource values in the Stein Valley first began in the 1970s. At the time the Stein Valley garnered considerable media interest, and in many respects the Stein became a harbinger for the war in the woods that was to develop on Vancouver Island and the BC coast. Advocacy and protests by a range of environmental groups, including several local organizations, eventually led to a two-year development moratorium. However, logging and mining concerns remained interested in the valley region, and while the recession of the early 1980s lessened resource development pressure on the area, the potential for development remained in the background.

Aboriginal interests came to the fore in the 1980s and early 1990s, notably with the tendering of a comprehensive land claim that included the Stein Valley. With advocacy groups the Lillooet Tribal Council and Lytton and Mt Currie First Nations sponsored the annual *Voices for the Wilderness Stein Festivals*. These events were highly successful in popularizing the appeal to protect the Stein, the zenith of which was the fifth festival at Mt Currie. Over 26,000 people attended the event (BC Parks Division, 2000). The Wilderness Advisory Committee recommended in 1986 that no access roads be developed in the Stein without formal agreement between the Lytton Indian Band and the provincial government (BC Parks Division, 2000).

In 1987, the Lytton and Mt Currie bands released their joint Stein Declaration; 'we will maintain the Stein Valley as a wilderness in perpetuity for the enjoyment and enlightenment of all peoples and the enhancement of the slender life thread of the planet' (BC Parks Division, 2000). Then, in 1989, the Lytton and Mt Currie First Nations, with no legal authority under BC law, declared the area a park, and renamed the watershed the 'Stein Valley Tribal Heritage Park: A Living Museum of Cultural and Natural History'.

One of the characteristics of the campaigns to protect the Stein was the ability of shifting and disparate environmental coalitions, composed of both Aboriginal and non-Aboriginal elements, to 'get the message out' and recruit public support, all at a time when the larger economic and social milieu was still quite development-friendly. It was a message that was being broadcast not just locally or regionally, but across the province and Canada, and external organizations were also showing interest. The campaigns to preserve the Stein were to act as models for subsequent movements, and certainly informed events evolving at Clayoquot Sound on Vancouver Island.

The Stein campaigns began a process of public discourse where traditional approaches to development and public land use were being seriously questioned. The BC compact between government, industry, and resource-based communities also began to fray. After a while there seemed to be little appetite on the part of the province and most development interests for a protracted fight in the Stein, especially since rather than abating, the conflict was gaining strength. The wilderness value of the Stein watershed was formally recognized with the designation of the Upper Stein and Lower Stein Wilderness Areas by *Order in Council* in 1987, with the proviso, however, that an access road for resource development could be constructed through the Lower Stein Wilderness Area (BC Parks Division, 2000). The Stein Valley was finally and permanently protected as a Class 'A' Park under the Parks Act of BC in November of 1995, becoming Stein Nlaka'pamux Heritage Park. The area is currently managed cooperatively by the BC government and the Lytton First Nation.

Clayoquot Sound

It is hard to underestimate the impact that land use conflict on Vancouver Island has had on resource management policy in BC. Clayoquot Sound covers an area of about 3,500 square kilometres of the western side of Vancouver Island. The physiography of Clayoquot Sound is very rugged and diverse, to say the least; it consists of coastal plain, islands, inlets and fjords, mountain slopes, lakes of variable size, and small rivers. The marine environment is also varied and includes shallow- and deep-water environments (BC Parks Division, 2003e); exposed and sheltered waters; numerous small islets and rocks; and a number of small estuaries. Other features of note include well-expressed karst landforms and geothermal hot springs (BC Parks Division, 2003b). The Clayoquot area is under hypermaritime influence, with all the wet climate characteristics of BC's outer coast (Pojar and Meidinger, 1991). Biogeoclimatic zones in the area are typical of wetter, exposed areas of BC's south coast – Coastal Western Hemlock Very Wet Southern Hypermaritime; Submontane Very Wet Maritime and Montane Very Wet Maritime variants; Mountain Hemlock Windward Moist Maritime variant and Alpine Tundra. The Windward Islands Mountain and Vancouver Island Shelf Marine Ecosections are also found in the area (BC Parks Division, 2003a–n). Ecosystems in the area extend from marine, shoreline, and foreshore environments through lowland, submontane, montane, and subalpine forests to coastal alpine environments.

Clayoquot's forest cover ranges from stands of low-elevation sitka spruce and western red cedar-western hemlock to mid-elevation stands dominated by western hemlock and amabilis fir to high-elevation forests of mountain hemlock, amabilis fir, and yellow cedar. Seral environments dominated by red alder and alluvial black cottonwood are also found in the area. At least 29 rare plant species have been recorded in the area, the most significant of which is dwarf trillium (BC Parks Division, 2003a–n). High-quality spawning habitat for Sockeye, Chinook, and Coho salmon is found in the area, as is habitat for Dolly Varden Char, Steelhead, Rainbow, and Coastal Cutthroat trout (BC Parks Division, 2003a–n).

The Sound's marine habitats also support very diverse faunal communities, including mammals, such as sea otter, harbour seals, and grey whales (BC Parks Division, 2003e). The region supports some of the densest populations of marbled murrelets and bald eagles in the Pacific Northwest. Wolves, black bear, elk, cougar, and coastal black-tailed deer are also found. Amphibians, such as north-western, clouded, and western red-backed salamanders, are also found in the area (BC Parks Division, 2003i). The area is large and complex and, in general, it can be said that the Clayoquot region not only represents outer coastal habitats along the south coast of BC, it also defines them.

Conflict over forest practices in the Clayoquot Sound area constituted one of the highest-profile environmental campaigns in Canadian history. Campaigning began by the Friends of Clayoquot Sound to restrict logging in the 1970s (Friends of Clayoquot Sound, 2006). By the early 1980s, residents of the town of Tofino and nearby Nuu-chah-nulth First Nations began expressing concern about logging in the area, particularly plans to log Meares Island, which is very close to both Tofino and the First Nations settlement of Ahousaht. Activism and good press coverage resulted in the relatively rapid convening of a land use planning consultation process. The result was a recommendation that called for excluding logging from that portion of Meares Island most visible from the town of Tofino. However, McMillan Bloedel, which at the time held the forest licence, did not agree and eventually the provincial Cabinet also rejected the outcome (Spaces for Nature). This was to have a reverberating effect, one that heralded a period of protracted and particularly acrimonious conflict.

The resentment generated in local communities by the government and industry responses led to escalating confrontations. Some of the first blockades of forest operations in BC history came about in 1984 and 1985, when McMillan Bloedel attempted to begin logging on Meares Island. The Nuu-chah-nulth eventually obtained an injunction against logging activity on the island which resulted in a permanent suspension of logging on the island (Spaces for Nature). Relations remained strained between pro- and anti-development groups. By the late 1980s environmentalists had broadened their area of interest to include the whole of Clayoquot Sound. This resulted in more blockades and ever more extensive publicity for the issue and the organizations that were seeking to end the forest industry in the area. In response, government initiated a land use planning process for the whole Clayoquot Sound area. Even the town of Port Alberni, about 75 kilometres away from Clayoquot, but a place where most of the logs from the Clayoquot area went for processing, was included in this process. But environmental groups regarded the process as being biased and rejected the results – they saw the process as flawed, unbalanced, and favouring industry (Spaces for Nature).

In the early 1990s more blockades followed, and by 1993 confrontation had widened just as a new provincial government led by the New Democratic Party (NDP) attempted to implement the findings of the land use planning panel. During the summer of 1993, large-scale protests occurred and approximately 850 people were arrested; some claim this was the largest act of civil disobedience in Canadian history (Friends of Clayoquot Sound). Throughout this period protests and arrests occurred on a daily basis. Some estimates hold that up to 12,000 people protested. Then Greenpeace International became directly involved. This led to an extensive national and international media campaign that came to include a boycott of products from forest companies operating in the area. Eventually, the major licensee, McMillan Bloedel (MB), supported the findings of an independent scientific panel appointed by government to study the issue of Clayoquot Sound (Spaces for Nature). By 1995, much of the area was protected and restrictions on logging were implemented. But, for some, this was not enough. People representing the Friends of Clayoquot Sound and Greenpeace blockaded and 'took over' a McMillan Bloedel logging camp at Rankin Sound in 1996 (Friends of Clayoquot Sound). A clear-cut near the highway to the western side of the Island was marketed as the 'Black Hole' by environmentalists, and touted as an example of the 'horrors of commercial logging'. Eventually, MB withdrew altogether from operating in the Sound, maintaining a token presence as part of the joint venture with Iisaak Forest Resources, a small logging company run in conjunction with the Nuu-chah-nulth First Nation.

The campaigns to protect Clayoquot Sound resulted in profound changes in land use policy by the government of British Columbia and forest companies. In 1979, before active campaigning began, protected areas within the Sound consisted of: two small provincial parks (Maquinna and Gibson Marine); a number of small ecological reserves; and one significant national park (Long Beach Unit of Pacific Rim National Park Reserve). By the end of 1995, a total of 14 provincial parks encompassing some 36,000 hectares were designated (BC Parks Division, 2003a–n). Most of these new parks were created in 1995, although the Megin Valley and adjacent areas were designated earlier as a park in the controversial 1993 land use decision implementation (BC Parks Division, 2003j, n). Maquinna Provincial Park was significantly enlarged in 1995. Currently, Hesquiat Peninsula Provincial Park with an area of 7,888 hectares is the largest provincial park in the Sound. When the area of the Long Beach unit of Pacific Rim National Park Reserve is considered (13,715 hectares), the amount of directly protected area in Clayoquot Sound is 14.23 per cent of its total area. However, in addition to directly protecting areas, land use decisions have profoundly changed logging practices in the Sound.

The recommendations of the 1995 Scientific Panel essentially led to the elimination of clear-cut logging practices. MB abandoned clear-cut logging in all of its operations in favour of retention management in 1998. International Forest Products has also switched from clear-cutting to 'retention' management and in recent years has logged only on Kennedy Flats. MB closed its operations in the Sound late in the 1990s, including its Tofino office, and today it no longer exists as a company. Its successor remains as a partner in Iisaak Forest Resources. International Forest Products has greatly reduced its logging activity in the Sound and has closed its office in Ucluelet. It is rumoured that International Forest Products will soon withdraw from the Sound altogether. Iisaak will likely be the only company operating in the Sound. It is difficult to identify the Black Hole now, it is covered with vigorous young conifers.

Carmanah and Walbran valleys

The Carmanah and Walbran valleys lie on the western side of Vancouver Island, just south of the Clayoquot Sound area. They are remote, rugged, and difficult to access. Each includes biophysical features that are significant at local, regional, and continental scales. Although public attention was initially drawn to the record-sized sitka spruce trees found in the area (Carmanah Valley Forest Management Advisory Committee, 1992), the valleys are perhaps more significant for its inclusion of an entire medium-sized watershed, Carmanah Creek and its tributaries, and completes preservation of the topographic gradient for lower-elevation areas on the windward west coast of Vancouver Island. Carmanah/Walbran Provincial Park is likely significant for its preservation of a representative landscape rather than for preservation of specific species or biophysical features.

Three biogeoclimatic zones are found in the two valleys: the Southern Very Wet Hypermaritime Coastal Western Hemlock variant is found in the lowest elevation sections, nearest the ocean; the Submontane Very Wet Maritime Coastal Western Hemlock variant is above; and the Montane Very Wet Maritime Coastal Western Hemlock variant sits at the height of the land. Tree species in the valleys include western hemlock, western red cedar, amabilis fir and, in immediate beach areas and riparian zones, sitka spruce. Salal, deer fern, red huckleberry, and evergreen huckleberry are the most common understory species. Leavers (1996) writes that the understory in the area vegetation is quite diverse; noting that 50 species of vascular plants were recorded in a 1990 survey, in addition to many species of mosses, liverworts, and lichen. While bear and cougar are found in the area, the heavily forested terrain limits the mammal populations, and no species of mammal has been noted to have an extraordinary distribution in the Carmanah or Walbran valleys (Leavers, 1996). But some 36 bird species, including two endangered species, the marbled murrelet and northern pygmy owl, have been identified and recorded (Leavers, 1996).

In the spring of 1988 the Carmanah Valley joined Clayoquot Sound as a conflict place. Environmentalist Randy Stoltmann hiked through the valley and discovered what is likely Canada's largest tree, a 400-year-old sitka spruce standing about 95 metres high, soon called the Carmanah Giant. Stoltmann wrote and spoke about the valley and the trees he found and brought the valley into BC's now volatile forest debate (Wilson, 1998). The valley was within MB's timber tenure (TFL 44), and they had plans to log parts of the area. This led to a conflict that would not only result in park designation for the Carmanah and part of the neighbouring Walbran Valley, but would also influence the larger protected areas policy discourse. Tension eased when MB announced the initiation of a Management and Working Plan (MWP) process, which initially projected harvesting in the valley to begin in 2003. This seemed to remove any immediate threat to the sitka stands (Wilson, 1998). But once the province approved the plan, MB shifted timber quotas from other divisions into the Carmanah area, in essence accelerating the logging plan (Wilson, 1998).

Stoltmann and organizations such as the Carmanah Forestry Society and the Sierra Club held rallies to protest the harvesting proposal at the provincial legislature and on the logging roads being punched into the valley. Environmental organizations developed an information campaign centred on two themes. First, by portraying the protests as grassroots environmentalism against corporate interests, environmental organizations sought to link the industrial use of 'special' places such as Carmanah to policies which, it seemed, could no longer sustain jobs or the economies that relied on them. Second, the valley was called the last unlogged watershed between Barkley Sound and Victoria. By tying it to representations of lost landscapes and natural systems, the resulting image created the sense that the valley was unique and the last of its kind. Environmentalists offered Carmanah as a place with values other than as a source of wood fibre.

Conflict moved from the forest to the courts and the political level, with industry seeking injunctions against the protesters, and the environmental organizations working to apply political pressure. After the media campaign began, which targeted MB's clients, the company stopped construction of the access road. In June 1988 MB proposed the creation of two recreation sites to protect the Carmanah Giant and nearby stands of giant spruce. The environmental movement had been successful at achieving what Wilson (2001a) characterizes as an alternative problem definition – a refocusing of attention on forestry practices, landscape change, and the need for what they termed sustainable practices.

One month later, the BC Forest Service requested that MB prepare a Special Management Plan for the Carmanah – a plan that would be publicly reviewed (Carmanah Valley Forest Management Advisory Committee, 1992). The Forest Service saw little potential for diffusing the conflict without some form of park designation larger than MB's proposed recreation site. The provincial government passed the Carmanah Pacific Park Act, which removed the lower valley from TFL 44, but maintained MB's logging rights in the upper valley. This split valley objective was to placate conservationists without significantly changing the larger resource management setting.

The Carmanah Valley Forest Management Advisory Committee was created to advise the Forest Service on best practices for the area, specifically on options for balancing uses, in essence, mitigating conflict. The Committee was composed of industry, community, government, and environmental representatives. By early 1992 the Advisory Committee's report was complete; it emphasized multiple uses, but was to remain essentially unimplemented. It contained ecosystem inventory information which, although not particularly complex, spurred further conflict by heightening concern about the split valley nature of the management plan proposed for the watershed. Throughout the deliberations the issue of the upper valley remained. For environmental interests, the notion of preserving the bottom half of a valley while logging the upper seemed to beg downstream impacts and negate the integrity of the valley's natural systems. In 1991 new protests began when another timber company, Fletcher Challenge, announced plans to log in the neighbouring Walbran

Valley. Once again protesters were arrested for blocking access to loggers and road-building crews, injunctions were served, and the government was lobbied. In an attempt to defuse conflict and develop a compromise approach, limited placation was tried again, but the policy of containment failed. Conflict escalated and the split valley approach did not succeed in addressing demands from environmental organizations, industry, or forest workers (Cashore et al., 2000: 98). Then there was a change in provincial government.

By 1992, when the NDP won the election and formed the government, the first part of Carmanah Park had already been designated, but they took office in the middle of new conflicts over the upper Carmanah and Walbran valleys, and there were protests emerging at Clayoquot Sound. With this new government came changes to the nature of resource politics and the approach to planning at places of contention on Vancouver Island. The key linkage had been between government and industry, with labour as a proxy player. Industry acted in an indirect sense to represent employment interests since the view was that what was good for the forest industry was good for those it employed. Environmental organizations had links to the NDP. But things had already begun to change. In the years leading up to the change in government, pressure from environmental organizations and corresponding change in public opinion led to a gradual, albeit limited, shift to accommodate some demands for new protected areas (Wilson, 2001a: 39). Labour organizations and the environmental movement were important NDP constituents and Vancouver Island was a traditional NDP stronghold. Dealing with forest conflicts such as Carmanah Walbran was now important to the political health of government members. They initiated a process of macro policy change, beginning with the Commission of Resources and Environment, which we describe in detail below. One of the actions of the Commission was to initiate the Vancouver Island Land Use Plan, which became central to the Carmanah Valley: it recommended an expanded park. This became Carmanah Walbran Park which now includes the entire Carmanah Valley and most of the lower portion of the Walbran Valley. The upper Walbran remains open to logging. Events in the Stein and on Vancouver Island had already changed land use planning in BC, and this would help provide the process foundation for addressing conflicts on BC's north and central coasts.

The 'Great Bear Rainforest'

The combined area of the British Columbia central and north coasts totals 64,000 square kilometres. This is approximately twice the size of Belgium (Ministry of Agriculture and Lands, 2006c). BC's central and north coasts have one of the most extensive fjord coastlines of any global region. This area extends from the Bute Inlet area in the south, roughly the same latitude as northern Vancouver Island, up the coast to the Alaska Panhandle and from the height of the Coast Mountains to tidewater. This is essentially the BC coast north of Vancouver Island, with the exception of the Queen Charlotte Islands. This region includes the very rugged topography of the Coast Mountains and the more subdued terrain of the Hecate Lowlands (Holland, 1964). This area is regarded as the largest remaining natural stretch of temperate rainforest on the planet. The World Wildlife Fund considers the coast region as globally outstanding and as one of the most important places on Earth for biodiversity conservation (Office of the Premier, 2006b). The area in question has a cool, moist climate that is strongly maritime in character but ranges from outer coastal hypermaritime, through maritime and submaritime, to subcontinental in nature (Pojar and Meidinger, 1991). Generally speaking, this region is wetter and cooler than the Clayoquot Sound area. Biogeoclimatic units include variants of the coastal western hemlock, montane and alpine tundra but also the occasional intrusion of transitional variants of interior zones, such as the interior Douglas fir, Engelmann spruce subalpine fir, and sub-boreal spruce.

The area is heavily forested with stand types ranging from extensive areas of low-elevation, low-productivity western red cedar-western hemlock types that have more or less of a bog character, to more productive stands consisting of western hemlock, western red cedar and amabilis fir to subalpine forests of mountain hemlock, amabilis fir, Engelmann spruce, and yellow cedar. Much of the forest growth is of low productivity and under strong bedrock control, due to the thin soils. The more productive forests are associated with alluvial systems and fluvial fans. Seral forests with a greater presence of red alder and cottonwood are associated with watercourses.

Approximately 350 species of birds have been noted in the area. All five species of Pacific salmon spawn in the numerous watercourses and lakes of this area. More than 250 endangered or threatened terrestrial animal, plant, and marine species have been noted in the area (Office of the Premier, 2006b). 'Charismatic' species, such as grizzly bear, wolves, mountain goats, and black bear, are found in abundance here. But perhaps the most important animal found in the area is the Kermode bear, a white-coated natural mutation of the black bear. Although found as far east as Minnesota, the largest concentration of Kermode bears is seen on the central and north coasts, particularly in the Princess Royal Island area (Office of the Premier, 2006a). As part of their campaign to keep the area from being logged, environmental organizations christened the Kermode the 'Spirit Bear' and the area was marketed as the 'Great Bear Rainforest'.

Environmental organizations first became interested in the central coast in the early 1990s. It was also at this time the area was branded as the 'Great Bear Rainforest' and the campaigns began to attract international interest in preserving the area. The conflict was addressed to forest companies, the provincial government, and international customers of forest product manufacturers, notably in Europe. The BC government initiated a multi-party land and resource management planning process (LRMP) for the central coast in 1997 (Office of the Premier, 2006b). A similar process was initiated in 2001 for the north coast. Environmental groups initially refused to participate in the LRMP process. They regarded it as a status quo vehicle, and referred to the initiative as 'talk and log'. The international campaign aimed at discrediting forest companies and influencing their customers was intensified in 1997. Forest products customers in Europe condemned logging practices and arrests and began boycotting products from companies involved in the dispute.

Once again the familiar round of blockades, arrests, boycotts, and media blitzes dominated the discourse. In 1999 Home Depot, a major purchaser of forest products, committed to no longer purchasing wood products from coastal BC old-growth forests. A Greenpeace-sponsored delegation convinced German business interests to cancel existing contracts for BC wood products unless forest practices changed. In response to the increasing pressure, forest companies entered into negotiations with environmental groups about how to resolve the situation. These negotiations led to a suspension of media and boycotting campaigns and allowed limited logging to continue in agreed upon areas, while further discussions continued. In 2001 the Joint Solutions Project (JSP) allowed companies and environmentalist groups to resolve issues and to provide input into the LRMP process (Office of the Premier, 2006b). The process not only marked a change in industry attitudes, but it also saw the emergence of compromise by some environmental organizations. The JSP and the provincial government jointly funded the Coast Information Team, to provide scientific analysis of the area and technical assistance to the LRMPs. First Nation groups have played an increasingly significant role in the process, and have emerged as major players in regional land use planning.

Resolution of the north and central coast LRMPs was announced in February of 2006, when the provincial government agreed to implement land use decisions from the planning tables. Significantly, 180,000 square kilometres, or 28 per cent, of the area in question will be preserved as protected areas, including parks, where no further commercial resource extraction will occur (Ministry of Agriculture and Lands, 2006b). It is planned that 100 new protected areas will be established in the area. Existing protected areas cover 6,000 square kilometres and another 120,000 square kilometres will be added (Ministry of Agriculture and Lands, 2006b). The amount of area protected will be approximately twice the size of Ireland. This addition will increase the size of BC's protected areas to 14 per cent of its land base. The agreement also seeks to implement an Ecosystem-Based Management (EBM) approach to land use planning and management. This will be based on techniques that balance social needs with ecosystem protection (Ministry of Agriculture and Lands, 2006b). The details of this are yet to be finalized but in addition to the protected areas, two other land use zones are defined for the area. Biodiversity Areas will allow for tourism and mining activity but no other types of development and will exclude commercial forestry. Biodiversity Areas account for three per cent and ten per cent of the central and north coast areas, respectively.

Ecosystem-Based Management Operating Areas will ideally allow for a full range of economic activity, while using the EBM approach to balance economic activity with conservation (Ministry of Agriculture and Lands, 2006b). This is a tall order, but bears promise for a very different approach to resource management. These areas account for 66 per cent and 65 per cent, respectively, of the north and central coast areas (Ministry of Agriculture and Lands, 2006b). A very significant feature of the process has been the role of First Nations. A total of 25 First Nations were consulted during the LRMP process and played a pivotal role in the land use decisions made. It is anticipated that 18 of these First Nations will sign government-to-government land use agreements with the BC government (Ministry of Agriculture and Lands, 2006a).

The influence of specific places on larger policy

In BC the pervasive influence of land use conflict – centred on demands for protected areas designation – has led to a wholesale reassessment of public land use planning. New planning processes were initiated to address the conflicts represented in the cases just outlined; three were particularly large in scale: the Commission on Resources and Environment, the Protected Areas Strategy, and the ongoing Land and Resource Management Plans. Each undoubtedly had conflict mitigation as a core objective, and each emerged as a response to a setting where conflict had become commonplace.

The Commission on Resources and Environment and Vancouver Island Land Use Planning (CORE)

The Commission's mandate was to give communities and other stakeholders a voice in regional land use planning, and to recommend lands for protection within some of BC's more controversial areas (Owen, 1998; Tollefson, 1998). CORE was to be the forum for dispute resolution and making regional land use decisions across BC to avoid the valley-by-valley protests that plagued the policy setting (Owen, 1998; Burrows, 2000: 212). The Protected Areas Strategy emerged from CORE deliberations.

CORE's creation was, in no small part, a response to the events at Clayoquot Sound and Carmanah Walbran. Its first activity was to initiate a land use planning process for the Island based on a negotiative multi-stakeholder process with representation from preservationists, industry, labour, fishing, and a range of other social and economic interests. Under CORE's mandate, in late 1992 the Vancouver Island Negotiation Process began, with the central objective of producing a Vancouver Island Land Use Plan (VILUP). The VILUP process was given one year to produce a consensus-based plan – a tall order given the conflictual context. The diversity of issues brought to the table by participants posed organizational challenges, and the prioritization of interests and values became an ongoing aspect of contention not only within the VILUP, but also in other CORE deliberations (Tollefson, 1998).

Central to the VILUP were principles of equal representation, process efficiency, and feasibility – which can be translated as the likelihood of achieving a stable and lasting agreement (Burrows, 2000). But the VILUP process faced an inherent difficulty. The agenda was seen by some to accord preference to forest use, while for others forest preservation was the goal. Perhaps most difficult for some environmental organizations was that any talk of protection for Clayoquot Sound had been excluded from the agenda. The CORE/VILUP rationale was that Clayoquot Sound was being addressed by a concurrent planning process devoted to that area alone. Only six environmental organizations chose to formally participate, including the Carmanah Forestry Society and the Sierra Club of Western Canada, while another 46 groups who did not participate directly indicated loose support for those who chose to (Burrows, 2000). Some groups returned to conflict tactics based on various forms of civil protest. After a year, the VILUP process did produce a plan, but without consensus. The results can best be likened to a broad, multiple-use plan, with some protected areas designations, which emerged as one of its more specific recommendations.

The CORE process ended in 1996, and the province ceded much of the regional land use and resource planning to agency-based processes. Burrows (2000) suggests CORE was stopped in no small part because it lost credibility with those who participated, and the recommendations CORE made to the government seemed increasingly remote from what was being said at community-level consultation. After CORE ended, the LRMPs assumed the role of developing regional land use plans, which are implemented through regionally based multi-stakeholder round tables that operate on a smaller scale than CORE (Wilson, 2001b; Scudder, 2003).

Land Resource Management Plans

The defeat of the NDP and the election of a more ideologically conservative government (the BC Liberal Party) has led to a range of changes in forest and land use management, but the LRMP process, so influential in the 'Great Bear Rainforest' case, has remained an important planning process in BC.

The LRMPs are regional integrated resource management plans that seek to create a publicly defined vision for use and management of public provincial lands and resources. They ideally involve a broad range of interests and values. Their development requires involvement of people representing a wide range of interests and values. LRMPs generally provide: broad land use zones defined on a map; objectives that guide management of natural resources in each zone; strategies for achieving the objectives; and a socio-economic and environmental assessment that evaluates the plan. Decisions are intended to reflect social choice, they are negotiated, and diverse values are considered and debated within the LRMP setting.

In practice the LRMP process has emerged as highly participatory – communities express a profound sense of ownership over the outcomes, and this has ensured the longevity of the LRMP process. While, arguably, work on the LRMPs has been hesitant, progress has been made. Implementation and monitoring of approved LRMPs continues across BC, and the LRMPs for the central coast and north coast regions have advanced with identification of ecosystem-based management as the desired framework for land use planning and management. BC's Protected Areas Strategy is implemented in part through LRMPs, which serve as settings where such areas are identified. The LRMPs are, more importantly, the setting where boundary and conditions of designation are debated, thus natural area designations can be weakened by the dominance of other values.

The Protected Areas Strategy

The Protected Areas Strategy (PAS) emerged in 1993, just as CORE was expanding to other BC regions. The VILUP had a direct role in creating the boundary of Carmanah Walbran, and was particularly instrumental in the creation of the PAS. The government's experience at Carmanah, and other Vancouver Island conflicts, pointed to the need for a strategic approach to protected areas designation. The PAS was also BC's response to the United Nations 12 per cent protected areas goal, an objective to which BC legally committed itself (Scudder, 1999). Of course, such an objective might have little ecological meaning (e.g. Soulé and Sanjayan, 1998; Woodwell, 2002). The objective may have been to establish protected areas that represent BC's diverse natural systems, but doing this requires knowing more about natural areas. In practice, the process has been very much about balancing two dominant competing perspectives in the protected areas debate; economic growth versus environmental protection.

The PAS became a venue for the analysis of options and the determination of protected places. This was achieved largely by using the ecosystem classification systems. The problem was that regardless of how complete or extensive such inventories might be, some will question their comprehensiveness, and other dynamics will also play unacknowledged roles. While the challenges of mediating the conflicting agendas of environmental groups and forestry companies and other users remained, the government moved forward with the PAS, and by 2000 it had achieved the 12 per cent goal (Harding and McCallum, 1994: 367). But, as we commented above, this will be exceeded with the addition of new protected areas in the central and north coasts. While the PAS is a conflict resolution process, it has, nevertheless, set the stage for clashes with environmental groups. While environmental organizations believe it should be about preserving ecosystems and biodiversity, in practice a more complex range of values is considered.

Even if the primary goal of the PAS is the protection of natural diversity, implementation depends on a more complex definition, where the measure of diversity was based on more than natural values, and irrespective of the rhetoric, land tenure is an important conflict factor in PAS. Public lands are easier to designate and there has been no appetite for applying the strategy to private lands. Diversity in this process is not the same as biological diversity (Scudder, 2003). Put another way, designation of protected areas through PAS yields what has been called a 'negotiated product' (e.g. Yearley, 1996; Latour, 1999; Song and M'Gonigle, 2001).

The risk inherent in a process driven primarily by the conflict reduction objective is that it will result in an imbalance in ecosystem representation and a disproportionate allocation of lands to protected status that are either not economically valuable or are inaccessible to resource companies. There is a familiar refrain on the west coast 'there's a lot of rock and ice in BC's new parks' (e.g. Binkley, 2000: 181). Indeed, Scudder's (2003) work suggests that the PAS has yielded a remarkable lack of coincidence between biodiversity rarity and richness and protected areas, and that despite the growth of inventory information there are significant gaps in understanding the regional distribution and frequency of endangered species. Recent LRMP developments may weaken this criticism. Despite remaining a formal policy, the PAS has become tangential to the LRMP process.

Discussion and conclusions

From each of the cases described here, and the response of industry and government, we can distinguish four conflict stages: attempted containment, limited negotiation, the breakdown of negotiation, and, finally, protected area designation. In the first stage there were efforts to contain and diffuse conflict through a policy of containment (Wilson, 2001a). This began as ignoring protests, closing ranks, then questioning the legitimacy of those who challenge established practice (Burrows, 2000; Wilson 2001a, b), and, ultimately, hoping conflict would go away. In BC this did not work. Conflict grew and moved beyond localized action into a realm where transnational environmental organizations became involved. In the second stage, government and industry responded with a combination of limited negotiation, and legal actions such as arresting protesters and seeking injunctions against those who were actively blocking logging activities. At this stage negotiation can assume different forms and achieve varying success. Negotiation among opposing perspectives can be genuine in that it seeks to incorporate alternative views into management and practice; or it may be intended to co-opt those views and limit their impact through protracted consultation and consensus-building exercises; or negotiation may simply be a way of stalling, again hoping that conflict will subside. But negotiative processes can produce compromise, and this might include some protection for a place of controversy. A third stage occurs when negotiations break down. Environmental organizations return to protest, often with a new fervour and sometimes with new coalitions. And, fourth, depending on the effectiveness of protest actions and the political expediency with which the government wants to resolve the issue, an acquiescence response to protest may ultimately dictate the outcome (Hessing and Howlett, 1997; Cortner and Moote, 1999).

Arguably, conflict was generated by aesthetics, supported by the perception that the area had a particular ecosystem value, and the last-of-its-kind syndrome where arguments focused on the rationale that each place must be preserved because it was the last of its kind within a region or possessed unique characteristics. The risk inherent in a conflict-based policy informed by imprecise knowledge is that as reserves are designated in attempts to appease environmental movements, communities, and industry, the results could be based on political expediency rather than ecological significance or integrity.

The BC experience shows that those outside traditional decision-making can develop a powerful capacity to advance protected area designation. Local conflicts can also assume the power to transcend the local, and influence policy change at a larger scale. Did conflict improve overall land use decision-making processes in BC? The short answer is yes. A range of new integrative and agency-based land use planning programmes were a direct response to the 'war in the woods'. The issue now is how lasting such change really is. In BC the impacts have been variable, and while some programmes have remained in place, others have ended, all have been altered, and in time these alterations might re-ignite conflict. But as the 'Great Bear Rainforest' example suggests, there is reason for cautious optimism about the role of new BC approaches to large-scale land use planning, embodied in the LRMPs.

Protected areas designation not uncommonly comes about from conflict inspired by the power of place, through perceptions of an area as unique, rare, and threatened, or having ecosystem value. Flexibility for planning and use is possible after an area has been protected, but if a landscape is greatly changed by human actions such opportunities are lost. The most lasting legacy of BC's recent conservation and parks policy initiatives may well be the creation of a new setting where strategic land use planning is not only the new norm, it is now expected by the public.

Note

The opinions expressed in this chapter are those of the authors and do not represent those of their respective agencies.

Literature cited

BC Parks Division (2000). *Management Plan for Stein Valley Nlaka'pamux Heritage Park*. Victoria: British Columbia Ministry of Environment, Lands and Parks.
—— (2003a). *Clayoquot Arm Provincial Park Purpose Statement and Zoning Plan*. Victoria: British Columbia Ministry of Environment, Lands and Parks.
—— (2003b). *Clayoquot Plateau Provincial Park Purpose Statement and Zoning Plan*. Victoria: British Columbia Ministry of Environment, Lands and Parks.
—— (2003c). *Dawley Passage Provincial Park Purpose Statement and Zoning Plan*. Victoria: British Columbia Ministry of Environment, Lands and Parks.

—— (2003d). *Flores Island Provincial Park Purpose Statement and Zoning Plan*. Victoria: British Columbia Ministry of Environment, Lands and Parks.
—— (2003e). *Gibson Marine Provincial Park Purpose Statement and Zoning Plan*. Victoria: British Columbia Ministry of Environment, Lands and Parks.
—— (2003f). *Hesquiat Provincial Park Purpose Statement and Zoning Plan*. Victoria: British Columbia Ministry of Environment, Lands and Parks.
—— (2003g). *Hesquiat Peninsula Provincial Park Purpose Statement and Zoning Plan*. Victoria: British Columbia Ministry of Environment, Lands and Parks.
—— (2003h). *Kennedy Lake Provincial Park Purpose Statement and Zoning Plan*. Victoria: British Columbia Ministry of Environment, Lands and Parks.
—— (2003i). *Kennedy River Bog Provincial Park Purpose Statement and Zoning Plan*. Victoria: British Columbia Ministry of Environment, Lands and Parks.
—— (2003j). *Maquinna Provincial Park Purpose Statement and Zoning Plan*. Victoria: British Columbia Ministry of Environment, Lands and Parks.
—— (2003k). *Sulphur Passage Provincial Park Purpose Statement and Zoning Plan*. Victoria: British Columbia Ministry of Environment, Lands and Parks.
—— (2003l). *Sydney Inlet Provincial Park Purpose Statement and Zoning Plan*. Victoria: British Columbia Ministry of Environment, Lands and Parks.
—— (2003m). *Tranquil Creek Provincial Park Purpose Statement and Zoning Plan*. Victoria: British Columbia Ministry of Environment, Lands and Parks.
—— (2003n). *Vargas Island Provincial Park Purpose Statement and Zoning Plan*. Victoria: British Columbia Ministry of Environment, Lands and Parks.
Bean, M.J. (1997). A policy perspective on biodiversity protection and ecosystem management. In S.T.A. Pickett, R.S. Ostfeld, M. Shachak, and G.E. Likens (eds), *The Ecological Basis of Conservation*, pp. 174–192. New York: Chapman and Hall.
Beaty, T. (2000). Preserving biodiversity: challenges for planners. *Journal of the American Planning Association* 66 (1), 5–20.
Binkley, C. (2000). A crossroad in the forest: the path to a sustainable forest sector in British Columbia. In C.D. Salazar and D.K. Alper (eds), *Sustaining the Forests of the Pacific Coast: forging truces in the war in the woods*, pp. 174–192. Vancouver: UBC Press.
British Columbia (1993). *Protected Areas Strategy*. Victoria: Government of British Columbia.
British Columbia Ministry of Forests (1999). *Managing Identified Wildlife: procedures and measures*, Volume 1. Victoria: Queen's Printer.
Burrows, M. (2000). Multistakeholder processes: activist containment versus grassroots mobilization. In C.D. Salazar and D.K. Alper (eds), *Sustaining the Forests of the Pacific Coast: forging truces in the war in the woods*, pp. 209–228. Vancouver: UBC Press.
Carmanah Valley Forest Management Advisory Committee (1992). *Forest Management in the Upper Carmanah Valley: recommendations of the public advisory committee*. Victoria: British Columbia Ministry of Forests.
Cashore, B., Hoberg, G., Howlett, M., Rayner, J., and Wilson, J. (2000). Change and stability in B.C. forest policy. In M. Howlett, J. Wilson, B. Cashore, G. Hoberg, and J. Rayner (eds), *In Search of Sustainability: British Columbia forest policy in the 1990s*, pp. 234–257. Vancouver: UBC Press.
CORE Commission on Resources and Environment (1994). *Vancouver Island Land Use Plan*, Volume I. Victoria: Crown Publications.
Cortner, H.J. (2000). Making science relevant to environmental policy. *Environmental Science and Policy* 3, 21–30.
—— and Moote, M. (1999). *The Politics of Ecosystem Management*. Washington, DC: Island Press.
Dearden, P. and Rollins, R. (2002). *Parks and Protected Areas in Canada* (2nd edn). Oxford and Toronto: Oxford University Press.
Demarchi, D.A., Marsh, R.D., Harcombe, A.P., and Lea, E.C. (1990). The environment. In R.W. Campbell, N.C. Dawe, I. McTaggart-Cowan, J.M. Cooper, G.W. Kaiser, and M.C.E. McNall (eds), *The Birds of British Columbia*, Volume I, pp. 55–142. Victoria: Royal British Columbia Museum.
Eden, S. (1998). Environmental issues: knowledge, uncertainty and the environment. *Progress in Human Geography* 22, 425–432.
Friends of Clayoquot Sound. *History*. www.focs.ca/about/history.asp. Accessed 1 May 2006.

Greenpeace. *A History of Greenpeace Canada's Rainforest Campaign*. www.greenpeace.ca/e/campaign/forests/greatbear/background/history.php. Accessed on 1 May 2006.

Grumbine, R.E. (1997). Reflections on 'what is ecosystem management?'. *Conservation Biology* 11, 41–47.

Harding, Lee E. and McCullum, E. (1994). *Biodiversity in British Columbia*. Ottawa: Ministry of Supply and Services.

Hessing, M. and Howlett, M. (1997). *Canadian Natural Resource and Environmental Policy*. Vancouver: UBC Press.

Holland, Stuart S. (1964). *Landforms of British Columbia A Physiographic Outline*. Victoria: British Columbia Department of Mines and Petroleum Resources, Bulletin No. 48.

Lanskail, D.A.S. (1984). Resource use conflicts: the forest industry perspective. In P.J. Dooling (ed.), *Parks in British Columbia: emerging realities*, pp. 45–55. Vancouver: UBC Press.

Latour, B. (1999). *Pandora's Hope: essays on the reality of science studies*. Cambridge, MA: Harvard Press.

Leavers, D. (1996). *Carmanah/Walbran Provincial Park Master Plan and Background Report*. Victoria: BC Parks, Ministry of Environment, Land and Parks.

Lippke, B. and Bishop, J. (1999). The economic perspective. In M.L. Hunter Jr (ed.), *Maintaining Biodiversity in Forest Ecosystems*, pp. 597–638. Cambridge: Cambridge University Press.

Margules, C.R. and Pressey, R.L. (2000). Systematic conservation planning. *Nature* 405, 243–253.

M'Gonigle, M. (1997). Behind the green curtain. *Alternatives* 23 (4), 16–21.

—— (1999). Ecological economics and political ecology: towards a necessary synthesis. *Ecological Economics* 28, 11–26.

Ministry of Agriculture and Lands (2006a). *Backgrounder: First Nation consultation key to land use decisions*. Victoria: British Columbia Ministry of Agriculture and Lands and Office of the Premier, 2006AL0002–000066.

—— (2006b). *Backgrounder: province announces a new vision for coastal B.C.* Victoria: British Columbia Ministry of Agriculture and Lands and Office of the Premier, 2006AL0002–000066.

—— (2006c). *News Release: province announces a new vision for coastal B.C.* Victoria: British Columbia Ministry of Agriculture and Lands and Office of the Premier, 2006AL0002–000066.

Norton, D. 1999. Forest reserves. In M.L. Hunter (ed.), *Maintaining Biodiversity in Forest Ecosystems*, pp. 525–555. Cambridge: Cambridge University Press.

Noss, R.F. and Scott, J.M. (1997). Ecosystem protection and restoration: the core of ecosystem management. In M.A. Boyce and A.W. Haney (eds), *Ecosystem Management: concepts and methods*, pp. 239–264. New Haven, CT: Yale University Press.

Office of the Premier (2006a). *Factsheet: Spirit Bear facts*. Victoria: British Columbia Office of the Premier and Ministry of Agriculture and Lands, 2006AL0002–000066.

—— (2006b). *Coastal Land Use Agreements*. Victoria: British Columbia Office of the Premier and Ministry of Agriculture and Lands, 2006AL0002–000066.

Owen, S. (1998). Land use planning in the nineties: CORE lessons. *Environments* 14 (2), 14–26.

Perry, D.A. (1998). The scientific basis of forestry. *Annual Review of Ecology and Systematics* 29, 435–466.

Pojar, J. and Meidinger, D. (1991). British Columbia: the environmental setting. In D. Meidinger and J. Pojar (eds), *Ecosystems of British Columbia*, pp. 39–68. Victoria: British Columbia Ministry of Forests, Special Report Series 6.

Priddle, G.B. (1982). Parks and land use planning in Northern Ontario. *Environments* 14, 47–51.

Satterfield, T. (2002). *Anatomy of a Conflict Identity, Knowledge, and Emotion in Old-growth Forests*. Vancouver: UBC Press.

Scudder, G.G.E. (1999). Endangered species protection in Canada. *Conservation Biology* 13, 963–965.

—— (2003). *Biodiversity Conservation and Protected Areas in British Columbia*. Vancouver: Department of Zoology and Centre for Biodiversity Research, The University of British Columbia.

Shafer, C.L. (1995). Values and shortcomings of small reserves. *BioScience* 45, 80–88.

Song, S.J. and M'Gonigle, R.M. (2001). Science, power, and system dynamics: the political economy of conservation biology. *Conservation Biology* 15, 980–989.

Soulé, M.E. (1991). Conservation: tactics for a constant crisis. *Science* 253, 744–750.

—— and Sanjayan, M.A. (1998). Conservation targets: do they help? *Science* 279, 2060–2061.
Spaces for Nature. *Clayoquot Sound, Flores and Vargas Islands Provincial Parks*. www.spacesfornature.org/greatspaces/clayoquot.html. Accessed 1 May 2006.
Tollefson, C. (1998). *The Wealth of Forests: markets, regulation and sustainable forestry.* Vancouver: UBC Press.
Wilson, J. (1998). *Talk and Log: wilderness politics in British Columbia*. Vancouver: UBC Press.
—— (2001a). Experimentation on a leash: forest land use planning in the 1990s. In M. Howlett, J. Wilson, B. Cashore, G. Hoberg, and J. Rayner (eds), *In Search of Sustainability: British Columbia forest policy in the 1990s*, pp. 31–60. Vancouver: UBC Press.
—— (2001b). Talking the talk and walking the walk: reflections on the early influence of ecosystem management ideas. In M. Howlett (ed.), *Canadian Forest Policy*, pp. 94–126. Toronto: University of Toronto Press.
Woodwell, G.M. (2002). On purpose in science, conservation and government. *Ambio* 31 (5), 432–436.
Yearley, S. (1996). Nature's advocates: putting science to work in environmental organizations. In A. Irwin and B. Wynne (eds), *Misunderstanding Science: the public reconstruction of science and technology*, pp. 172–190. Cambridge, UK: Cambridge University Press.

Chapter 8
Deconstructing ecological integrity policy in Canadian national parks

Douglas A. Clark, Shaun Fluker, and Lee Risby

Introduction

Criticisms of the traditional preservationist 'fortress park' model for protected areas policy are numerous and well documented (Janzen, 1989; Pretty, 1997; Neumann, 1998; Adams and Hulme, 2001; Pimbert and Risby, 2002; Bengtsson et al., 2003; GEF, 2005), and ecological integrity has been promoted as a more enlightened principle for conservation policy formation (e.g. Woodley et al., 1993). Numerous definitions of ecological integrity exist (Karr, 1991; Kay, 1991; Woodley et al., 1993; Westra, 1994) and though some commentators question the adequacy of ecological integrity as a policy goal (Wicklum and Davies, 1995), relatively few authors have examined what this plurality of definitions means for practice (DeLeo and Levin, 1997; Manuel-Navarette, 2003). This chapter describes the transformation of ecological integrity theory into policy and practice in the Canadian national park system, and examines specific park management problems in Canada that illustrate difficulties arising from applying a narrow interpretation of ecological integrity to protected area management.

An overview of ecological integrity

In his literature review of ethical and scientific work concerning ecological integrity, David Manuel-Navarette (2003) conceptualizes ecological integrity into three discourses based on existing literature, and suggests a fourth discourse. His typology provides a useful framework for examining different interpretations of the concept and the worldviews upon which these interpretations are based:

1 *Wilderness-normative*: this discourse views ecological integrity as reflecting a pristine state of nature that does not include humans. According to this discourse, humans are destined to destroy nature. Therefore, entire ecosystems (or significant components of them) must be protected from human influence, and the resulting unimpaired ecological structure and function will lead to a high (and objectively measurable) degree of ecological integrity. Protection of integrity is a moral choice, which is informed by positivistic biological science and implemented by professional managers. This discourse has received dedicated attention from philosophers who identify ecological integrity as a locus for intrinsic value in pristine wilderness (Westra, 1994) and is implicitly promoted by much of the conservation biology and wilderness advocacy literature (e.g. Brandon et al., 1998; Terborgh, 1999; Foreman, 2004).

2 *Systemic-normative*: this discourse is based on an understanding of ecosystems as dynamic complex systems whose integrity is a reflection of their resilience in the face of change – usually human-induced stresses. This discourse presents itself as more pragmatic than wilderness-normative (i.e. recognizing there are no 'pristine' ecosystems), but still holds that integrity is an empirically measurable objective that represents the wellness of ecological systems apart from human influence (Westra, 1994), and also reinforces the roles of science and managerial implementation. It incorporates the main principles of the literature on adaptive management (e.g. Holling, 1978; Walters, 1986) and ecosystem-based management (e.g. Slocombe, 1993; Grumbine, 1994).

3 *Ecosystemic-pluralistic*: this discourse builds on the systemic-normative understanding of complex social-ecological systems, but, importantly, it incorporates consideration of diverse social values and perspectives. By doing so, the discourse fundamentally changes the decision-making process from expert-based to participatory and transparent, and transcends the human–nature dualism followed by the previous two discourses. The role of science becomes more interdisciplinary and collaborative rather than predictive or directive, and the power structure is more horizontal than vertical. This discourse includes the ideas of post-normal science (Funtowicz and Ravetz, 1994), and many of the concepts of the community-based management and adaptive co-management literature (see Berkes, 2004). This discourse, however, remains within the worldview that environmental issues tend to be problems stemming from conflict. Proponents of the two normative discourses criticize this discourse for assigning instrumental value to ecological integrity.

4 *Transpersonal collaborative*: this discourse departs radically from the previous three by rejecting Western liberal ideology and its solitary definition of 'self' and conflict between individuals in favour of an ideology that recognizes interdependence of individuals with surrounding social and ecological systems.[1] Ecological integrity is less of a management tool used to solve specific problems between conflicting management goals (i.e. economic development versus environmental protection) or conflicting human value judgements, and more of an ongoing, internalized exercise within individuals as they create meaning. This discourse incorporates insights from systems theory as well as trans-scientific knowledge systems, especially those of non-Western cultures; all of which are a priori considered equally valid. This discourse suggests that the need to assert the primacy of maintaining ecological integrity is, itself, a product of dominant narratives in Western culture: there is no need to manage for integrity as individuals engaging in collaborative learning collectively produce ecological integrity.

Management for ecological integrity in protected areas is usually rooted in the wilderness-normative discourse, and in application is distinguishable from the preservationist model only by its more contemporary ecological terminology; avoiding terms such as 'natural' in favour of more 'objective' characteristics. The reasons for this are plentiful and likely include the prospect of institutional 'inertia' within park management organizations and the appropriation of new terms to re-label their own dominant prevailing norms and behaviours, rather than adapting to new demands. For example, Risby (2002) showed that protected area managers and planners in Uganda adopt the discourse of 'community involvement' (transpersonal collaborative) while still practically employing wilderness-normative legal/policy and management framework(s) to secure the hegemony of the wilderness-normative discourse.

The dominant Western narrative views nature as a collection of resources for humans, and individuals exploit those resources in accordance with laws enacted by social power structures. Wilderness-normative ecological integrity is the product of an environmental narrative seeking to flip this dualism by valuing nature above humans; or at least devaluing the human role within nature. Ecological integrity, or at least those geographic areas maintaining it, is superior to humans whose activities are inherently destructive and inferior. This environmental narrative, however, is arguably the dominant Western narrative in new clothing; it retains the dualism that assigns nature as an 'other', distinct from humans (Rose, 1994; Cronon, 1996; Morito, 1999; Merchant, 2004). It also retains the assumption that ecological integrity is an objectively measurable state that can be managed for in accordance with a legal and governmental system that dictates substantive management results (Manuel-Navarette, 2003). The foregoing suggests that normative discourses, more so than the non-normative, are readily assimilated into Western legal systems because they do not challenge the fundamentals of Western culture.

Two main implications for protected area management result. First, ecological integrity has lost its original complex systems perspective as the concept is applied in its more simplistic normative discourse(s) (Ascher, 2001). As such, ecological integrity policy fails to recognize the dynamic, self-organizing nature of ecosystems, leading to insufficient consideration of change (human induced and otherwise), multiple stable states, and scale. Moreover, the wilderness-normative discourse renders ecological integrity too rigid to address land use issues in all but the most remote protected areas that see little or no human activity (Fluker, 2003).

Second, the wilderness-normative discourse perpetuates a hierarchical, bureaucratic approach to natural resource management, to which has been applied such labels as 'command-and-control' (Holling and Meffe, 1996), and 'scientific management' (Brunner et al., 2005). The wilderness-normative discourse impedes progress towards the participatory 'new paradigm' for protected areas (Beresford and Philips, 2000; Risby, 2002). Where agencies are also moving towards that paradigm, e.g. through community-based management approaches, their policy objectives can conflict. Reproduction and reinforcement of preservationist models in developing countries have an embedded history and longevity that can impede resolution of community-based management problems (GEF, 2005).

Ecological integrity policy in Canadian national parks

The Parks Canada Agency[2] (PCA) began to investigate and apply principles of ecosystem management (Agee and Johnson, 1988; Slocombe, 1993; Grumbine, 1994) in the 1980s and early 1990s (Henry and Lieff, 1992; Canadian Parks Service, 1992). In 1994 ecosystem management was adopted as an explicit national policy approach (Canadian Parks Service, 1994) and some of its concepts, such as using ecological boundaries and improving inter-agency cooperation, found ready application to existing issues in many parks (e.g. Herrero et al., 2001). Curiously, though, the term 'ecosystem management' is conspicuously absent from the Agency's more recent high-level policy documents. By the late 1990s, discussion of ecosystem management had been largely overshadowed by ecological integrity (PCA, 2000, 2001a, b). One reason for this, perhaps, is because a productive author and advocate of ecological integrity is employed by the Agency. It is also possible that a policy of maintaining ecological integrity – then defined very much as a wilderness-normative concept – posed less threat to established organizational interests than ecosystem management, in which sharing power is more explicit.

During the late 1990s, the minister responsible for Canada's national parks commissioned a panel investigation to assess the ecological integrity of the parks (PCA, 2000). Among many recommendations concerning action to restore or preserve park ecological integrity, the *Report of the Panel on the Ecological Integrity of Canada's National Parks* (the Panel Report) suggested legislative amendments to ensure the maintenance or restoration of ecological integrity is the overriding priority in parks management. As a result, the 2000 Canada National Parks Act (CNPA) elevated ecological integrity to the first priority in the management of Canadian national parks with a legislated definition[3] and Section 8 (2) which states the '[m]aintenance or restoration of ecological integrity, through the protection of natural resources and natural processes, shall be the first priority of the Minister when considering all aspects of the management of parks'.

The CNPA ecological integrity provisions must be placed within the traditional dual mandate of 'use' and 'preservation' in Canada's national parks (Bella, 1987). Section 4 (1) of the legislation states:

> [t]he national parks of Canada are hereby dedicated to the people of Canada for their benefit, education and enjoyment . . . and the parks shall be maintained and made use of so as to leave them unimpaired for the enjoyment of future generations.

The Panel Report was unequivocally critical of the trend in parks management to over-emphasize 'use' in the parks. The Panel, therefore, recommended the enactment of the CNPA ecological integrity provisions to re-affirm preservation (apart from human influence) as the first priority in parks management.

The last 15 years have been a turbulent time in Canada's national parks: budgets and staff have been significantly reduced, and the Agency has been repeatedly reorganized (Van Sickle and Eagles, 1998; Wright and Rollins, 2002; Dearden and Dempsey, 2004). The Panel Report stated that the ecological integrity of most national parks was in jeopardy, and recommended both substantially increased funding and organizational reform in order for the Agency to meet its mandate to maintain the parks' ecological integrity (PCA, 2000). Nevertheless, implementation of the Panel's recommendations has been severely constrained by the absence of such new funding (PCA, 2001b; Dearden and Dempsey, 2004), although by 2004 some funding was available for individual projects. Achieving ecological integrity in Canadian national parks has now become a highly symbolic goal which, according to some observers, could remedy an urgent environmental crisis (Searle, 2000).

Considered in this social and political context, it is not surprising that Parks Canada's search for a simple, defensible, and universally applicable definition of ecological integrity (PCA, 2000) has substantially altered and shaped understanding of the term itself. In practice, ecological integrity is represented not only as a management objective (Canadian Parks Service, 1994; Woodley, 2002), but also as a single-equilibrium continuous variable – less technically, an inherent state of nature that can be lost or recovered (PCA, 2000). This conflation of goal and metric has produced a linear, deterministic management model that presupposes that a park is a stable ecosystem that has high integrity unless stressed by human activities (Figure 8.1). Such stressors reduce the ecosystem's integrity by a measurable quantity that can subsequently be restored by active intervention.

Not only does this model fail to account for nonlinear ecosystem behaviour and multiple stable states (Holling, 1973), but also it simplistically categorizes all social-ecological interactions as either stressors or management interventions. The normative dichotomy of that categorization implies superiority of the intervening Agency over those who stress ecosystems, justifying a top-down style of management. Hermer (2002) calls this institutional phenomenon 'emparkment'. Emparkment is an outcome of serious concern to many stakeholders and local inhabitants, especially to Aboriginal peoples, who do not view their own activities as stressors on park ecosystems (e.g. Lewis, 1989; Budke, 1998; Weitzner and Manseau, 2001) and whose own identity is inseparable from the land itself (Berkes, 1999; Brody, 2000; Nadasdy, 2003). Here, the Panel Report sets up a paradox, advocating ecological integrity in a wilderness-normative sense that excludes people, while at the same time promoting stronger relationships between Parks Canada and Aboriginal peoples.

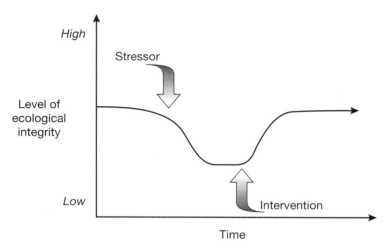

Figure 8.1 Managing ecological integrity: single equilibrium model (after PCA, 2000)

Wilderness-normative ecological integrity, as defined in the CNPA, leads to ineffective resolution of conservation problems that might be conceptually straightforward in ecological terms (Caughley, 1994), but whose solutions are rarely simple to implement. The magnitude of this mismatch is illustrated by the reintroduction of wolves (*Canis lupus*) to Yellowstone National Park in the US. The actual reintroduction took only two years to accomplish, but was preceded by 20 years of planning and controversy (Clark and Gillesberg, 2001). The full dimensions of conservation problems in protected areas rarely fit within the wilderness-normative discourse of ecological integrity, and there are consequences to ignoring the dimensions that do not. Three examples of such problems in Canadian national parks are summarized below. In different ways, each case shows that the strictly ecological definitions of system integrity and simplistic, unitary goals of wilderness-normative conceptions of ecological integrity hinder effective responses to conservation problems, which are often of a polycentric and socio-political character.

Snow geese in Wapusk National Park

Long-term research on lesser snow geese (*Chen caerulescens*) nesting in Wapusk National Park, Manitoba has documented the causal links between a larger goose population, their foraging behaviour, and the widespread conversion of coastal salt marshes around Hudson Bay to hypersaline mud-flats (Jefferies et al., 2004; Abraham et al., 2005). The dramatic increase in goose numbers was apparently due to higher over-winter adult survival as geese fed on agricultural wastes and crops in the US midwest and gulf states.

This case demonstrates the rapid, widespread change of an ecosystem to an alternative stable state with significantly reduced productivity and species diversity (Figure 8.2). Issues of scale characterize management responses so far. Deliberately increased hunting in the US and Canada has reduced this goose population on a continental scale (Jefferies et al., 2004). In contrast, the park is not contemplating any local-scale intervention (PCA, 2004), although the salt marshes' return to a productive state is judged unlikely even with stabilized goose numbers (Jefferies et al., 2004; Abraham et al., 2005). The wilderness-normative discourse has prevented the park from developing a comprehensive problem definition and hindered formation of a coherent response. Essentially, the park will monitor the situation and accept liberalized goose hunting outside the park, but not inside it, even for Aboriginal

Deconstructing ecological integrity policy

Figure 8.2 Former salt marsh ecosystem overgrazed by lesser snow geese, La Pérouse Bay, Wapusk National Park. Note the dead willow, and the patches of moss, not the formerly abundant sedge or grass

people (Clark, 2001; PCA, 2004). The complexity and multi-scalar nature of the problem appear to have prevented the park from comprehensively assessing any alternative strategies (Clark, 2001; PCA, 2004). Finally, the park expresses a strong desire to maintain control in a situation where it is manifestly infeasible: 'Within Wapusk National Park of Canada there will be no sport hunting, no encouragement of special hunts for Aboriginal people . . . furthermore, any hunt or management action on park lands will be under Park control' (Clark, 2001: B3.3).

Abalone, urchins, and sea otters in Haida Gwaii

Sloan (2004) presents the complex case of two listed species-at-risk: northern abalone (*Haliotis kamtschatkana*) and their predator, sea otters (*Enhydra lutris*), in the Haida Gwaii Archipelago (Queen Charlotte Islands), British Columbia. The region includes Gwaii Haanas National Park Reserve and Haida Heritage Site, as well as a proposed national marine conservation area. Sea otters were extirpated there over 100 years ago, and kelp forests diminished as populations of the herbivorous abalone and red sea urchins (*Strongylocentrotus franciscanus*) were released from predation by otters (Figure 8.3). The central issue is that a re-established otter population would probably depress abalone and urchin populations which, in the absence of otters, came to support important commercial and Aboriginal subsistence fisheries. In this situation, determining the ecologically optimal strategy for abalone or otter recovery is a very different question than asking what sort of a coastal ecosystem Haida Gwaii's inhabitants want to have. Answering the latter question involves an explicitly non-scientific choice about desired futures that is based on a wide range of cultural and commercial values as well as wilderness ideals (Levin, 1988; Sloan, 2004).

This particular situation demands a choice between alternative, mutually exclusive ecosystem states that provide different societal and ecological benefits: kelp forests with otters and few urchins or abalone, versus 'urchin barrens' that have less biodiversity but support high-value fisheries (Fanshawe *et al.*, 2003). The wilderness-normative discourse allows little room for such context-specific judgements though, by giving automatic precedence to the less 'impaired' system state, kelp forests. Nevertheless, Sloan (2004) points out that this discourse has failed to gain local support. Further, empowered by a federally negotiated co-management regime, the Haida people have put forward an alternative discourse based on their own complex relationship with coastal ecosystems, including abalone harvest.

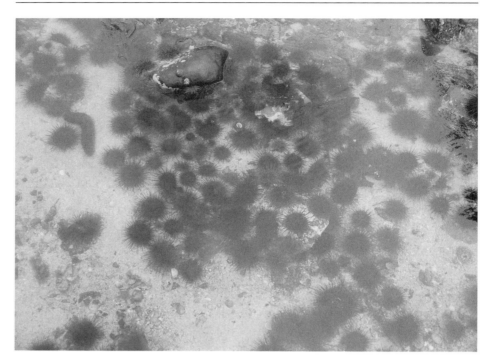

Figure 8.3 Red sea urchins in the intertidal zone, Gwaii Haanas Haida Heritage Site and National Park Reserve

Judicial decisions: road construction in Wood Buffalo and water diversion in Banff

In May 2001, the minister responsible for Canada's national parks approved the construction of a winter road traversing the remote Wood Buffalo National Park in northern Alberta. Despite the CNPA ecological integrity provisions and the release of the Panel Report only a year earlier, the minister approved the road without even mentioning ecological integrity in her decision. In subsequent judicial review, the Federal Court of Canada upheld the minister's decision, finding that she had taken ecological integrity into account, notwithstanding the absence of express consideration of the concept (*Canadian Parks and Wilderness Society* v. *Canada (Minister of Canadian Heritage)* (2003), 1 CELR (3d) 20 (Fed. CA)). In the Court's view, the maintenance of ecological integrity need not be the determinative factor in parks management despite the express requirement in the CNPA. Indeed, the Court held that in some cases the interests of humans can override those of ecological integrity where impairment is minimized to ensure the park can be enjoyed by future generations.

In 2002, the minister renewed a water permit that allows the Chateau Lake Louise in Banff National Park to drain water from Lake Louise to sustain hospitality operations. An environmental group challenged the renewal on the grounds that it was prohibited by the CNPA ecological integrity provisions. The Federal Court of Canada denied the challenge, upholding the minister's renewal as striking a reasonable balance between human activity and the preservation of ecological integrity (*Mountain Parks Watershed Assn.* v. *Chateau Lake Louise*, (2004) FC 1222). As in the 2001 Wood Buffalo decision noted above, the Court had no reservations about trumping the ecological integrity management priority in favour of human interests.

Deconstructing ecological integrity policy

Implicit in these two judicial decisions is the view that wilderness-normative ecological integrity leads to park management decisions that exclude humans from parks. This result is simply unacceptable for most, if not all, of Canada's national parks. The wilderness-normative discourse (humans versus nature) excludes the consideration of human values in assessing ecological integrity. One commentator on the Wood Buffalo decision suggests the wilderness-normative discourse facilitates a conflict between southern preservationists seeking to preserve 'pristine' northern wilderness and the local communities that rely on the park to provide sustenance and other cultural needs (Scott, 2003).

Another commentator argues the normative rigidity of wilderness-normative ecological integrity impairs the ability of legal decision-makers to apply the concept (Bagg, 2005). As a result, legal decision-makers applying the CNPA ecological integrity provisions to adjudicate human value conflicts concerning the Canadian parks have either read down the normative weight of wilderness-normative ecological integrity or have dismissed it altogether (Fluker, 2003; Bagg, 2005), and the maintenance of ecological integrity remains policy rhetoric in Canada's national parks despite its legislated status.

Prospects for alternative ecological integrity discourses

Fluker (2003) asserts that Parks Canada's wilderness-normative definition of ecological integrity – which he terms 'natural ecological integrity' – is rigid and ultimately self-defeating. The worldview underlying the wilderness-normative discourse tells a story of conflict between economic development and environmental preservation; this dualistic story is too simple to resolve difficulties facing protected areas. As an alternative to natural ecological integrity policy in Canada's national parks, Fluker (2003) argues for 'socio-ecological integrity', which is closely aligned with the non-normative ecological integrity discourses (particularly the ecosystemic-pluralistic) in which social and ecological systems are conceived as mutually dependent, self-organizing entities. Socio-ecological integrity depends on human values about a particular ecosystem and interactions with it; as such it cannot be considered 'objectively' as a single quantifiable entity independent of those values. Theoretically, this concept resolves the competition in protected areas between orthodox ecological integrity and societal concerns, and it has been advanced in some developing countries as noted below.

Socio-ecological integrity incorporates elements of resilience theory, suggesting a new model for protected area management (Figure 8.4). Social-ecological systems are dynamic in this model, and have no 'pristine' or 'baseline' ecological states. When a system enters

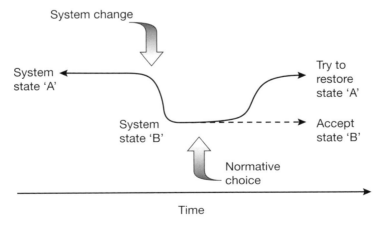

Figure 8.4 Conceptual model for managing socio-ecological integrity for multiple equilibria and human values

(or is predicted to enter) a new state, an explicitly normative decision – informed, but not determined, by science – must be made about which state is more desirable to manage for.[4] If a new system state is undesirable, adaptations to it can be learned and communicated. Decision-makers must recognize that state-changing interventions will likely be costly, and could be impossible because the system might not possess sufficient capital or connectivity to be restored to its previous state (Kay and Schneider, 1994), or the system's whole context might have changed (Walker et al., 2004). Managing for socio-ecological integrity would be demanding since it would require tremendous flexibility, resources, openness at multiple institutional and geographic scales, patience, and clarity about all participants' standpoints and values (Kay et al., 1999; Walker et al., 2002; Waltner-Toews et al., 2003).

Primacy is no guarantee that an existing ecosystem is the only one to have integrity in a particular situation (Kay and Schneider, 1994). Consequently, human relationships with existing or past ecosystem states may have more influence than science does in pronouncing a particular state to have greater ecological integrity than another. Letting go of attachments to past systems is particularly difficult in parks, as demonstrated by the persistence of management goals such as Leopold et al.'s (1963) 'vignettes of primitive America' and Cole's (1971) natural regulation (Wright, 1999). However, South Africa's century-old Kruger National Park appears to have implemented management strategies based on an understanding of ecosystem heterogeneity and change, so it is possible (Carruthers, 1995; DuToit et al., 2003).[5] Nevertheless, accepting change in principle does not mean that every change is inevitable or even acceptable. The slide into relativism feared by some conservationists (Soulé and Lease, 1995) needs to be examined and re-examined through applied research and, where possible, monitoring systems to assist managers in tracking and anticipating changes in ecosystems and causal factors.

Socio-ecological integrity represents a desirable relationship between social and ecological systems (Fluker, 2003). Choices must, therefore, be made about which system states to manage protected areas for, and such value-based choices will require criteria for determining the desirability of specific changes in systems. There are constitutive issues to such decisions: how they will be made and who will make them are probably more important considerations than their actual content (Clark, 2002). Decisions about preferred system states will likely be made in contexts of high uncertainty about the system at hand and with high stakes for the system's varied, and often competing, human interests and livelihoods. Involving the people who interact directly with the system not only can provide instrumentally useful information, but also is the most ethical course of action since those are the people who will be most affected by any decisions (Funtowicz and Ravetz, 1994).

The polycentric character inherent within socio-ecological integrity, or the ecosystemic-pluralistic discourse, complements existing judicial interpretations of the CNPA ecological integrity provisions that require Parks Canada to undertake multi-factored analysis of park management decisions (Bagg, 2005). Notwithstanding that the CNPA ecological integrity provisions failed to compel the application of wilderness-normative ecological integrity in the 2001 Wood Buffalo decision, socio-ecological integrity has, arguably, influenced the result. In 2005, the Supreme Court of Canada quashed the road approval on grounds that Parks Canada and its minister failed to adequately consult with the local Mikisew Cree First Nation concerning implications the road would have on the First Nation's treaty hunting and trapping rights (*Mikisew Cree First Nation v. Canada (Minister of Canadian Heritage)*, (2005) SCC 69). The Court's focus on the absence of local participation in the road decision has parallels with socio-ecological integrity and the ecosystemic-pluralistic discourse. This judicial decision suggests that the ecosystemic-pluralistic discourse may be capable of resolving the difficulties of 'emparkment'. The decision also suggests that the principle of ecological integrity has more influence on legal decision-makers by mandating a process than specifying the substantive result; the form by which the principle becomes a legal rule remains contested ground.

Socio-ecological integrity potentially resolves the competition in protected areas between normative ecological integrity and societal concerns. In practice, this discourse has been advanced on a wide scale in many developing countries through Integrated Conservation and Development Projects (ICDPs) since the early 1990s. A key underlying justification has been the prospect and need for operationalizing sustainable development as espoused in Agenda 21 and the Convention on Biological Diversity (CBD) and more recently packaged within the Millennium Development Goals (MDGs) (United Nations, 2006). However, successful ICDPs in terms of delivering 'win-win outcomes' for conservation and development have been hard to achieve in reality, often due to biases toward either conservation or development in project design and implementation, in addition to a multitude of common design faults such as short timescales, overly complex objectives and activities, unrealistic conservation and development incentives, and poor monitoring and evaluation[6] (see Wells and McShane, 2004).

The outcome of the failure of ICDPs has been disenchantment among some practitioners and a call to return to basic wilderness-normative discursive principles with respect to protected area management (e.g. Brandon et al., 1998; Terborgh, 1999; Terborgh et al., 2002 and responses by Chapin, 2004; Redford et al., 2006). Others are now calling for a more nuanced design and implementation of ICDPs (Wilshusen et al., 2002) and decentralized governance of resources (Borrini-Feyerabend, 2003; Borrini-Feyerabend et al., 2005; Hardin and Remis, this volume) within the context of poverty–conservation linkages to contribute towards the MDGs. In such ecosystemic-pluralistic discourses science is no longer the determining voice of reason or justification, but one viewpoint within the socio-politics of conservation. Appreciating and clarifying the trade-offs between social development demands and ecosystems in a move away from 'win-win' assumptions associated with specific choices is thus a particularly important contribution of ecological science (Kay and Schneider, 1994; Wilshusen et al., 2002; Wells and McShane, 2004). Such an adaptive approach requires a new function for protected area regimes that differs from the behaviour of a typical centralized bureaucracy. Regimes must empower individual protected areas to innovate and adapt, conserve accumulated knowledge and memory of successful adaptations, and facilitate learning by spreading information about those innovations throughout the system (Folke et al., 1998, 2003; Brunner et al., 2005).

Conclusions

The ecological integrity provisions contained in Canadian national park legislation have locked the Parks Canada Agency into a rigidity trap, where high connectedness and efficient control mechanisms reinforce a system's governing paradigm, but at the cost of adaptive capacity (Holling et al., 2002). The legislation provides a strong defence against challenges to the Agency's normative assumptions and privileged position in the wilderness-normative discourse. Ironically, the legislation has little normative influence over park management decisions. Moreover, while the legislation purports to ensure the primacy of ecological integrity in parks management, it has failed to compel legal decision-makers to implement it as such. As a result, the maintenance or restoration of ecological integrity in Canada's national parks remains firmly entrenched as policy rhetoric, but has little influence in the field. Judicial interpretations of the CNPA ecological integrity provisions suggest that legal decision-makers are heavily influenced by the non-normative ecological integrity discourses. Arguably, this was the case in Wood Buffalo National Park, and in many other northern parks where Aboriginal land claims have granted specific rights to peoples (Slocombe, 1996; Weitzner and Manseau, 2001).

Simplistic calls for either strictly 'top-down' or 'bottom-up' solutions to this rigidity-trap dilemma fail to recognize a potentially wide range of mixed-approach solutions that would probably better reflect the social and biophysical complexities of most nations' protected areas (Redford, 2006). A policy environment oriented more towards supporting local-scale participation initiatives and adaptive governance (Brunner et al., 2005) would be helpful for the development of such alternative and parallel discourses. In Canadian national parks, for example, the co-management models developed in the northern parks could be considered as prototypes whose wider diffusion and adaptation might prove useful elsewhere in the system; but whose utility would have to be determined on a case-by-case basis. Mechanisms for managers and stakeholders to learn about successful innovations in other parks are also lacking, and creating them would be a particularly constructive 'top-down' function (Berkes and Folke, 1998). Ultimately, de-emphasizing the expert-privileging normative discourses and dealing with the proximate residual effects of emparkment will require leadership and commitment sustained over time at the local scale, combined with trust, respect, and institutions capable of sanctioning any excesses of power without preventing innovation and adaptation.

Notes

1. The policy emphasis on poverty–environment (including conservation within and outside PAs) linkages within the context of the Millennium Development Goals is illustrative of transpersonal collaborative discourses.
2. Previous names for the Parks Canada Agency such as the Canadian Parks Service and Parks Canada are still commonly used, so we use them here as well.
3. Section 2 (1) of the CNPA states:

    'Ecological integrity' means, with respect to a park, a condition that is determined to be characteristic of its natural region and likely to persist, including abiotic components and the composition and abundance of native species and biological communities, rates of change and supporting processes.

4. In practice, this is often dependent on the manager formulating and testing a set of assumptions about the ecosystem and having a monitoring system in place to detect changes and therefore decide what it is he/she is managing for.
5. For example, elephant culling has been part of the Kruger National Park management strategy. Although, large-scale active management of national park habitats dates back to the 1950s when hippopotamus and elephant culling was practised in the Queen Elizabeth and Murchison Falls National Parks in Uganda to reduce overgrazing and deforestation pressures (see Bere, 1959, 1972). Managers in Uganda and now in South Africa are managing in a sense to 'combat' change and maintain a perceived sense of 'ecological equilibrium'.
6. Making it difficult to know the reasons for success and failure during implementation and thus to employ adaptive management.

Literature cited

Abraham, K.F., Jefferies, R.L., and Rockwell, R.F. (2005). Goose-induced changes in vegetation and land cover between 1976 and 1997 in an Arctic coastal marsh. *Arctic, Antarctic and Alpine Research* 37, 269–275.

Adams, W.A. and Hulme, D. (2001). Conservation & community: changing narratives, policies & practices in African conservation. In D. Hulme and M. Murphree (eds), *African Wildlife and Livelihoods: the promise and performance of community conservation*, pp. 9–23. Oxford: James Currey.

Agee, J.K. and Johnson, D.R. (eds) (1988). *Ecosystem Management for Parks and Wilderness Areas*. Seattle, WA: University of Washington Press.

Ascher, W. (2001). Coping with complexity and organizational interests in natural resources management. *Ecosystems* 4, 742–757.

Bagg, S. (2005). Don't be fooled by the fancy language: ecological integrity is not a value (in the eyes of the law). Unpublished. Calgary: University of Calgary.

Bella, L. (1987). *Parks for Profit*. Montreal: Harvest House.
Bengtsson, J., Angelstam, P., Elmqvist, T., Emanuelsson, U., Folke, C., Ihse, M., Moberg, F., and Nyström, M. (2003). Reserves, resilience, and dynamic landscapes. *Ambio* 32, 389–396.
Bere, R. (1959). Queen Elizabeth National Park, Uganda: the hippopotamus problem and experiment. *Oryx* 5, 116–124.
—— (1972). *A History of Uganda National Parks*. Unpublished manuscript.
Beresford, M. and Philips, A. (2000). Protected landscapes: a conservation model for the 21st century. *George Wright Forum* 17, 15–26.
Berkes, F. (1999). *Sacred Ecology: traditional ecological knowledge and resource management*. Philadelphia, PA: Taylor & Francis.
—— (2004). Rethinking community-based conservation. *Conservation Biology* 18, 621–630.
—— and Folke, C. (eds) (1998). *Linking Social and Ecological Systems: management practices and social mechanisms for building resilience*. Cambridge: Cambridge University Press.
Borrini-Feyerabend, G. (2003). Governance of protected areas – innovation in the air. *Policy Matters* 12, 92–101.
——, Pimbert, M., Farvar, T., and Kothari, A. (2005) *Sharing Power: learning by doing in co-management of natural resources throughout the world*. Gland: CEESP and IUCN.
Brandon, K., Redford, K.H., and Sanderson, S.E. (eds) (1998). *Parks in Peril: people, politics, and protected areas*. Washington, DC: Island Press.
Brody, H. (2000). *The Other Side of Eden: hunters, farmers, and the shaping of the world*. Vancouver: Douglas & McIntyre.
Brunner, R., Steelman, T.A., Coe-Juell, L., Cromley, C.M., Edwards, C.M., and Tucker, D.W. (2005). *Adaptive Governance: integrating science, policy, and decision-making*. New York: Columbia University Press.
Budke, I. (1998). *A Review of Cooperative Management Arrangements and Economic Opportunities for Aboriginal People in Canadian National Parks*. Vancouver: Parks Canada.
Canadian Parks Service (1992). *Toward Sustainable Ecosystems: a Canadian Parks Service strategy to enhance ecological integrity. Final report of the Ecosystem Management Task Force, Canadian Parks Service, Western Region*. Calgary: Canadian Parks Service.
—— (1994). *Guiding Principles and Operational Policies*. Ottawa: Minister of Public Works and Government Services.
Carruthers, J. (1995). *The Kruger National Park: a social and political history*. Pietermaritzburg: University of Natal Press.
Caughley, G. (1994). Directions in conservation biology. *Journal of Animal Ecology* 63, 215–244.
Chapin, M. (2004). A challenge to conservationists. *WorldWatch* (November/December), 17–31.
Clark, D. (2001). Lesser snow geese in Wapusk National Park of Canada: a case study. In Parks Canada, *An Orientation Program on Ecological Integrity – a call to action. Participant's workbook version 1.0*, pp. B3.1–B3.4. Ottawa: Parks Canada.
Clark, T.W. (2002). *The Policy Process: a practical guide for natural resource professionals*. New Haven, CT: Yale University Press.
—— and Gillesberg, A. (2001). Lessons from wolf restoration in Greater Yellowstone. In V.A. Sharp, B. Norton, and S. Donnelley (eds), *Wolves and Human Communities: biology, politics, and ethics*, pp. 135–149. Washington, DC: Island Press.
Cole, G.F. (1971). An ecological rationale for the natural or artificial regulation of native ungulates in parks. *Transactions North American Wildlife and Natural Resources Conference* 36, 417–425.
Cronon, W. (1996). *Uncommon Ground: rethinking the human place in nature*. New York: W.W. Norton.
Dearden, P. and Dempsey, J. (2004). Protected areas in Canada: decade of change. *The Canadian Geographer* 48, 225–239.
DeLeo, G. and Levin, S. (1997). The multifaceted aspects of ecosystem integrity. *Conservation Ecology* 1 (1), 3. Retrieved 26 November 2002 from www.consecol.org/vol/iss1/art3.
Du Toit, J.T., Rogers, K.H., and Biggs, H.C. (2003). *The Kruger Experience: ecology and management of savanna heterogeneity*. Washington, DC: Island Press.
Fanshawe, S., Vanblaricom, G.R., and Shelley, A.A. (2003). Restored top predators as detriments to the performance of marine protected areas intended for fisheries sustainability: a case study with red abalones and sea otters. *Conservation Biology* 17, 273–283.

Fluker, S. (2003). 'Maintaining ecological integrity is our first priority' – Policy Rhetoric or Practical Reality in Canada's National Parks? A case comment on *Canadian Parks and Wilderness Society v. Canada (Minister of Canadian Heritage)*. *Journal of Environmental Law and Practice* 13, 131–146.

Folke, C., Berkes, F., and Colding, J. (1998). Ecological practices and social mechanisms for building resilience and stability. In F. Berkes and C. Folke (eds), *Linking Social and Ecological Systems: management practices and social mechanisms for building resilience*, pp. 414–436. Cambridge: Cambridge University Press.

——, Colding, J., and Berkes, F. (2003). Synthesis: building resilience and adaptive capacity in social-ecological systems. In F. Berkes, J.F. Colding, and C. Folke (eds), *Navigating Nature's Dynamics: building resilience for complexity and change*, pp. 352–387. New York: Cambridge University Press.

Foreman, D. (2004). *Rewilding North America: a vision for conservation in the 21st century*. Washington, DC: Island Press.

Funtowicz, S.O. and Ravetz, J. (1994). Uncertainty, complexity and post-normal science. *Environmental Toxicology and Chemistry* 13/12, 1881–1885.

Global Environmental Facility. (2005). *The Role of Local Benefits in Global Environmental Programs*. Washington, DC: GEF Office of Monitoring and Evaluation.

Grumbine, R.E. (1994). What is ecosystem management? *Conservation Biology* 8, 27–38.

Henry, J.D. and Lieff, B. (1992). *Ecosystem Management of National Parks, Western Region, Canadian Parks Service: seminar proceedings*. Calgary: Canadian Parks Service.

Hermer, J. (2002). *Regulating Eden: the nature of order in North American parks*. Toronto: University of Toronto Press.

Herrero, S., Roulet, J., and Gibeau, M. (2001). Banff National Park: science and policy in grizzly bear management. *Ursus* 12, 161–168.

Holling, C.S. (1973). Resilience and stability of ecological systems. *Annual Review of Ecology and Systematics* 4, 1–23.

—— (1978). *Adaptive Environmental Assessment and Management*. New York: John Wiley and Sons.

—— and Meffe, G.K. (1996). Command and control and the pathology of natural resource management. *Conservation Biology* 10, 328–337.

——, Gunderson, L.H., and Peterson, G.D. (2002). Sustainability and panarchies. In L. Gunderson and C.S. Holling (eds) 2002. *Panarchy: understanding transformations in human and natural systems*, pp. 63–102. Washington, DC: Island Press.

Janzen, D.H. (1989). The evolutionary biology of national parks. *Conservation Biology* 3, 109–110.

Jefferies, R.L., Rockwell, R.F., and Abraham, K.F. (2004). The embarrassment of riches: agricultural food subsidies, high goose numbers, and loss of Arctic wetlands – a continuing saga. *Environmental Reviews* 11, 193–232.

Karr, J.R. (1991). Biological integrity: a long-neglected aspect of water resource management. *Ecological Applications* 1, 66–84.

Kay, J.J. (1991). A nonequilibrium thermodynamic framework for discussing ecosystem integrity. *Environmental Management* 15, 483–495.

—— and Schneider, E. (1994). Embracing complexity: the challenge of the ecosystem approach. *Alternatives* 20, 32–39.

——, Regier, H.A., Boyle, M., and Francis, G. (1999). An ecosystem approach for sustainability: addressing the challenge of complexity. *Futures* 31, 721–742.

Leopold, A.S., Cain, S.A., Cottam, C.M., and Gabrielson, I. (1963). Wildlife management in the national parks. *Transactions North American Wildlife and Natural Resources Conference* 28, 28–45.

Levin, S.A. (1988). Sea otters and nearshore benthic communities: a theoretical perspective. In G.R. Vanblaricom and J.A. Estes (eds), *The Community Ecology of Sea Otters*, pp. 202–209. New York: Springer-Verlag.

Lewis, H.T. (1989). Ecological and technological knowledge of fire: Aborigines versus park rangers in northern Australia. *American Anthropologist* 91, 940–961.

Manuel-Navarette, D. (2003). *Ecological Integrity Discourses as a Heuristic Tool for Understanding Conservation Initiatives: the case of the Maya Rainforest in the Usmacinta Watershed*. Ph.D. dissertation. Waterloo: University of Waterloo.

Merchant, C. (2004). *Reinventing Eden: the fate of nature in western culture*. New York: Routledge.

Morito, B. (1999). Examining ecosystem integrity as a primary mode of recognizing the autonomy of nature. *Environmental Ethics* 21, 59–73.

Nadasdy, P. (2003). *Hunters and Bureaucrats: power, knowledge, and aboriginal-state relations in the southwest Yukon*. Vancouver: University of British Columbia Press.

Neumann, R. (1998). *Imposing Wilderness: struggles over livelihood and nature preservation in Africa*. Berkeley, CA: University of California Press.

Parks Canada Agency (2000).*Unimpaired for Future Generations? Conserving ecological integrity with Canada's national parks. Report of the Panel on the Ecological Integrity of Canada's National Parks*. Ottawa: Minister of Public Works and Government Services.

—— (2001a). *An Orientation Program on Ecological Integrity – a call to action. Participant's workbook version 1.0*. Ottawa: Parks Canada Agency.

—— (2001b). *First Priority: progress report on implementation of the recommendations of the panel on the ecological integrity of Canada's national parks*. Ottawa: Minister of Public Works and Government Services.

—— (2004). *Wapusk National Park of Canada: draft management plan*. Churchill: Parks Canada Agency.

Pimbert, M.P. and Pretty, J.N. (1997). Parks, people and professionals: putting 'participation' into protected area management. In K.B. Ghimire and M.P. Pimbert (eds), *Social Change and Conservation*, pp. 297–330. London: Earthscan.

Redford, K.H., Robinson, J.G., and Adams, W.M. (2006). Parks as shibboleths. *Conservation Biology* 2, 1–2.

Risby, L.A. (2002). *Defining Landscape, Power and Participation: an examination of a national park planning process for Queen Elizabeth National Park, Uganda*. Ph.D. thesis. Cambridge: Cambridge University.

Rose, C. (1994). Given-ness and gift: property and the quest for environmental ethics. *Environmental Law* 24, 1–31.

Scott, D. (2003). Standing in the road: the battle for Wood Buffalo National Park. *Journal of Environmental Law and Practice* 13, 145–151.

Searle, R. (2000). *Phantom Parks: the struggle to save Canada's national parks*. Toronto: Key Porter Books.

Sloan, N.A. (2004). Northern abalone: using an invertebrate to focus marine conservation values. *Coastal Management* 32, 129–143.

Slocombe, D.S. (1993). Implementing ecosystem-based management. *BioScience* 43, 612–622.

—— (1996). Hinterlands, wilderness, and protected areas in northern Canada. In R.G. Wright (ed.), *National Parks and Protected Areas: their role in environmental protection*, pp. 369–387. New York: Blackwell Scientific.

Soulé, M.E. and Lease, G. (eds) (1995). *Reinventing Nature: responses to postmodern deconstruction*. Washington, DC: Island Press.

Terborgh, J. (1999). *Requiem for Nature*. Washington, DC: Island Press.

——, van Schaik, C., Davenport, L., and Rao, M. (eds) (2002). *Making Parks Work: strategies for preserving tropical nature*. Washington, DC: Island Press.

United Nations (2006). *The Millennium Development Goals Report 2006*. New York: United Nations Secretariat.

Van Sickle, K. and Eagles, P.F. (1998). Budgets, pricing policies, and user fees in Canadian parks' tourism. *Tourism Management* 19, 225–235.

Walker, B., Holling, C.S., Carpenter, S., and Kinzig, A. (2004). Resilience, adaptability and transformability in social-ecological systems. *Conservation Ecology* 9 (2), 5. Retrieved 24 September 2004 from www.consecol.org/vol9/iss2/art5.

——, Carpenter, S., Anderies, J., Abel, N., Cumming, G., *et al.* (2002). Resilience management in social-ecological systems: a working hypothesis for a participatory approach. *Conservation Ecology* 6 (1), 14. Retrieved 15 August 2002 from www.consecol.org/vol6/iss1/art14.

Walters, C.J. (1986). *Adaptive Management of Renewable Resources*. New York: McGraw Hill.

Waltner-Toews, D., Kay, J.J., Neudorffer, C., and Gitau, T. (2003). Perspective changes everything: managing ecosystems from the inside out. *Frontiers in Ecology* 1, 23–30.

Weitzner, V. and Manseau, M. (2001). Taking the pulse of collaborative management in Canada's national parks and national park reserves: voices from the field. In D. Harmon (ed.), *Crossing Boundaries* in *Park Management: proceedings of the 11th conference on research and resource management in parks and on public lands*, pp. 253–259. Hancock: George Wright Society.

Wells, M. and McShane, T. (2004). *Getting Biodiversity Projects to Work: towards more effective conservation and development*. New York: Columbia University Press.

Westra, L. (1994). *The Principle of Integrity: an environmental proposal for ethics*. Landham: Rowman & Littlefield.

Wicklum, D. and Davies, R.W. (1995). Ecosystem health and integrity? *Canadian Journal of Botany* 73, 997–1000.

Wilshusen, P.R., Brechin, S.R., Fortwangler, C.L., and West, P.C. (2002). Reinventing a square wheel: critique of a resurgent 'protection paradigm' in international biodiversity conservation. *Society & Natural Resources* 15, 17–40.

Woodley, S. (2002). Planning and managing for ecological integrity. In P. Dearden and R. Rollins (eds), *Parks and Protected Areas in Canada: planning and management* (2nd edn), pp. 97–114. Don Mills: Oxford University Press.

——, Kay, J., and Francis, G. (eds) (1993). *Ecological Integrity and the Management of Ecosystems*. Boca Raton, FL: CRC Press.

Wright, P. and Rollins, R. (2002). Managing the National Parks. In P. Dearden and R. Rollins (eds), *Parks and Protected Areas in Canada: planning and management* (2nd edn), pp. 207–239. Don Mills: Oxford University Press.

Wright, R.G. (1999). Wildlife management in the national parks: questions in search of answers. *Ecological Applications* 9, 30–36.

Chapter 9
The science and management interface in national parks

R. Gerald Wright

Introduction

Managing the natural resources of parks and protected areas is a difficult and complicated task. Management decisions must take into account the dynamics of natural cycles in plant and animal populations, the variation in natural ecological, hydrological, and geological processes, and the increasing importance of anthropogenic-induced environmental changes caused by climate warming and air and water pollution. Management decisions must also take into account other, more pragmatic factors. These include political considerations, economic limitations, existing laws, and the necessity of satisfying the needs and perceptions of the public that visit and/or cherish these areas. The degree to which each of these considerations influences the management decision-making process depends on the situation and, clearly, some influences such as public opinion can be more powerful than others. It is well acknowledged that management decisions which ignore the views of political leaders, violate laws, are economically unsound, or ignore public sentiment, do not normally last.

This chapter is, however, concerned only with one of the many influences on resource management decisions in parks and protected areas – that of scientific knowledge and the informed opinions of scientists. It is written from the perspective of experience gained from a long tenure as a research scientist with the United States National Park Service (USNPS).

Origin and purpose of the US National Park Service

The USNPS was established in 1916 although several national parks were in existence prior to that time. From the very beginning, the underlying purpose for most natural area parks administered by the agency has been debated. Were they primarily established to preserve unique landscapes, resources, or processes, or were they established to provide an environment for recreation and education? The language of the Organic Act which founded the USNPS only confounds this dilemma, stating that the purpose of parks is

> to conserve the scenery and the natural and historic objects and wildlife therein and to provide for the enjoyment of the same in such a manner and by such means as will leave them unimpaired for the enjoyment of future generations.
> (Ch 408, 39 [US] stat. 535 [1916])

The Act is especially ambiguous because it does not define terms such as 'natural' or 'unimpaired'. This has left the terms and the Act open to a variety of interpretations, none of which has been totally satisfactory (see Sellars, 1997 for a full account of this dilemma). There have also been few judicial interpretations of the USNPS Organic Act to indicate the extent that parks should be managed for preservation as opposed to use and enjoyment (Lemons and Stout, 1984).

Early national park natural resource management efforts

In its early years, the USNPS struggled to build a constituency that would support the new and relatively unknown agency. To that end, the emphasis was on attracting visitors and providing for them in comfortable accommodations. The scenery of the existing parks was the selling point – be it the geysers of Yellowstone, the grandeur of Yosemite Valley, the big trees at Sequoia, the beautiful lake at Crater Lake, or the rugged mountains and lakes of Glacier (Sellars, 1997). Wildlife was viewed as being particularly important to visitor satisfaction, and, because of that, natural resources were managed to provide visitors with neat, tidy landscapes and panoramas with opportunities to see wildlife close up (Wright, 1992). Sellars (1997) has described this practice as 'facade management'. Of special importance were the so-called 'charismatic species' that included elk (*Cervus elaphus*), deer (*Odocoileus* sp.), mountain sheep (*Ovis canadensis*), and black bear (*Ursus Americanus*).

This type of management viewed park ecosystems as static, isolated, independent landscapes in which any change in the natural system was seen as detrimental. Managers *believed* they knew what was best and right for the park ecosystems and managed accordingly (Davis and Halvorson, 1996). Park managers thought that predators, fire, native insects, and plant succession were bad influences and, therefore, they removed wolves (*Canis lupis*), coyotes (*Canis latrans*), and cougars (*Felis concolor*) from parks to protect the good animals, primarily the large herbivores. Fires were suppressed, native insect infestations were eradicated, and successional influences were curtailed (Wright, 1992).

The emergence of science and its effect on natural resource management

The archaic attitudes of the first two decades began to change in the early 1930s when a young biologist, George Wright, and his colleagues began the first series of scientific investigations in parks and challenged many of the belief-based concepts. Their first publication, *A Preliminary Survey of Faunal Relations in National Parks* (Wright et al., 1933), represented a landmark in applying scientific knowledge to park resource management, and most of the policies it advocated are still valid today. As a result of this publication, the USNPS established the Wildlife Division at the Washington level, headed by Wright. For a few years under George Wright's charismatic leadership, the programme flourished. However, the success was short-lived, caused in part by the untimely death of Wright in 1936 as well as the advent of the Second World War a few years later. The pattern of short-lived research programmes followed by their demise would be repeated in the future (Wright, 1992).

Although some of the more egregious actions, such as predator control, ceased by the end of the 1930s, other management actions based on preconceived notions of 'right' and 'wrong' biota and static conditions in parks continued in the parks for the next 30 years. Thus, in the 1960s, the USNPS was culling large numbers of elk and mule deer in western parks to prevent them from 'destroying' their habitat, spraying native forest insects to protect desirable trees, maintaining open-pit garbage dumps that were the scene of bear shows for visitors, and planting fish in naturally fishless lakes to provide recreational fishing opportunities (Wright, 1992). As a result of these types of management policies, opposition to the belief- and opinion-based policies of park managers began to surface within the scientific community, and scientists began to press the agency to adopt a more rigorous scientific approach to park management. It should be noted that as early as 1916, Grinnell and Storer argued in a paper in the journal *Science* that human manipulation of park environments should be minimized, and they suggested that a park's greatest value was that it was an example of a natural ecosystem as it existed before the advent of European invasion. A 1921 resolution by the American Association for the Advance of Science (AAAS) also argued for minimal management interference, suggesting that

> the maintenance of the national parks in an absolute natural condition is of upmost importance ... The national parks are rich fields for the natural sciences ... because in them the native fauna and flora may be found more nearly undisturbed than anywhere else.
> (AAAS, 1922: 63)

Renewed efforts to bring science to bear on natural resource management

The appointment of Stewart Udall as Secretary of the Interior in 1962 was an important turning point in the push for more science in parks. Udall, responding to growing criticisms of USNPS wildlife management policies (particularly the killing of elk at Yellowstone and Rocky Mountain National Parks), commissioned a review of wildlife management policies headed by noted wildlife biologist A. Starker Leopold. Udall also requested a National Academy of Science review of the USNPS research programme. The report of the committee, headed by Leopold (Leopold *et al.*, 1963), was a comprehensive examination of park wildlife-management goals and of the policies and methods that would be most appropriate in achieving those goals. It strongly endorsed an applied research programme in the parks. In what has become the most widely quoted passage of the report, the authors concluded that the primary goal of park management was that 'biotic associations within each park be maintained, or where necessary recreated, as nearly as possible in the condition that prevailed when the area was first visited by white man. A national park should represent a vignette of primitive America' (Leopold *et al.*, 1963: 32). The report concluded that the successful habitat management in national parks was dependent on maintaining the ecological processes producing the habitat. Its lasting legacy was the influence the report's ecological orientation had on USNPS resource management policy (Wright, 1992).

The report of the National Academy of Sciences (Robbins *et al.*, 1963) was a blunt condemnation of the past and existing USNPS research programme. The authors concluded that 'its status has been and is one of many reports, numerous recommendations, vacillations in policy and little action, [or] financial support (Robbins *et al.*, 1963: 24). It further concluded that USNPS research

> lacked direction, has been fragmented ... has been applied piecemeal and has suffered because of a failure to recognize the distinctions between research and administrative decision making, and has failed to insure the implementation of the results of research in operational management.
> (Robbins *et al.*, 1963: 31)

While both of these reports were endorsed by USNPS administrators, specific recommendations for changing the research programme of the agency were only partially enacted, and then often for only short time periods. The result of this inaction was that over the next 30 years, nine additional advisory boards or panels where established to review the USNPS science programme and recommend solutions to improve the effectiveness of science in park management and to address problems between scientists and managers (e.g. Darling and Eichhorn, 1967; Conservation Foundation, 1979; National Parks and Conservation Association, 1988; National Research Council, 1992; Risser and Lubchenco, 1992). As with the first two reviews described above, these additional reviews provided both general and specific recommendations for strengthening science in support of the parks. Many of the suggested improvements were recommended repeatedly, and all reviews were virtually unanimous in their conclusions that administrative structural changes were needed to solve the ultimate problem. Such a change typically involved the creation of an expanded and independent science organization with adequate funding and supervision independent of park management. Other recommendations included substantive changes of the training of scientists and managers and provision of a clear understanding of the types of research that fall within the science/management relationship. However, despite the repeated admonitions from these reviews, the importance of a strong science programme never received agency-wide support. In the view of many of the reviewers, as well as scientists in the agency, managerial decisions in the agency continued to be made at variance with the judgements of scientists, ignored sound scientific data, were not supported by scientific research, or were made in situations where science was brought to bear far too late in the management process (Risser and Lubchenco, 1992). As a result, through the mid-1990s, relations between many scientists and managers remained tense, and often led to mutual frustration and outright verbal or written conflict. Scientists continued to feel that placing research personnel under the authority of superintendents resulted in a research programme characterized by crisis management and research priorities that reflected the park managers' personal preferences. The transfer of park managers often resulted in changes of research direction and emphasis and the termination of projects already under way. Park managers, on the other hand, felt that placing scientists under supervisory authority outside the park allowed scientists to work on projects of interest to them rather than park priorities, and therefore jeopardized field-level managers' commitment to research (Kitchell and Nichols, 1987).

The basis of the dichotomy between park managers and scientists

It is pertinent to explore some of the specific reasons why agency management failed to adopt an approach that better integrated scientific knowledge and judgement into natural resource management decisions. The frustrations and conflict referenced above frequently have, as their basis, the fact that there is often a very different mind-set held by the two parties, (i.e. the expectations that protected areas managers have of scientists and the expectations that scientists have of managers). These expectations are derived from the differences in background, training, and experience of both parties, and, in the case of management, the culture of the agency. These expectations also influence the way scientific knowledge is typically communicated to managers and, as an end result, how effective it actually is. Referring again to the underlying purpose of parks, the extent to which parks are considered to be natural preserves, unaltered by consumptive activities and anthropogenic processes as opposed to recreational areas, has clearly influenced the questions asked by, and of, scientists and, therefore, the character of the research that has taken place in these parks. In recent decades, a not uncommon view of the large US national parks has been that they are one of the few remaining areas in the country containing unaltered habitat where the natural forces of nature can take place unimpeded by human management. They are also considered to be repositories of biodiversity often lost in the rest of the country (Stein et al., 2000). This view, abetted by the increasing loss of habitat and biodiversity in other areas of the country, clearly enhances their research potential. This idea that parks are valued venues for research has, in fact, long been recognized by the scientific community (Wright and Hayward, 1985; Parsons, 1989).

Scientists' and managers' views of appropriate research

The pristine nature of parks tends to lend itself to a variety of types of research studies and their legal protection makes them particularly appropriate for long-term studies, such as the single predator–prey system at Isle Royale (Peterson, 1999), multiple predator–prey systems in Yellowstone National Park (Yellowstone National Park, 1997), Denali National Park and Preserve (Singer and Dalle-Molle, 1985), climate and vegetation change at Olympic (Peterson et al., 1997), climate and glacier changes at North Cascades and Glacier National Parks (Leovy and Sarachik, 1991), climate change and sea level rise at the barrier island seashores (Halley and Curry, 1993), wildfire and insect infestation at Rocky Mountain, Glacier, and Yellowstone (Wuerthner, 1988), long-term changes in air quality at the Sierra Nevada parks, Shenandoah, and Great Smoky Mountains (Meixner et al., 2002), and the maintenance of cutthroat trout (*Salmo clarki bouvieri*) populations at Yellowstone Lake (Varley and Schullery, 1996). The list is long. Such research contributes to resource management by (1) providing information on the baseline conditions of natural resources, (2) providing the foundation for a long-term monitoring programme of those resources, and (3) improving our scientific understanding of the way natural systems and their processes work (Peterson, 1996).

Suffice to say, that while these types of topics are of intense interest to natural scientists and the scientific community, they might not register highly with a park manager faced with growing numbers of visitors and a limited and outdated physical infrastructure, or one who must deal with the potential impacts caused by overabundant animal species (Porter and Underwood, 1999). The problems and research questions faced by a park manager whose park is viewed by millions of visitors as a pleasuring ground and who want to see charismatic animals up close are quite different from those the scientist might wish to study, but no less important (Wright, 1998).

By virtue of their training and experience as well as their professional goals, scientists and managers clearly look at natural resource management issues and problems quite differently. Scientists working in parks are normally motivated by the need to understand the underlying causes of a given issue and to look for unique or heretofore unexplored relationships involved with the situation; they are stimulated by the excitement of conducting original research, particularly in the beautiful undisturbed environment of parks. Often in this process they become involved in relationships tangential to the original problem – a point of frustration to managers. Scientists are normally motivated by the goal of actually discovering something unique and publishing these findings in prestigious journals, thereby further enhancing their status among peers and their ability to compete for additional research funds (Chalmers, 1982). Solving the given resource management problem that spawned the research, particularly in a timely manner, can become a secondary concern to conducting the research itself.

Managers are, conversely, motivated primarily by the need to solve a given problem as quickly and efficiently as possible. Managers are often less concerned with the need for scientists to attend professional meetings and to seek recognition by other scientists (Kitchell and Nichols, 1987). What managers want of scientists are the data to guide management which can be used to derive guidelines for action (Peterson, 1996). This is particularly imperative for high-profile resource management problems that have a high degree of visibility to the visitors, political consequences, or that impair the ability of management to undertake what are considered to be necessary physical developments.

Case example of the debate: Crater Lake National Park

One example of this phenomenon occurred at Crater Lake National Park. This park was established by the US Congress in 1902. Crater Lake was recognized to be the primary resource of the park, noted for its deep-blue colour and extreme water clarity. There was thus considerable concern when results of limnological studies conducted between 1978 and 1981 suggested that lake clarity had declined relative to the results of intermittent studies conducted earlier (Larson, 1996). In 1982, the USNPS conveyed a panel of limnologists to evaluate the existing data for Crater Lake. The panel concluded that the database was insufficient to determine if lake clarity had changed, and recommended a limnological study of the lake. At the same time, Congress passed a law that authorized the Secretary of the Interior to promptly initiate studies on the status and trends of changes of lake water quality for ten years and to implement immediately such actions necessary to ensure retention of the lake's pristine water quality. In hindsight, there seems little doubt that the Congressional action produced a much greater impetus for a long-term research programme than did the recommendations of a panel of scientists (Larson, 1996). From a management perspective, the factor of primary importance in this study was to determine if long-term changes in lake conditions had occurred and, if so, were they human related and, if they were, how could they be mitigated. Scientists were more interested in developing a better understanding of the physical, chemical, and biological characteristics and processes of the lake.

The ten-year limnological research programme was completed in 1992. The results from the numerous studies clearly demonstrated that some components of the lake system exhibited little inter-annual change, whereas other components exhibited considerable long-term cyclic changes. A key challenge to the researchers was to separate the natural dynamics of the lake system from any changes caused by human activities. Nonetheless, the lake was considered to be pristine, with the exception of impacts from introduced fish (Larson, 1996).

How scientists and managers view research

Except in those cases where research is needed to deal with an explicit management concern or is legally mandated, managers are often reluctant to endorse a particular project, particularly if, in their view, it can be conducted outside the park. Thus they are often leery of 'hobby research' (i.e. research on the pet subject of a given researcher and research that may have little bearing on pressing park issues (Wright, 1987)). They also might not be supportive of research if it is seen to adversely influence public perceptions or use of the resource. Thus, managers are often reluctant to permit research in parks that involves the consumptive use of resources or that may involve the handling and/or marking of animals or permanent plots. Policies are sometimes based on personal feelings, established only because managers do not want visitors seeing marked animals; but others are motivated by a true concern for the welfare of the resources (Wright, 1988). All park research is, therefore, heavily scrutinized, and an increasingly bureaucratic review and permitting process is required to conduct research in parks.

Scientists, on the other hand, are often insensitive to the needs of managers – failing to follow park research guidelines or regulations, such as working only in permitted areas and/or keeping park managers informed about all of their actions. Scientists also fail, in many instances, to produce much information of any real utility to park managers. Therefore, despite the fact that many parks have a long history of research, park managers are often frustrated by the lack of basic data on natural resources that can be used to guide management and to quickly make informed decisions in their park, as the case example of Crater Lake illustrated. This can occur when interpretations of the research conducted and the data collected are typically not reported or made available in a medium understandable to managers. Research findings communicated only in graduate theses and journal articles are usually insufficient for managers. In other cases, information and data may be provided but are quickly lost in park files and/or overlooked because of frequent changes in resource management personnel. Information management has long been one of the weakest links in the USNPS resource management programme and efforts to correct it have only recently been inaugurated (Ostergren and Wright, 1998).

In other cases, research might have been terminated because of the transfer of a scientist or a cut-off of funding. In other cases, no definitive results were ever produced or reported. In still other cases, one research project might spawn another and another with no closing point. In these cases, a scientist might be reluctant to provide a definitive answer for a certain issue, again frustrating management. Managers are, in turn, often unaware of the scientific method that guides research through a framework of challenge and questioning designed to continually redefine ideas and objectives as new results are obtained in order to develop objective knowledge (Ford, 2000). In the eyes of managers, the scientific method is often an endless process producing little of immediate benefit to management. This attitude leads to an oft-repeated comment I have heard from managers throughout my career – 'haven't we researched that enough already – just tell me what I need to know'.

The above comment reflects a common view of protected area managers and administrators that all adverse situations in their areas are 'problems' – whether they be the results of natural processes or anthropogenic in origin. And whereas science is based on uncertainty (e.g. the idea that nothing can really be proved, only disproved), management longs for certainty. Management, in fact, views research as the mechanism to solve 'problems' with certainty (Wright, 1992). Reducing research to a 'problem-solving mechanism' is, however, clearly at variance with the above description of the nature of research, and scientists see this viewpoint as compromising their abilities to engage in long-term research programmes, to be able to look for subtle interactions or nuances in the resource they are studying, and restricting their ability to obtain results meaningful enough to publish. The view of managers also influences the way research in parks tends to be funded. Funds are allocated only to the extent that they can be used to solve a given problem rather than provide an understanding of its underlying causes. This system has, in turn, led to a common criticism of research in the parks; it has been reactive rather than proactive, although in many cases, reactive management is often unavoidable (Wood, 1988). The above discussion touches on only a few of the reasons why managers and scientists view research in parks differently. However, it should be emphasized that both parties do recognize the immense value of research in parks and it can probably be said that in a perfect world with unlimited funds for research and research personnel, these differences would melt away.

Patterns of scientific research in parks

Historically the research patterns of almost all parks consisted of many individual projects without overriding themes or long-term planning. This was because most projects were directed by scientists working independently with funding from diverse sources. Much of this research produced useful scientific information on natural resources without impacting park budgets or personnel (Peterson, 1996). However, it may also have had little impact on solving issues that management felt were most important. In large part, this last statement can be attributed to the fact that parks, particularly in past years, have often had surprisingly little oversight of individual research projects on their lands. As a result, park managers were often not involved in developing the objectives of a given project, a process that could help assure a more mutually acceptable product.

The exception to the scattered or disorganized pattern of research in parks are those instances where one or more research scientists were stationed at, or worked in, a park for a long period of time. However, because long-term funding for research in parks is generally difficult to obtain, these exceptions are relatively few in number. Table 9.1 lists some of the major parks that have had long-term research programmes on the listed resources.

These research programmes were, in general, highly productive and produced much useful information about the resources studied in the individual parks. However, they also became somewhat entrenched over the years because the time period that the scientists worked in a specific park almost always exceeded the tenure of various park managers who supervised

Table 9.1 Long-term research projects that have occurred in US national parks over the past three decades

Park	Research subject	Personnel affiliation
Channel Islands	Marine resources	USNPS
Crater Lake	Limnology	USNPS
Eastern Barrier Islands	Coastal geomorphology	University
Everglades	Hydrology	USGS
Everglades	Water birds	USNPS
Great Smokey Mts	Black bears	University
Isle Royale	Wolves, moose	University
Isle Royale	Lake watersheds	USNPS
Hawaii volcanoes	Exotic species invasions	USNPS
Mammoth cave	Karst, hydrogeological	University
Saguaro National Monument	Saguaro cactus dynamics	Institution
Sequoia, Yosemite	Prescribed fire	USNPS
Yellowstone	Grizzly bear, elk, bison, cutthroat trout	USNPS

them (park managers are routinely transferred). Therefore, as management priorities changed, it was often difficult to change the focus of the research. Today, because of retirements and the transfer of scientific personnel out of the agency as described below, it is far less common to have scientists devoted to a single park research programme for long time periods.

Case example of long-term research: Isle Royale National Park

The most significant long-term research programme in the USNPS, that is also an exception to the last statement, has been the wolf–moose (*Cervus canadensis*) research programme at Isle Royale National Park, an archipelago consisting of one large island and a complex of many small islands lying in northern Lake Superior, 24–29 kilometres from the Ontario, Canada shore. A study of wolf predation patterns, wolf behaviour and ecology, and moose population dynamics began in the park in 1958 and has continued uninterrupted to the present day. One of the most striking features of this programme has been the continuity of the researchers in charge of the project. The first phase was directed by Durward L. Allen of Purdue University from 1958 through 1975. The second phase (1975 to the present) has been directed by Rolf O. Peterson based at Michigan Technological University in Houghton. Peterson was one of Allen's Ph.D. students (Wright, 1996). Another unique aspect of the research programme was its longstanding policy that none of the living subjects were to be handled except in extreme circumstances (see Peterson and Krumenaker, 1989; and Oelfke and Wright, 2000 for details) and all forms of human disturbance were kept to a minimum.

Park use in the winter, the primary time when research activities occur, has always been very difficult because of access problems. During this period, the long-term lack of human disturbance on the island has allowed researchers to view wolf and moose behaviours rarely witnessed elsewhere (Peterson and Page, 1983; Peterson, 1999). To avoid disruption of the winter research programme, the NPS has legally closed the park between 1 November and 15 April. This is the only action of its kind in the USNPS. This factor is made easier due to the overall low visitation at the park (less than 20,000 per year).

The long-term research at Isle Royale has provided major scientific benefits, including a better understanding of the predator–prey process. However, the real lesson that this research teaches is that events in nature should always be interpreted cautiously and that any conclusion is subject to change. Through the now 48 years of the study, there have been many instances in which interpretations about a phenomenon made during one period were later altered (Wright, 1996). The programme also provides insight into those factors that serve to successfully maintain a long-term research programme. First, it is clearly a benefit that wolves are truly a charismatic species enjoying great public support in the US (Fisher, 1995). In part because of this resource, the programme has also enjoyed strong administrative support from USNPS administrators at park, regional, and national levels. The long-term success of the programme is also clearly due to the continuity of the research personnel and the fact that they were affiliated with universities rather than directly with the agency. This has allowed them a certain amount of independence, as long as personality conflicts do not arise. It has also given the researchers the ability to more effectively seek outside funding, which has been essential in the programme's success. Finally, the isolation of the park and the small number of visitors insulates it from political concerns that often dominate resource management policies in most other parks.

All of the above factors that have contributed to the success of Isle Royale research programme and its strong science/management interface are unique to that situation and some are simply fortuitous. Therefore, while this study can teach us much, it is difficult to see this situation serving as a model for the science/management interface for parks elsewhere. Most parks have far greater visitor-use pressures, are beset by threats outside their boundary, have a more complex set of resources, and may be far more prominent on the political radar screen.

Search for a solution of the science/management interface

In 1994, problems with the science/management interface not only in the USNPS, but also in the US Fish and Wildlife Service and Bureau of Land Management, were again recognized at the federal level. Many of these problems were those presumed to occur when federal regulations, such as the Endangered Species Act, were not based on good science. A structural solution was seen as the answer and in 1994 then Secretary of the Interior Bruce Babbit created by administrative feat in the Department of the Interior a new agency, the National Biological Survey (NBS). The name was chosen to parallel the United States Geological Survey, the other Interior science agency which had, in the Secretary's opinion, already distinguished itself as a world-class scientific institution (National Research Council, 1993; Pulliam, 1995). The new agency was established in recognition of the critical importance of science to natural resource decision-making, and its goal was to develop an anticipatory, proactive biological science programme that would enable land and resource managers at federal, state, and local levels to develop comprehensive ecosystem management strategies and to respond to resource issues in a timely and efficient manner (National Research Council, 1993). The NBS was designed to be an independent science bureau which would not advocate positions on resource management issues and it had no regulatory authority.

All research scientists in the three above-listed land management agencies within the Department of the Interior were transferred into this new agency in 1994. The transfer of the USNPS research scientists removed what little legislative authority the agency had to conduct research on its own lands. In many ways, at least on paper, the creation of the NBS effectively addressed many of the recommendations made in the reviews of the NPS science programme discussed above. Scientists were now administered by scientists. Professional evaluations and advancement were based on scientific achievement, and scientists no longer were pressured to be an advocate for a preferred management position.

Almost immediately however, problems emerged. The NBS did not have the sanction of the US Congress and many members, therefore, objected to its creation, feeling it usurped its authority. Equally important among the members of Congress and various special interest groups was the belief that the 'survey' aspects of the NBS were a backdoor approach to taking over private lands in the country. Renaming the agency the National Biological Service did little to quell this concern, and the agency was threatened with dissolution (Pulliam, 1998a, b). For various reasons, the ultimate political solution was to dissolve the agency and move the personnel and functions into a newly created division, the Biological Research Division (BRD), of the existing US Geological Survey in 1995.

The BRD: a retrospective

In the 11 years that have followed the creation of the division and the resulting shifts in personnel among variously named entities, there has been little objective analysis as to whether the radical change in administrative structure has effectively addressed many of the fundamental problems between science and management considered in the early reviews. A survey conducted of all USNPS research scientists just after their transfer to the NBS revealed a high degree of dissatisfaction with the move and a great deal of scepticism as to whether the formulation of a new agency would solve the science/management problems within the agency or would benefit the science programme of the USNPS (Van Riper, 1994). The events of the intervening years have shown these concerns to be well founded. Some former USNPS research scientists opted to return to their parent agency under the guise of newly created positions of 'science administrators', which are essentially research contractors. Others managed to retain a good working relationship with their former parent agency and continued to serve its science needs. However, the majority appeared to be drawn off to new research endeavours not connected with the former parent agency, where better funding existed. In this case, the USNPS most assuredly lost needed scientific expertise.

In retrospect, it appears (to this writer) that the creation of the NBS and now the BRD of the US Geological Survey was a good idea that came at the wrong time politically and was poorly implemented. From the NPS perspective, it did not take into account the culture of the agency, the long working relationship of NPS scientists with the agency, and the dependence of the agency on in-house expertise rather than going to universities or other federal agencies. Unlike research scientists associated with other Department of the Interior land management agencies, USNPS scientists worked almost exclusively on research associated with that agency. Most were long-term employees who understood both the mission and the culture of the agency. Thus, although there were problems, they were still able to function at the science/management interface.

Conclusions and recommendations

It is somewhat ironic that, although an administrative separation of scientists from managers was long advocated in the many reviews of the science programme, a structural solution, at least on the magnitude that resulted with the creation of the NBS/BRD, does not seem to be the answer. The solution enacted, if anything, further weakened the links between management and science. In the long run, we may find that little will be accomplished if the science is good but the managers do not use it. As discussed earlier, one reason for the longstanding gap between researchers and resource managers has been contrasting expectations in terms of goals and objectives (Huenneke, 1995). Structural changes alone will not solve this problem and could actually exacerbate them.

Therefore, rather than structural changes, there appears to be a real need for scientific personnel that exhibit special qualities. These are those individuals who understand and appreciate the value of parks and park managers, who are willing to work on the applied problems that management confronts, and who will explain their findings in an understandable manner. While individuals with these qualities are not unique, they are, in my opinion, uncommon in today's crop of graduate students who attend natural resource colleges in the US. It also does not appear that the employment qualifications for scientists within the BRD, as limited as these opportunities are, reflect these types of characteristics. Therefore, the NPS needs to make its own efforts to obtain needed scientific expertise, even if these individuals cannot be classified as research scientists. These hires should be very selective of individuals who reflect the qualities listed above. Further, rather than being permanently placed in one location, these new hires should be initially placed in an apprentice programme that would allow them to be exposed to a variety of parks and resource problems as well as differences in management styles. Some 15 years ago the NPS had a cadre of about 110 research scientists almost all with long career tenure with the agency. The great majority are now retired or very soon will be. They need to be replaced by a similar cadre of scientifically trained individuals who thoroughly understand the mission and culture of the agency. However, this change cannot occur in isolation. There is also a strong need for resource managers who have better scientific backgrounds and technical training (Peterson, 1996), a situation that is, in fact, slowly occurring. The ultimate melding of the science/management interface will occur when the distinction between management-oriented scientists and science-oriented managers no longer exists.

Literature cited

American Association for the Advance of Science (1922). Resolution on the maintenance of natural conditions in national and state parks. *Science* 55, 63.

Chalmers, A.F. (1982). *What is This Thing Called Science?* Milton Keynes; Open University Press.

Conservation Foundation (1979). *Federal Resource Lands and their Neighbors*. Washington, DC: The Conservation Foundation.

Darling, F.F. and Eichorn, N.D. (1967). *Man and Nature in the National Parks*. Washington, DC: Conservation Foundation.

Davis, G.E. and Halvorson, W.L. (1996). Long-term research in national parks: from beliefs to knowledge. In W.L. Halvorson and G.E. Davis (eds), *Science and Ecosystem Management in the National Parks*, pp. 3–10. Tucson, AZ: The University of Arizona Press.

Fischer, H. (1995). *Wolf Wars*. Helena, MT: Falcon Press Publishing Co.

Ford, E.D. (2000). *Scientific Method for Ecological Research*. Cambridge: Cambridge University Press.

Grinnell, J.B. and Storer, T. (1916). Animal life as an asset of national parks. *Science* 44, 375–380.

Halley, R.B. and Curry, R. (1993). Coastal geology and national parklands. *Park Science* 13 (1), 4–6.

Huenneke, L.F. (1995). Involving academic scientists in conservation research: perspectives of a plant ecologist, *Ecological Applications* 5 (2), 209–214.

Huff, D.E. and Varley, J.D. (1999). Natural regulation in Yellowstone National Park's northern range. *Ecological Applications* 9 (1), 17–29.

Kitchell, K. and Nichols, R. (1987). Scientists, superintendents differ on researchers' role in RM Region. *Park Science* 7 (2), 4–5.

Larson, G.L. (1996). Exploring the dynamics of Crater Lake, Crater Lake National Park. In R.G. Wright (ed.), *National Parks and Protected Areas: their role in environmental protection*, pp. 435–449. Cambridge, MA: Blackwell Science.

Lemons, J. and Stout, D. (1984). A reinterpretation of national park legislation. *Environmental Law* 15 (1), 41–65.

Leopold, A.S., Cain, S.A., Cottom, S., Gabrielson, I.N, and Kimball, T. (1963). Wildlife management in the national parks. *North American Wildlife and Natural Resources Conference, Transactions* 28, 28–45.

Leovy, C. and Sarachik, E.S. (1991). Predicting climate change for the Pacific Northwest. *The Northwest Environmental Journal* 7 (2), 169–201.

Meixner, T., Allen, E., and Tonnessen, K. (2002). Atmospheric nitrogen deposition: implications for managers of US parks. *Park Science* 21 (2), 30–33.

National Parks and Conservation Association (1988). *Research in the Parks: an assessment of needs*. Volume 2. *Investing in Park Futures: a blueprint for tomorrow*. Washington, DC: National Parks and Conservation Association.

National Research Council (1992). *Science and the National Parks*. Washington, DC: National Academy Press.

——— (1993). *A Biological Survey for the Nation*. Washington, DC: National Academy Press.

Oelfke, J.G. and Wright, R.G. (2000). How long do we keep handling wolves in the Isle Royale wilderness? *Park Science* 20 (2), 14–19.

Ostergren, M. and Wright, R.G. (1998). Creating a bibliographic database for a widely distributed collection. *Information Outlook* 21 (1), 27–30.
Parsons, D.J. (1989). Evaluating parks as sites for long-term studies. In G.E. Likens (ed.), *Longterm Studies in Ecology*, pp. 171–173. New York: Springer-Verlag.
Peterson, D.L. (1996). Research in parks and protected areas: forging the link between science and management. In R.G. Wright (ed.), *National Parks and Protected Areas: their role in environmental protection*, pp. 417–434. Cambridge, MA: Blackwell Science.
——, Schriener, E.G., and Buckingham, N. (1997). Gradients, vegetation, and climate: spatial and temporal dynamics in the Olympic Mountains, USA. *Global Ecology and Biogeography Letters* 6 (1), 6–17.
Peterson, R.O. (1999). Wolf–moose interactions on Isle Royale: the end of natural regulation? *Ecological Applications* 9 (1), 10–16.
—— and Page, R.E. (1983). Moose wolf fluctuations at Isle Royale National Park, Michigan USA. *Acta Zoology Fennica* 174 (3), 251–253.
—— and Krumenaker, R.J. (1989). Wolves approach extinction on Isle Royale: a biological and policy conundrum. *George Wright Forum* 6 (4), 10–16.
Porter, W.F. and Underwood, H.B. (1999). Of elephants and blind men: deer interactions in the US national parks. *Ecological Applications* 9 (1), 3–9.
Pulliam, H.R. (1995). The birth of a federal research agency. *BioScience* 45 (13), S1–S6.
—— (1998a). The political education of a biologist Part I. *Wildlife Society Bulletin* 26 (2), 199–202.
—— (1998b.) The political education of a biologist Part II. *Wildlife Society Bulletin* 26 (3), 499–503.
Risser, P.G. and Lubchenco, J. (eds) (1992). *Report of a Workshop for a National Park Service Ecological Research Program*. Washington, DC: National Park Service Ecological Study Program Report.
Robbins, W.J. et al. (1963). *A Report by the Advisory Committee to the National Park Service on Research*. Washington, DC: National Academy of Sciences.
Sellars, R.W. (1997). *Preserving Nature in the National Parks*. New Haven, CT: Yale University Press.
Singer, F.J. and Dalle-Molle, J. (1985). The Denali ungulate-predator system. *Alces* 21 (4), 339–358.
Stein, B.A., Kutner, L., and Adams, J. (eds) (2000). *Precious Heritage: the status of biodiversity in the United States*. Oxford: Oxford University Press.
Van Riper, C. (1994). *Ombudsman Committee Report: solutions to problems faced by former NPS scientists transferred to NBS*. A Report to the director of the USDI, National Biological Service. Washington DC: National Biological Service.
Varley, J.D. and Schullery, P. (1996). Yellowstone Lake and its cutthroat trout. In W.L. Halvorson and G.E. Davis (eds), *Science and Ecosystem Management in the National Parks*, pp. 49–73. Tucson, AZ: The University of Arizona Press.
Wood, J. (1988). Improving the role of science in the National Park Service. *Park Science* 8 (2), 22–23.
Wright, R.G. (1987). Improving the science program in the National Park Service: a rejoinder. *Park Science* 6 (2), 14–15.
—— (1988). Wildlife issues in national parks. In W. Chandler (ed.), *Audubon Wildlife Report 1988/1989*, pp. 169–196. San Diego, CA: Academic Press.
—— (1992). *Wildlife Research and Management in the National Parks*. Urbana, IL: University of Illinois Press.
—— (1996). Wolf and moose populations in Isle Royale National Park. In W.L. Halvorson and G.E. Davis (eds), *Science and Ecosystem Management in the National Parks*, pp. 74–95. Tucson, AZ: The University of Arizona Press.
—— (1998). A review of the relationships between visitors and ungulates in national parks. *Wildlife Society Bulletin* 26 (4), 471–476.
—— (1999). Wildlife management in the national parks: questions in search of answers. *Ecological Applications* 9 (1), 30–36.
—— and Hayward, P. (1985). National parks as research areas with a focus on Glacier National Park, Montana. *Bulletin of the Ecological Society of America* 66 (4), 354–357.
——, Dixon, J., and Thompson, B. (1933). *A Preliminary Survey of Faunal Relations in National Parks*. Washington, DC: National Park Service Fauna Series, no. 1.
Wuerthner, G. (1988). *Yellowstone and the Fires of Change*. Salt Lake City, UT: Haggis House Publications.
Yellowstone National Park (1997). *Yellowstone's Northern Range: complexity and change in a wildland ecosystem*. Mammoth Hot Springs, WY: National Park Service.

Chapter 10
Indigenous peoples and protected heritage areas:
acknowledging cultural pluralism

David Neufeld

> 'Just smell that breeze,' Dad said as we rounded a little grove. He inhaled deeply and I did the same. The warm air was both sweet and sharp: a delightful mixture of wild honeysuckle, roses, wild sweet peas, green grass, sap, tall slough plants, rich brown earth, and the yeasty odour of the silver wolf-willows. 'It's fair wonderful, isn't it?' Dad said as we jogged along again. 'Just like God's own garden.' We came to the top of a little rise and Dad let Nelly stop. Darkie stopped too, and we sat there for a while and looked at the beauty around us: at the poplars and willows both silver and green, and at the roses, wild mint, and harebells that were everywhere.
> 'Take a good look at it, Mary,' Dad said quietly. 'You'll never see it this way again.'
> I did as I was told. I looked at the tall grass and the peavine and the soft green silk of the wild barley, but the sad note in Dad's voice puzzled me. How could the prairie change? I wondered. I did not realize then what an instrument of change a plough is.
> The trees and willows are gone now, grubbed out and burned, and the roses and wild mint have been ploughed under. Wheat now grows where the chook-cherries and the violets bloomed. The wind is still sweet, but there is no wildness in it and it no longer seems to have wandered a great way over grass and trees and flowers. It now smells of dry straw and bread. The keen wild fragrance the wind knew in those days has gone forever.
>
> (Heimstra, 1955)

Introduction

Mary Heimstra's (1955) pang of loss is one of the primary impetuses behind publicly protected heritage areas (PHA) in North America. In Saskatchewan, where Heimstra's family settled in 1904, it was the children of the original settlers who sharply felt this loss. As they reached the end of their active lives, they undertook the rituals of their age – burying parents and remembering their own initiation to the place they learned to call home. And they, and their children, took action to remember and honour their home and its creators. The family picnic sites, berry-picking patches, community rodeo grounds, swimming holes, the beaches on the fish-stocked reservoirs created by the federal Prairie Farm Rehabilitation Act (PFRA), these remnants of 'God's own garden' and those human-created contributions to it, were made into regional parks. Unguided by any national or even provincial organization, local communities identified these special places that spoke to the achievements of their pioneer forebears – the transformation of a wild place to a productive home. The children and grandchildren made sure there were places of memory and reflection on their good life.

These regional parks[1] were also to preserve tiny pieces of that original natural world their parents entered as newcomers. Partly to allow a nostalgic glimpse of the land before its transformation and partly to allow every visitor the sensation of being a pioneer, the park experience included a chance to be the first one in this wild place, to be in a time before time. Palliser Regional Park in south-central Saskatchewan, named after the leader of a British scientific expedition which first reported on the area, includes a 'buffalo rubbing stone' as part of its heritage display. It is said to be a piece of a much larger transformer rock, a physical manifestation of the sacred Plains Cree oral tradition. Recognized as an Indigenous spiritual item, and thus a monument to the long-ago past, the stone lay in the valley to be flooded by the Lake Diefenbaker reservoir in the mid-1960s. Plans to remove the rock up to the new lake shore as a monument ultimately resulted in its being blown to pieces. One large fragment of rock, not likely part of the original, was then removed to the park (Herriot, 2000: 69ff.). In many ways the purposes of Palliser Regional Park and others across Saskatchewan reflect the culturally entrenched values and interests expressed in PHAs across Canada. At the same time they also illustrate the tragically limited understanding of Indigenous peoples typical of the policies and governance shaping management of these PHAs.

This chapter forwards the idea that the PHAs of Canada and, by association, those developed and supported by the West around the world, are culturally entrenched tools of State power. They are designed to strengthen the State through fostering citizen identity with the State and to gain citizen acknowledgement of the State's responsibility to represent them in the world. A review of some of Parks Canada's experiences with Indigenous peoples related to the management of PHAs highlights challenges raised by First Nations and notes the resulting policy responses from Parks Canada. Finally, the chapter considers the present forms of recognition that PHAs extend to Indigenous peoples and suggests the consideration of significant revisions to our notions of governance to ensure that both policy and the context for policy application are conducive to the desire to more effectively and meaningfully address the interests forwarded by Indigenous peoples.

Constructing the nation-state

From our preface we can understand that the network of PHAs in Canada is an elaborate set of cultural constructions reflecting the interests, values, and aspirations of the people, and their governments, who created and maintain them. Although the larger systems of PHAs, managed by the Canadian, provincial, and territorial governments, are now largely understood as representative elements of the various ecosystems that make up Canada and the history of its settlement, this understanding is founded upon the mainstream societal values reaching back to the origins of parks and protected areas.

Modern protected areas have their origin in North America. Alfred Runte, historian of the American national parks, suggests that the idea for national parks arose as part of the process of building the republic. Although nominally free of the social hierarchy and wars of the old world, the settler societies of the new world shared the desire for, and faced the same challenges in creating, a modern nation-state. The nation-state, a political entity representing ethnic or cultural groups, as the primary element in the international order evolved in Europe from the mid-seventeenth century. German political philosophers forwarded a set of rational criteria defining the nation-state – common language, race, and shared traditions, emphasizing the cultural unity of the nation. Ernest Renan, a French Orientalist, challenged this rigid focus on culture. From the more culturally diverse background of France he forwarded instead a definition of the nation-state built upon the idea of a 'willfulness to live together' expressed through continuing consent, common memories, and the will to exploit a common inheritance (Wikipedia, n.d.; Webber, 1976: 112). This latter idea also more accurately reflected the values of the diverse immigrant populations making up the nation-states of the new world. The founders of the American states developed governing structures to ensure continuing consent. However, the challenge of establishing common memories and identifying a common inheritance proved more difficult. In Europe these memories and inheritances were written into the landscape as venerable cathedrals, ancient centres of learning, ruined castles, and a shared history of place. Americans originally felt at a loss in their lack of equivalent cultural achievement. What they had in abundance, however, was nature. Nature offered the shared experience of the frontier and the common interest in the material transformation of wilderness into farms and cities – into civilization.

And so the settlers turned to nature as the foundation of nation building. While the rationalist elements of the Enlightenment predominated in this process of absorbing nature, the Romantic response to the open frontier also played a role in developing a national feeling among citizens. The emotive responses to wilderness and the home they carved from its wildness shaped both the Euro-American cultural views of nature and the social character of nation building through the nineteenth century. In the fine arts the appreciation of the sublime – the fearful majesty and power of the natural forces shaping human life – inspired music, literature, and painting about place. The more ordered discipline of history was similarly shaped by the frontier, imaging it as a beacon of freedom drawing settlers westward. The sciences, acting through the western explorations that noted and measured the continent, also contributed to the sense of a common future of development and prosperity. Finally, the idea of progress, the idea that time had both a direction and a destination, underlying the expression of the frontier experience, also incorporated a universalist notion of the perfect state of man. There was a belief in the perfectability of human society through material wealth. These visceral and intellectual responses to nature were the foundational elements of nation building in the United States and Canada – the impressing of individuals with their shared experience and common future as marks of their citizenship of a nation. National parks and, later, historic sites thus became powerful tools in the business of constructing the State.

This approach from the European intellectual tradition of the Enlightenment culminated in a Modernity seeking the emancipation of man, through passage to his highest and best form – western European civilization – and the control of nature, by bringing the resources of the State into the ordered and efficient service of man. The consequence of this reduction of nature to a platform for human agency was the complete separation of culture from nature. The consequences of this bifurcated universalist approach to the world were especially hard on the Indigenous peoples of North America.

Both Canada and the United States have worked diligently through the past two centuries to construct themselves as modern nation-states. The governments of both countries drew upon many different resources to create a citizen community that would identify with their new nations. This process, coloured by both the romanticism of the arts and the rational appropriation of nature for development, included such elements as a common public education, national military service, standard weights and measures, transport and commerce linkages, and a shared vision of a national community (Webber, 1976).

These ideas were formalized into distinct intellectual frameworks that both justified the State and forwarded a shared national vision of a future. In the United States, nature was incorporated into the State through Frederick Jackson Turner's Frontier thesis. According to Turner the power and vitality of the American republic grew from the vastness of the continental United States and the opportunities arising from the 'free lands' beyond the frontier. Turner developed his ideas in the late nineteenth century during the intellectual crisis spawned when the frontier was officially closed, that is, development had consumed all of the open free lands. National parks were thus established to preserve elements of this primal force in the creation of America. Ted Catton, historian of national parks, suggests that Denali National Park in Alaska was, in part, established to commemorate the time of the pioneers, in fact to preserve the opportunity of experiencing the frontier in its raw state. Thus, the national park is a geographical relic of the land settlement process that made the American republic (Catton, 1997: 105).

In Canada, historical interpretation of the development of Canada similarly relied upon the State's expansion across the continent. Harold Innis' sophisticated economic development model sought to justify the existence of Canada, emphasizing both the trans-Atlantic cultural ties to Britain and France and the nation's difference from the United States. His case rested upon the 'natural' boundaries of Canada, that is, the network of transport and commercial linkages expanding from the St Lawrence River valley and integrating them into a nation. The resulting Laurentian thesis was the unchallenged framework for understanding Canada as a nation well into the 1960s. Canadian historic sites reflect these interests through the preservation of military forts from the French and English wars, fur trade posts, and sites related to the expansion of settlement and economic and administrative development, that is, the process of nation building. National parks, likewise, played an important role in constructing the idea of the State (Neufeld, 2002).

The natural wonders, especially the spectacular examples in the west that formed the first national parks in both countries, were the sublime emotive elements reminding visitors, and the viewers of the many art works of these places, of the power of the Christian God that created the world and provided the new world to the newcomers for their use. The national parks and the slightly later historic sites became the manifest symbols of God's blessing of the newcomers' settlement and development project. The western Christian significance of this revelation as a foundation for the State meant there was broad public support for the preservation of the prominent elements of this original pre-Columbian, prelapsarian really, landscape for the spiritual renewal of its citizens. In the same way, historic sites were recognized as mythic markers of the successful transformation of God's largesse into productive land and stable, well-provided-for communities. This transformation similarly represented the highest order of a rational world. The application of reason, through science and technology, led to an obvious improvement in the material welfare of mankind. The dual goals of the Enlightenment – the conquest of nature and the emancipation of man – were both represented by the PHAs and, through them, integrated into the character of the State itself. All of this was based upon a belief in progress, that is, the idea that through the application of reason, that distinctive element of humanity, the world can be made a better place.

The implications of the idea of progress in the modern world, and for the management and direction of PHAs, are significant. The assumption that time has a direction fosters a belief, particular to the West, in the gradual but incremental increase of knowledge, an increase leading to the improvement of the human condition. The corollary of this path through time was destination, that is, an assumption of an eventual convergence of the diversity of humanity into a single, well-adjusted pattern. The diversity of cultures, opinions, and thoughts about the past and the future were simply personal opinions or antiquated superstitions individuals decided to adopt in their ignorance. John Gray, Professor of European Thought at the London School of Economics, cautions:

> We have inherited the faith that as the world becomes more modern it will become more reasonable, more enlightened and more balanced. We expect that, as modern habits of thinking advance across the world, people everywhere will become more like us – or at least as we imagine ourselves to be.
>
> (Gray, 2004: 17)

Edward Said, the Arab critic of Western colonialism, notes that such a denial of other histories is an imperial tool to gain control over, and attribute meaning to, a foreign region. This creation of a past gives control over the present, that is, it creates a friendly cultural space; a friendly cultural space that is cemented in citizens' minds and made sacred by the identification and establishment of PHAs. Thus, by denying Indigenous peoples their histories, PHAs are a potent expression of a belief system creating and maintaining a vision of the new world as empty land to be developed, a vision noting Indigenous peoples only in contrast to the strength and vigour of newcomers, a vision regularizing this state of affairs as the norm for its citizens. Under this belief system Indigenous peoples in the present were effectively rendered invisible (Said, 1978: 66, 108–109).

With this thought in mind we can approach a different, perhaps broader, understanding of the roles played by PHAs in Canada. PHAs were, and continue to be, created as part of the State-building process, reflecting its needs and fulfilling its purposes. The Canadian State accomplishes this by using national parks and historical commemorations to establish a national cultural space. Such a national cultural space highlights values, it establishes the boundaries of the national community and it articulates a modernist vision of the ideal future. This space is an expression of cultural power, it reminds us of who we are and what we value. And it misrepresents all people in the region who resist inclusion in the identified cultural boundaries.

As the PHAs in Canada are clearly culturally entrenched entities, it follows that the policies directing their management and the governance shaping their purpose are expressions of the same modernist vision of the State and its purposes. That is, both policy and governance of PHAs are integrated into a comprehensive cultural narrative which, by recognizing only one culture, makes all culture irrelevant to order and purpose. Identity and values differing from the mainstream are simply choices practised by individuals that do not affect the gradual accumulation of knowledge leading humanity to a final convergence of order. This belief in progress denies any legitimacy to other perspectives on the world, effectively barring them from a role in society. This belief, currently challenged as outlined below, is the basis for the colonization of the world by the West. The addressing of this belief is a requirement for the decolonization of our Western understanding of landscape and place and the revision of the policies and governance guiding PHAs.

Contacts with Indigenous peoples

During the 1960s and 1970s, changing appreciations of social justice within the larger society supported the removal of barriers to political and legal activism among Indigenous peoples dissatisfied with their position in Canadian society. At the same time the complexities of environmental issues and the limits of related scientific knowledge were becoming more obvious. These social and environmental pressures affected Parks Canada and served to enhance the profile of Indigenous peoples in the strategic thinking of the organization's leadership. In 1985 the Historic Sites and Monuments Board of Canada (HSMBC), the federal body mandated to sanction places, events, and persons of national historic significance, acknowledged the cultural imbalance of the country's national historic sites and recommended consultations with First Nations to determine their interest in the national commemoration of their history.[2] Within national parks, the Panel for Ecological Integrity, a ministerial advisory committee struck in 1998, explored the possibilities of Indigenous 'naturalized knowledge' seemingly offering a complementary Indigenous approach to understanding the intricacies of ecosystems (Parks Canada Agency, 2000). The subsequent engagement of Indigenous peoples has challenged the cultural assumptions underlying the social and cultural purposes of PHAs in Canada and sparked a reconsideration of the policies and governance models guiding their management.

Parks Canada began direct consultations with Indigenous peoples in 1986. The primary objective of these and subsequent consultations was to more meaningfully include Indigenous peoples within Canada through appropriate forms of cultural recognition, that is, to identify, protect, and communicate their history and cultural values within a state programme of PHAs. The consultations were part of a broader public response to social justice issues raised by Indigenous groups through social activism, legal challenges, and public consultations from the 1960s onwards. This engagement of Indigenous peoples continues to significantly challenge and alter Parks Canada's understanding of its roles and programmes both within its mandated responsibilities and as an agent of a state government.

The Parks Canada consultations and subsequent activities with Indigenous peoples over the last 20 years have raised two interrelated questions that continue to complicate cooperative work between national government PHA programmes and the recently re-acknowledged sovereign Indigenous governments. The first relates to the practice of PHA management: what policies are needed to meaningfully and respectfully understand and include Indigenous cultural narratives within the existing culturally entrenched PHA system? The second tackles the larger issue of revisiting our understanding of the governance of existing PHAs: what are the changing responsibilities of the State to its citizens, both as individuals and as members of distinct and recognizable nations within Canada?

I started work with Parks Canada in 1986. Among my first assignments was to the team preparing the first management plan for Chilkoot Trail National Historic Site. The Chilkoot Trail is a passage connecting two distinct ecosystems – the mild Pacific coast rainforest of south-east Alaska and, separated by the rugged Coastal Mountains of north-western Canada, the drier but much colder boreal forest of the Yukon interior. Its long use as an Indigenous trading route is still visible in the family lineages joining communities. However, in the 1960s the trail was identified as a National Historic Site for its use during the Klondike gold rush of the late 1890s. Tens of thousands of gold-hungry Stampeders, mostly adventurous young men, moved across the trail leaving behind a colourful relict landscape of building remains and piles of abandoned tin cans and broken bottles.

Commemorating the Chilkoot Trail was part of a larger effort to recognize the gold rush as an important event in Canada's history. Following the Laurentian thesis, history began with the onset of regional Euro-American settlement and development, the incorporation of a far-flung corner of the country into the Laurentian network, and its economic contributions to the State's centre. Implicit in the commemorations of the gold rush was the recognition of the importance of the economic development of northern Canada. The celebration of the first large-scale exploitation of northern resources thus not only recognized the pioneers of the gold rush, it also gave a stamp of broad public approval to the mining and transportation improvements that opened the northern frontier regions to industrial development in the 1950s and 1960s (Neufeld, 2001).

This vision of economic development and settlement as progress had significant implications for Parks Canada's initial understanding of the historic role of the Carcross-Tagish First Nation along the Chilkoot Trail.[3] The three interpretive themes identified for this National Historic Site in the early 1980s were:

- life on the trail, including the experience of the Stampeders taken from their remains on the trail;
- transportation technology, noting the evolutionary progress of freight movement into the north; and
- national sovereignty, or the role of Canada's Mounted Police in extending social order and establishing the political boundary between Canada and the United States.

Cultural research by archaeologists and historians initially addressed the material culture on the trail and the rich lore found in the personal diaries and letters of Stampeders, later the operation of horse packing companies, aerial tramways, and, the railway were examined and, finally, the differences between stolid Canadian Victorian social values and the Wild West of the American republic were highlighted. This Laurentian analysis provided a clear and understandable story, at least to mainstream Canadians. The Indigenous people of the region, however, had only a limited role in this story. They were recognized in a transportation sub-theme as human pack animals, and thus effectively acted as a base line emphasizing the white man's more technologically advanced modes of transportation.

As a result of the government's attention to Indigenous activism in the mid-1980s, however, Parks Canada attempted to make the national story more inclusive. The 'Indian side' of the Chilkoot Trail story was identified as a research priority. Historical research, especially in archival photo collections, offered some limited access to the Indigenous experience during the gold rush. However, it was soon clear that the primary source would be the stories and memories of the local First Nation people.

Negotiations for a community oral history project occurred within a context of volatile land claims politics that inevitably shaped the project's outcome. Identification of the significance of the Chilkoot as a historic site in 1969 pre-dated the federal government's acknowledgement of Indigenous claims and initially no consideration was given to Indigenous interests in the land set aside for the historic site. Parks Canada's first contacts with the Carcross-Tagish First Nation about an oral history project in 1986 followed the initiation of Yukon First Nation claim negotiations. Thus the project became linked to the community's demand for recognition of their government and the return of traditional lands. Recognizing the possibility of misunderstanding, Parks Canada established clear expectations with the First Nation for the oral history project. The project was to obtain the 'Indian' side of the story for presentation at the National Historic Site. Research design and control over products would remain with Parks Canada.

Not surprisingly, the Chilkoot Trail Oral History Project did not fulfil Parks Canada's initial expectations. The attempt simply to throw light on the previously unexplored 'Indian side' of the presumed national story was a failure. The Carcross-Tagish were quick to challenge the project's assumptions about the past. In one instance, after an extended set of interviews, the project anthropologist and a First Nation Elder were relaxing on a lake shore. The anthropologist found a stone hammer nearby and showed it to the Elder as proof of the Indigenous presence in the region. The Elder briefly examined the stone and then casually threw it back in the bushes, saying 'What have I been telling you all week?'[4] As the project progressed, we watched the First Nation similarly discard the Parks Canada notion of the project's objectives. It became clear there was no 'Indian side' of the Chilkoot Trail gold rush story; the stampede was seen simply as an annoying but brief interruption of their ongoing lives. Community oral tradition and continuing land use practices instead forwarded a distinctly different historical narrative describing their long use of the area and their connection to it as 'home'. These activities conveyed a significant message to Parks Canada about how the Carcross-Tagish used their traditional territory, parts of which were now absorbed into the Chilkoot Trail National Historic Site, to sustain their cultural identity. The First Nation also used the project to make powerful statements about their connection to this territory, thus returning to the main issues they wished to raise with the federal government – their distinct and different vision of the future and their desire to be free to fulfil it.

The Carcross-Tagish effectively used the oral history project as a platform to challenge a national understanding of the cultural significance of the Chilkoot Trail. The efforts to document the 'Indian side' of the gold rush story proved to be a dead end. Implicit in the counter-narrative offered by the Carcross-Tagish during the project was a direct challenge to the authority of Western knowledge and related management practices. The community questioned the 'truth' presented by academic perspectives on Canadian history. They challenged the authority and power of the government agencies relying on this history to manage 'their' lands. The Yukon First Nation's understanding of the past suggested alternative explanations of the world. The Carcross-Tagish challenged the assumed distribution of the social power inherent in the Western understanding of the past and they articulated a different vision of how the world was made. They challenged Parks Canada to consider another way of understanding who we are and where we are going as joint or parallel societies.

These results and other early consultations with Indigenous peoples across Canada highlighted the complexity of the conversation. To a great extent the Parks Canada expectation was to invite Indigenous people into the national story, thus correcting an earlier oversight. This approach was quickly found unacceptable to Indigenous peoples, however, and First Nations resisted attempts to include them in this way. In response, Parks Canada began developing new approaches to acknowledge the different ways Indigenous peoples understood and articulated their relationships to place and to the State.

Parks Canada's difficulties associated with this set of perplexing parallel narratives were ones shared by other government departments and the Canadian public at large. These questions, highlighted by Indigenous protests, political and legal actions, prompted the State to consider how to more fully recognize Indigenous people as citizens. In commemorating the national story, the HSMBC began discussions to address 'the challenge of designating subjects related to Aboriginal Peoples' history which do not conform to the traditional definition of national significance' (HSMBC, 1998). These latter concerns began to be addressed when the HSMBC accepted the concept of 'Aboriginal cultural landscape' in 1999 as a framework for the national recognition of Indigenous culture.

The development of new tools for cultural recognition allowed Parks Canada to more positively engage with Indigenous peoples. National PHAs were a modernist expression of a progressive narrative, the successful material transformation of empty wild land to a domesticated productive condition. Land was deemed a commodity whose effective stewardship was expressed in tangible forms such as buildings, transportation systems, and crops. Land was understood as a platform for the exercise of human agency (Ingold, 2000: 149). The commemoration of the Indigenous past founded on this presumption limited acknowledgement to the materiality of archaeological sites and stories of European explorers' helpers. Through the 1960s the HSMBC discussed the commemoration of the Indian peoples of Canada, eventually suggesting a statue for each tribe noting its location and their time of highest achievement be erected at the Montreal World Exhibition site, EXPO 67. The project, forwarding the Euro-Canadian created past for Indigenous peoples, foundered on the difficulties of inventorying the different tribes of Canada and their achievements. Indigenous peoples were represented at the fair by the far more controversial Indian Pavilion.[5]

Thirty years later the Aboriginal cultural landscape concept opened the door for a new way of understanding both place and the past. Defined as

> a place valued by an Aboriginal group (or groups) because of their long and complex relationship with that land [an Aboriginal cultural landscape] expresses their unity with the natural and spiritual environment. It embodies their traditional knowledge of spirits, places, land uses and ecology. Material remains of the association may be prominent, but will often be minimal or absent.[6]

Rather than considering the tangible proofs of transformation, the concept encourages the consideration of the intangible knowledge and skill sets, faith practices, and beliefs arising from relations among beings, both human and non-human, and place.

One of the first cultural commemorations of an Aboriginal cultural landscape was forwarded by the Gwichya Gwich'in of Tsiigehtchic, Northwest Territories. Nagwichoonjik National Historic Site, a 175-kilometre stretch of the Mackenzie River, was put forward to have their distinctive relationship to place acknowledged and understood by their children and visitors. While tangible elements of these relationships exist, such as fish camps, hunting trails, and resource sites, it was the intangibles, traditional knowledge, land use practices, language, and oral tradition,[7] that were deemed equally important. Thus, it was not the exploitation of resources, with the consequent transformation of land into a commodity that was emphasized, it was the web of ongoing connection that was presented as the concrete expression of the Gwichya Gwich'in cultural values and their continuing life in the present.

At the same time, Parks Canada sought to address the erosion of national park ecosystem health through a broader appreciation of both regional and cultural factors affecting the health of the land and animals in national parks. The concept of ecological integrity, the healthy functioning of an ecosystem within natural bounds, was identified as the goal of national park management, setting aside an older model based on an inviolate park boundary. This shift in mandate opened new possibilities in PHA management policies. A national panel on the ecological integrity of national parks reporting in February 2000 noted the importance of engaging Indigenous peoples in the management of national parks within their traditional lands. With an emphasis on the shared vision to protect these 'sacred places' there was also the hope that these examples would inspire other Canadians to acknowledge Indigenous peoples in Canada (Parks Canada, 2000: Vol. I, p. 15).

Significant elements in the Ecological Integrity Panel's report included a new emphasis on the importance of the 'naturalized knowledge'[8] in the management of national parks. This direction, perhaps recognizing clause 8 (j) of the 1992 International Convention on Biodiversity,[9] acknowledged the (possibility of a) special relationship between Indigenous peoples and place. These elements of the Panel's report consequently shaped our work in the description of what ecological integrity looks like for Kluane National Park and Reserve in south-west Yukon. This description was collated by the Park ecologist with submissions from biology colleagues, cultural researchers, and members of the Champagne and Aishihik First Nations. The resulting Ecological Integrity Statement was one of the first to explicitly identify the presence of Indigenous peoples, and their special relationship to place, as a necessary precondition for the health of a national park ecosystem (Box 10.1). The challenge now is to figure out how this combination or integration of the Aboriginal cultural landscape idea with the Western biological construct of ecological integrity can address the different cultural perceptions brought to management by both First Nation and the State. In 2004 Parks Canada provided $1.3 million to fund a five-year project, 'Healing Broken Connections' at Kluane National Park, to address this question.

As a result of these policy changes First Nations have become more comfortable that Parks Canada might recognize the existence of parallel paths in land management. However, this recognition is only the start of a complex learning process still underway. The acceptance of new policy tools, such as ecological integrity and the concept of the Aboriginal cultural landscape, indicate the possible direction of management change. It also creates new opportunities for working together with Indigenous peoples to search for changes that are both meaningful and effective in addressing the interests and concerns of Indigenous peoples.

As important as the revision and application of evolving policy is, these only gain currency and effectiveness when the cultural milieu of their application is altered by new understandings of governance. Governance is the determination of the roles of the State and its responsibilities to its citizens. Governance establishes the context for policy application. Changes in governance are a way of recognizing the cultural biases of the State's original formulation and its adaptation ensures the utility of the State to all members of the State. Canada has acknowledged its multicultural nature since the 1960s. The idea of the country as a cultural mosaic still resonates with many citizens. But in many ways the cultural mosaic does not challenge the culturally entrenched nature of the State and its purpose. While multiculturalism, set within the original rubric of the nation-state, promotes tolerance of cultural diversity, it offers neither validation nor recognition of cultural identity as a group activity. The consideration of this issue calls upon a rethinking of the State and its relationship and responsibilities to its members.

> **Box 10.1 Excerpt from the Ecological Integrity Statement for Kluane National Park and Reserve**
>
> **Public release draft version of 05/12/2000**
>
> *Theme #2: Cultural Reintegration*
>
> The Southern Tutchone have had a long-standing relationship with the greater Kluane ecosystem, having sustained healthy animal and plant populations through their harvesting and other cultural activities for thousands of years. The park forms part of their *cultural landscape* . . .
>
> These deep rooted connections between aboriginal people and place have been recognized as important elements in achieving and maintaining ecological integrity (cl. 38, UNESCO *World Heritage Convention* Operational Guidelines (1996)). The health and vibrancy of the Southern Tutchone relationships to their cultural landscape and its expression as traditional knowledge are integral elements of the park's ecological integrity.
>
> The gradual and eventually final exclusion of aboriginal people from a part of their traditional cultural landscape through this century has eroded the cultural connections between the Southern Tutchone and the lands now in the national park (Lotenberg, G. 1998. *Recognizing Diversity: An Historical Context for Co-managing Wildlife in the Kluane Region, 1890-present.* Mss., Parks Canada, Whitehorse, Yukon. 66 pp.). The weakening of these long-term linkages has significantly impaired the ecological integrity of KNP&R. It also has had negative consequences on Southern Tutchone culture. Without use of the park, knowledge of park lands and resources and their people's history in this area could not be passed on through community members, thereby limiting Southern Tutchone traditional knowledge. The health and vibrancy of regional traditional knowledge has suffered from this deterioration of the connections between aboriginal people and their cultural landscape.
>
> The sustainable relationship the Southern Tutchone have had with this part of their cultural landscape needs to be re-established and fostered. Activities that enhance and pass on Southern Tutchone traditional knowledge within the local First Nations communities must also be encouraged. The key actions designed to achieve these ends will strengthen the regional aboriginal cultural landscape and the contribution of traditional knowledge to ecosystem management.
>
> **Strategic Goal**
>
> To recognize the aboriginal cultural landscape as both an integral part of the Kluane region ecosystem and through the expression of Southern Tutchone traditional knowledge, a significant contributor to ecosystem management.
>
> **Objectives**
>
> - To re-establish KNP&R as part of the Southern Tutchone cultural landscape
> - To integrate the concept of cultural landscape into our understanding of the ecological integrity of the Kluane region, and First Nations' traditional knowledge in ecosystem management
> - To support activities that enhance and pass on Southern Tutchone traditional knowledge, especially land-based aspects of Southern Tutchone traditional knowledge, within local First Nations communities

- To support educational programs for members of local First Nations that focus on their history and heritage in the park area, and the management of the cultural resources
- To promote an understanding among Park staff, First Nations members, local residents and visitors to the park of the long-standing relationship of Champagne and Aishihik First Nations and Kluane First Nation with the Southern Tutchone cultural landscape

Key Actions

Action: Help members of local First Nations get to know and re-establish a sustainable relationship with park lands, i.e., renew ties with this part of the Southern Tutchone cultural landscape.

- Education and training programs designed to assist members of local First Nations in learning about the Southern Tutchone cultural relationships with plant and animal communities in the park have been implemented.
- Education programs that involve linking younger First Nation members with Elders to learn Southern Tutchone traditional knowledge have been implemented.
- Members of local First Nations are carrying out sustainable traditional harvesting activities within park boundaries.

Action: Improve understanding of the contribution of traditional knowledge and the aboriginal cultural landscape in the maintenance of ecological integrity.

- First Nations staff and membership understand the role of the aboriginal cultural landscape in contributing to the ecological integrity of the region, and the effects of harvesting activities.
- Park staff support local First Nations in offering cultural programs which contribute to ecological integrity.

Action: Improve the understanding of the Southern Tutchone cultural landscape.

- Park staff and First Nations have worked together to understand the character, qualities and values attributed to the Southern Tutchone cultural landscape.
- An inventory of First Nations' heritage features, such as trails, campsites, caches, cabins, wildlife harvesting areas and gathering sites within the park is completed.
- Aboriginal place names for features in the park have been documented and researched.
- Information regarding the Southern Tutchone cultural landscape within KNP&R has been appropriately secured for future reference and is shared between local First Nations and Parks Canada.

Action: Acknowledge and respect First Nations' cultural heritage in all aspects of park management.

- Traditional knowledge is used in setting management priorities and designing programs.
- First Nation cultural presence in the landscape is acknowledged through the use of aboriginal placenames.

Action: Encourage the development and delivery of educational and training programs that focus on the First Nations cultural legacy in the park.

- Public understanding and support for First Nations presence within the park has been achieved.
- The history and culture of First Nations in the park and surrounding area are being effectively communicated through appropriate media channels.
- First Nations are interpreting their traditional cultural landscape.
- The character, qualities and values of the Southern Tutchone cultural landscape as represented by the lands in KNP&R are communicated to the different groups with an interest in this matter.

Revising the understanding of the State and culture

A review of the international conventions addressing the question of cultural diversity over the last 60 years offers some insights into the nature of the changes in governance needed to make policies and practice more effective in addressing the interests of Indigenous peoples. These agreements also trace a trajectory of changing thought among States about culture and identity. The negotiation and acceptance of international agreements addressing human diversity have been influenced by four major, generally chronological, factors in the post-Second World War period (UNESCO, 2004). Immediately after the war there was a search for tools to promote and preserve peace. During the de-colonization period of the 1950s to the mid-1970s, newly independent nations were recognized as equal partners in the world community. Growing out of the economic difficulties faced by these new countries there was recognition of the links between culture and development. Finally, bringing us to the present, are agreements acknowledging linkages between culture and democracy, noting the 'need for tolerance not only between societies, but within them as well' (UNESCO, 2004: 3–4). These agreements indicate a growing global awareness of the significance of culture in intra-state governance, a factor highlighting, among other things, the relations between Indigenous peoples and PHAs.

In the waning days of the Second World War, planning for the United Nations (UN) was already underway among Allied governments. Although nation-states were to remain sovereign in this new international order, there was a shared desire to avoid the terrors of future wars driven by economic, racial, and political distinctions, and a recognition that peace was the necessary foundation for freedom (Bailey, n.d.). UN working groups quickly identified education and knowledge as the key to this peace. The work unfolded as a programme emphasizing the common humanity of the people of the world, and resulted in the 1948 acceptance of the Universal Declaration of Human Rights (UDHR).

An agreement with a noble purpose – the perfectability of a universal civilization – the Declaration is framed within modernist notions of the centrality of the individual. In the effort to prepare common ground for international understanding of shared humanity the UDHR confirms the ephemerality of culture, thus denying cultural identity as a significant factor in society. The document assumes that all people are not only equal but, at their core, the same. And in recognition of the sensitivity of nation-states to any infringements upon their sovereignty, the UDHR recognizes a citizen's duties to his or her government (Article 21). However, the only social organization above the individual recognized in the UDHR is the family (Article 16). The possibility of distinctions between peoples, that is, by cultural identity, are recognized, but only as free associations exercised by individuals (Article 27). The nation-state remains the sole arbiter of identity.

The UDHR develops a modernist vision of humanity as a collection of individuals with basic rights. The recognition of these rights regulates relations among individuals rather than understanding society as collections of communities seeking good for their members.

Differences between individuals are erased and the rights described are those of the liberal Western materialist vision of the world. John Gray suggests that:

> the Enlightenment project embodies a distinctive philosophical anthropology, for which cultural difference is an inessential, and . . . a transitory incident in human affairs. . . . distinctive cultural identities are seen as chosen lifestyles, whose proper place is in private life, or the sphere of voluntary association . . . [C]ultural difference is seen through the distorting lens of *choice*, as an epiphenomenon of personal life-plans, preferences and conceptions of the good.
>
> (Gray, 1995: 124)

This denial of culture as a state responsibility is the foundation for the continued refusal to acknowledge Indigenous peoples as having different interests in the State. The UDHR was the attitude, and the opportunity, that limited and then allowed the Indigenous voice to be heard.

The creation of new States through the third quarter of the twentieth century effectively de-frocked the European empires. However, these new nations, despite their often revolutionary liberation, posed little threat to the pervasive modernist notion of state citizenship as an individual, as opposed to a group, privilege. The International Covenant on Economic, Social, and Cultural Rights (1966/1976) provides insights into how these new nations and new nationalities were absorbed into the international system. The Covenant recognized the equality of the new States, but did so in terms of the Western progressive economic development model, already the foundation shaping the cultural purposes of US and Canadian PHAs. This fact was reinforced by their subsequent neocolonial relationship to international financial organizations. Further, the Covenant is largely a rewrite of the UDHR with an international committee established to report on how successful the new countries were in fulfilling the individual rights of their citizens, that is, how successfully they were assimilating modernist values of individual over community. Although innately hostile to valuation of culture, the recognition of new countries also spurred a broader understanding of multiple cultures in the world.

This broader understanding is reflected in the evolving expressions of the 1972 Convention Concerning the Protection of the World Cultural and Natural Heritage. The international recognition of the legacy of cultural and natural heritage around the world was a commitment by nation-states to understand and protect the cultural legacies within their boundaries. Although the original criteria under the Convention privileged Western science and aesthetics, the designations flowing from other parts of the world soon demonstrated the diversity of human works and peoples' different valuations of their place in the world. The designations became not only statements of national pride in an international forum but also markers of States' commitment to recognition of cultural diversity.[10]

Changes continued to be made as non-Western attitudes increasingly introduced nuances of cultural difference into the international discourse on culture and nature. One of the earliest examples is the 1992 International Convention on Biodiversity. The Convention, in its acknowledgement of the role Indigenous and local communities played in maintaining biodiversity, and relying upon it for their livelihood, recognized the distinct connections between some people and place. The playing out of this recognition continues to be contested. In 1998 the Indigenous working group of the Convention highlighted the divisions that still existed between their cultural perspective and the modernist structures of the international field. In an appeal to the parties of the Convention the working group noted:

> [R]eports to this [working group] point out that SBSTTA [the scientific committee] is highly political and not entirely scientific. These reports also point out that the reductionist method of western scientists do not adequately serve the holistic,

> biosphere approach to bio-diversity. Mr. Chairman, SBSTTA appears not to see the forest for the trees. The Indigenous and traditional perspective, that all life is related appears to be incomprehensible to SBSTTA.
>
> (United Nations, 1998)

This presentation symbolizes the challenges to the assumption of objective, value-free approaches to the environment and the past. It is representative of the many voices forwarding culturally centred narratives of meaning that have long been unheard through the heavy veil of Western cultural domination of the international cultural discourse.

In 2001 the Universal Declaration on Cultural Diversity advanced this argument recognizing the intra-state responsibility to identify and foster cultural diversity. These documents indicate a changing role for nation-states with regard to cultural heritage at the end of the twentieth century. The original intent of the nation-state was the expression of a single people's will and identity. In the early days of the twenty-first century these conventions and declarations highlight the State's responsibility to act as a regional steward of the diversity of human cultural expression. Contemporary discussions in Canada about the 2006 Convention for the Safeguarding of the Intangible Cultural Heritage[11] – an articulation of alternatives to the Western narratives of progress and materialism – reflect the difficulties States have in absorbing this significant change in the relations between and within States. James Tully (1995: 42–43) suggests that the identification of other levels of social organization within the State – Indigenous peoples, cultural minorities, gender – represents a radical transformation in constitutionalism, of the same order as the introduction of human-based principles over divine guidance through kings.

For some 300 years the modern liberal constitutional model of equal States and equal citizens has been developed in the West to govern human affairs. It has proved a flexible and adaptable system, absorbing change through contact and exchange with others, but also imposing its own values in its extension around the world. However, cultural resurgence in the post-imperial world has eroded the previously solid foundations of modernism and new demands are being made upon the previous order of the world. John Gray suggests there are multiple, and sometimes incommensurable, values present in human thought. Sometimes radical choices, that is, choices that do not support the idea of the inevitable progress in human affairs through the application of reason, need to be made. These choices arise out of the contact between these different cultural valuations of life in the world. This '[value-pluralism] renders the Enlightenment conception of the historical progress of the species meaningless or incoherent'. Gray concludes his appeal for a post-Enlightenment world with Heidegger's *Gelassenheit* noting we must 'wean ourselves from willing and open ourselves to letting things be . . . however, it is not openness to "Being" that is needed, but instead an openness to beings, to things of the earth, in all their contingency and mortality' (Gray, 1995: 69, 182).

Governance and cultural pluralism

Perspectives on time and place – history and environment – are developed by communities to bring a sense of order, membership, and purpose or meaning to their activities. The resulting worldview is the foundation of what we know as culture. Culture legitimizes the existence of a group to its members. It establishes governing institutions and guiding policies to advance the objectives of the group by coordinating activities and, through the projection of interests, neighbours.

At cultural contact points different worldviews try to make themselves understood and have their communities acknowledged. Where there is a large power differential between contacting cultures, the more powerful may deny the authority, the existence, of the less powerful and attempt to simply absorb or incorporate them into their worldview. By imposing their models of governance and ruling on the other, the more powerful establish a colonial regime with the concomitant denial and oppression of the other's worldview. In Canada the Western settler culture has, for at least a century and a half, imposed its worldview upon the Indigenous peoples resident here. To escape this colonial situation, with its costs for both the oppressed and the oppressor, this understanding of cultural imperialism needs to be consciously acknowledged and addressed (Alfred, 2005: 266).

In Canada, one of the primary contact points between Western and Indigenous cultures is land, especially those areas that are regarded as sacred or special by either or both cultures. Diverse perspectives on time and place meet here. The designation and management of state PHAs are places where Western culture clearly outlines its interests yet they are also places where place is respected and there is growing interest in the possibilities of using traditional knowledge in co-management with Indigenous peoples. For their part, Indigenous peoples in Canada struggle to make their cultural perspective understood in the management of PHAs which are part of their traditional territories. However, this conversation is neither easy nor straightforward.

To facilitate a fuller understanding of the Indigenous view, Canadians need to recognize that their PHAs are not neutral or objective examples of the environment in which they live. To come to this understanding, however, requires Canadians to acknowledge the effects of the national narrative, popularly still expressed through the Laurentian thesis. The Laurentian thesis was born of a distinct set of political and intellectual conditions that have shaped the entire warp and weave of contemporary Canadian social, political, and environmental, that is, cultural, understanding of Canada. Innis, his students, their students, and their students' students have sat as members of the HSMBC which identifies places of national significance. They have been the frontline staff, the managers, the administrators of Parks Canada, myself among them, and the bulk of the Canadian educated population. Although Parks Canada's responsibilities are broadly defined as protecting, presenting, celebrating, and serving Canadians using 'nationally significant examples of Canada's natural and cultural heritage', the policies and governance of the agency arise from the fabric of the pervasive unified national perspective. It is no simple matter to accommodate alternative or parallel narratives. To come to see PHAs, these special places, as particularly articulate expressions of their own cultural understanding of place and time is a start. Canadians can begin to free themselves of the colonial attitudes that have not allowed them to hear Indigenous voices or accept the existence and value of Indigenous cultural approaches to the world.

The root of the difficulties in reconciling Indigenous cultures as distinct from the unified national narrative appears to be the recognition of the existence of the different ways that cultures frame their worldview. John Gray's notion of 'value-pluralism, that ultimate human values are objective but irreducibly diverse, that they are conflicting and often uncombinable, and that sometimes when they come into conflict with one another they are incommensurable; that is, they are not comparable by any rational measure' (Banville, 2004), suggests the need for a recognition and acceptance of these multiple meanings. Rather than attempting to compare or integrate 'by any rational measure', perhaps we need to communicate differences and respect alternative visions of the future. For Parks Canada this means a broadened understanding of the roles played by PHAs.

The recognition of diversity and multiculturalism, however, does not address the deep-seated concerns of Indigenous peoples over their relationship to, or participation in, Canada. The notion of national identity as a bounded set of meanings has denied participation by others. As the country incorporates the other it must also accept a more complex, less linear story. The modernist notion of a unified nation-state progressing along a path to a perfect form has been shattered over the last century (Eksteins, 1999: 15–16). The recognition of many peoples, of many nations, within States, completely undermines earlier narratives that so diligently constructed a vision of a homogeneous nation with a single identity and single vision of the future.

Canada has moved to address this transformation of the realities of the State. In the 1960s, multiculturalism, the recognition of many cultures, changed the national sense of identity and broadened the country's membership to effectively include all newcomers within the boundaries of the national community and its progressive narrative of development. However, for First Nations, this was inadequate. They not only want to be recognized as people, they want to be acknowledged as cultures with different conceptions of the future. As a country, Canadians now face, with some apprehension, cultural pluralism – not only many cultures, but also many futures. James Tully (1995: 116) suggests that the acceptance of cultural pluralism means a State with distinct cultural groupings constantly negotiating with each other on the basis of mutual recognition, respecting the continuity of group traditions with governance rising from mutual consent. A culturally pluralistic Canada will be a State built not on exclusive cultural identities but, rather, on dynamic relationships that bind together different cultural groups.

These dynamic relationships do not refer to the individual battles waged in the acknowledgement of value-pluralism but, rather, indicate the continuing tension that will exist between different cultural communities. Gray (1995: 29) suggests that:

> Toleration is a virtue appropriate to people who acknowledge their imperfectability. . . . Rather than pursuing a delusive utopia in which all ways of life are given equal (and possibly unmerited) respect, they are content if they can manage to rub along together. In this they are recognizing a profound truth . . . that freedom presupposes peace . . . We are most likely to enjoy an enduring liberty if we moderate our demands on each other and learn to put up with our differences. We will then compromise when we cannot agree, and reach a settlement – always provisional, never final – rather than stand on our (in any case imaginary) human rights. Oddly enough, we will find that it is by tolerating our differences that we come to discover how much we have in common. It is in the give and take of politics, rather than the adjudications of the courts, that toleration is practised and the common life renewed.[12]

Joanne Barnaby, for a long time the Director of the Dene Cultural Institute, recently reflected upon the obligations for both government and cultural groups in a culturally pluralistic State:

> Being strong like two men . . . means that people need to draw from the strength of their culture and history to maintain a strong identity based on [their values], while also developing the capacity to interact and live effectively with other cultures and draw from their knowledge systems and their skills and abilities . . . It is the government's responsibility to foster values-based debate and to ensure that the policies that they establish reflect the values of the North. The people's responsibility is active participation, openness, honesty, sharing values, open debate about choices.
>
> (Tesar, 2006)

Acknowledgements

In the southern Yukon I am indebted to Carcross-Tagish First Nation Elders Winnie and William Atlin and Edna and Walter Helm for their hospitality and stories of Bennett; Clara Schinkel for her guidance during my learning time in Carcross and in my continuing work throughout the Yukon; and Doris McLean, who always introduced me as 'Parks Canada' in the early stages of the Chilkoot Trail Oral History Project, ensuring that I remembered who I was and challenging me to figure out what it was I was supposed to be doing.

Indigenous peoples and protected heritage areas

In the region of Kluane National Park and Reserve in the south-west Yukon I must acknowledge the citizens and staff of the Champagne and Aishihik First Nation and Kluane First Nation for their interest in and review of my work. I am especially indebted to Sarah Gaunt, who always brought intelligence and diligence to our work, Lawrence Joe, Craig McKinnon, Diane Strand, and Gerald Dickson whose dedication to their heritage ensures that the Southern Tutchone voice will always be heard.

The work on Nagwichoonjik National Historic Site was led by the capable and engaging staff of the Gwich'in Social and Cultural Institute. The opportunity to work with Alestine Andre and Ingrid Kritsch and other members of the community of Tsiigehtchic was always a great pleasure.

Parks Canada provides a supportive environment for thoughtful reflection upon its work. My many colleagues within the Agency have always been interested in reviewing, and often challenging, my work to make it better. In this regard I especially note Bob Coutts, Tom Elliot, Robert Lewis, Duane West, Anne Landry, Patricia Kell, Leslie Maitland, and David Henry. Conversations with Miche Genest of Canadian Heritage were likewise stimulating.

In the academy, the erudite work and thoughtful personal communications of Dr Doug Moggach of the University of Ottawa, Dr Julie Cruikshank and Dr Patrick Moore, both of the University of British Columbia, Dr Paul Nadasdy, University of Wisconsin – Madison, Dr Laura Peers, Oxford University, and Dr Michael Bravo, University of Cambridge, have always been appreciated. Parks Canada, Clare Hall, and the Scott Polar Research Institute, the latter both congenial and stimulating elements of Cambridge University, are also gratefully acknowledged for their support in the preparation of this piece.

Notes

1 The Saskatchewan Regional Parks Association, established in 1962 as an umbrella organization, presently includes over 100 regional parks. URL: www.parkitthere.ca/regional_parks.php. Accessed 21 November 2006.

2
>The Board agreed that the Program should move forward on . . . the commemoration of native history themes in the North . . . adopt a go-slow approach as considerable ill-will might be created if our efforts were tied too closely to the current government-wide . . . negotiations respecting native land claims in the North. It was emphasized that . . . meetings with native organizations might well be worthwhile . . . to . . . clarify the role of the Board and the Program with respect to the commemoration of native history, it was recognized that caution should be exercised in this regard, if any such meetings were to prove beneficial . . . The Cultural Pluralism Committee to examine possible strategies . . . to smooth the way for discussions amongst members of the Board, the Parks Service and northern natives respecting these matters.
>
>(Historic Sites and Monuments Board of Canada, *Minutes*, November, 1985, Parks Canada Intranet)

3 The author was a core member on the Chilkoot Trail planning team from 1986–1988 and subsequently managed cultural research for the national historic site until the late 1990s. This narrative draws from this personal experience.

4 Sheila Greer, personal communication.

5
>The Indians of Canada pavilion resembled a giant 100 foot high teepee. Inside the Indians introduced their exhibit with an accusation addressed to their countrymen. *You have stolen our native land, our culture, our soul . . . and yet, our traditions deserve to be appreciated, and those derived from an age-old harmony with nature even merited being adopted by you.*
>
>(Stanton, J. (1997) *Indians of Canada Pavilion at Expo 67*, URL: naid.sppsr. ucla.edu/expo67/map-docs/indianscanada.htm. Accessed February 2006)

6 This definition from Susan Buggey, *An Approach to the History of Aboriginal Peoples Through Commemoration of Cultural Landscapes*, available at www.pc.gc.ca/docs/r/pca-acl/index_e.asp. This was accepted by the HSMBC in July 1999.
7 *Nagwichoonjik National Historic Site, Commemorative Integrity Statement*, Draft 14 April 2004. Copy on file with author.

8   A naturalized knowledge system (also known to many non-Aboriginal people as 'traditional ecological knowledge') comprises four basic phases that roughly parallel an individual's growth throughout life:

- innate knowledge with which one is born;
- intuitive knowledge about how and why things 'are';
- empirical knowledge that is collected by experience and which might contest intuitive knowledge;
- harmonious or spiritual knowledge realized when conflict between empirical knowledge and intuitive knowledge is reconciled and better understanding is achieved.

Like naturalized knowledge, Western science is 'a way of knowing.' Using this knowledge system, people grope for better understanding of the world by testing intuitive knowledge (current, best understanding about why things 'are') with observations (new empirical information). The two often have to be reconciled, and are sometimes harmonized with previous knowledge. Western science is often represented by its fiercest proponents as more rigorous – and thus producing better knowledge – than other ways of knowing.

Both systems use the assimilation of new knowledge to improve understanding of the world – that is, learning. By recognizing this similarity, instead of emphasizing differences, Western and Aboriginal cultures may agree upon the shared goal of learning to improve responsibility for the natural world.

(from Parks Canada, 2000, Vol. II, p. 4-3)

The definition carefully notes the individual's knowledge and avoids the existence of culturally based knowledge sets.

9 8 (j)   Subject to its national legislation, respect, preserve and maintain knowledge, innovations and practices of indigenous and local communities embodying traditional lifestyles relevant for the conservation and sustainable use of biological diversity and promote their wider application with the approval and involvement of the holders of such knowledge, innovations and practices and encourage the equitable sharing of the benefits arising from the utilization of such knowledge, innovations and practices.

(from www.biodiv.org/convention/articles.shtml)

10 1972 Convention definitions:

- cultural heritage is *monuments*, archaeology, fine arts and architecture; *groups of buildings*, architecture or place in landscape; both from the perspective of history, art or science or *sites*, works of man or combined works of nature and man from the historical, aesthetic, ethnological or anthropological point of view.
- natural heritage is *natural features* from the aesthetic or scientific point of view; *areas* that are physical formations or habitat for threatened animals from science or conservation; *natural sites* from the view of science, conservation or natural beauty.

11 Ratified by 47 countries to April 2006. The ratifying countries include 16 from Europe, nine from Asia, nine from Africa, seven from Latin America, and six Arab states. Interestingly not a single settler society – United States, Canada, Argentina, Chile, Australia, or New Zealand – has yet signed on to this convention.
12 Consider Raz's statement:

> Conflict is endemic. . . . Tension is an inevitable concomitant of accepting the truth of value pluralism. And it is a tension without stability, without a definite resting-point

of reconciliation of the two perspectives, the one recognizing the validity of competing values and the one hostile to them. There is no point of equilibrium, no single balance which is forever correct and could prevail to bring the two perspectives together. One is forever moving from one to the other from time to time.

(Raz 1994: 165)

Literature cited

Alfred, Taiaiake (2005). *Wasáse: indigenous pathways of action and freedom*. Peterborough: Broadview Press.
Bailey, Peter (n.d.). The creation of the Universal Declaration of Human Rights. URL: www.universalrights.net/main/creation.htm. Retrieved September 2006.
Banville, John (2004). Review of John Gray, *Heresies: against progress and other illusions*. *Manchester Guardian*, 9 April.
Catton, T. (1997). *Inhabited Wilderness: Indians, Eskimos and national parks in Alaska*. Albuquerque, NM: University of New Mexico Press.
Eksteins, M. (1999) *Walking Since Daybreak*. Toronto: Key Porter Books.
Gray, J. (1995). *Enlightenment's Wake: politics and culture at the close of the modern age*. London: Routledge.
—— (2004). *Heresies: against progress and other Illusions*. London: Granta Books.
Heimstra, Mary (1955). Gully Farm. In K. Mitchell (ed.) (1997), *Horizon: writings of the Canadian Prairie*, p. 106. Toronto: Oxford University Press.
Herriot, T. (2000). *River in a Dry Land: a prairie passage*. Toronto: Stoddard.
HSMBC (Historic Sites and Monuments Board of Canada) (1998). *Minutes*. Parks Canada Intranet.
Ingold, Tim (2000). *The Perception of the Environment: essays in livelihood, dwelling and skill*. London: Routledge.
Latour, Bruno (1993). *We Have Never Been Modern*. New York: Harvester/Wheatsheaf.
Neufeld, D. (2001). Parks Canada and the commemoration of the North: history and heritage. In K. Abel and K. Coates (eds), *Northern Visions: new perspectives on the north in Canadian history*, pp. 45–79. Peterborough, Ontario: Broadview Press.
—— (2002). The commemoration of northern Aboriginal peoples by the Canadian government. *The George Wright Forum* 19 (3), 22–33.
—— (in press). Parks Canada, the commemoration of Canada, and northern Aboriginal oral history. *Oral History and Public Memories*. Philadelphia, PA: Temple University Press.
Parks Canada Agency (2000). *'Unimpaired for Future Generations?' Protecting Ecological Integrity with Canada's National Parks: Report of the Panel on the Ecological Integrity of Canada's National Parks, Vol. I 'A Call to Action', Vol. II 'Setting a New Direction for Canada's National Parks'*. Ottawa: Minister of Public Works and Government Services.
Raz, J. (1994). Multi-culturalism: a liberal perspective. In J. Raz, *Ethics in the Public Domain: essays in the morality of law and politics*, pp. 170–192. Oxford: Oxford University Press.
Runte, A. (1997). *National Parks: the American experience*. Lincoln, NE: Nebraska University Press.
Said, E. (1978). *Orientalism*. New York: Pantheon.
Tesar, C. (2006). Joanne Barnaby on 'Preserving, Revitalizing and Promoting Culture and Identity'. *Northern Perspectives* 30 (1), 5–6 (Winter). Also available at: www.carc.org/northern_perspectives.php#2006.
Tully, J. (1995). *Strange Multiplicity: constitutionalism in an age of diversity*. Cambridge: Cambridge University Press.
UNESCO, Division of Cultural Policies and Intercultural Dialogue (September 2004) *UNESCO and the Issue of Cultural Diversity: Review and Strategy, 1946–2004*.
United Nations (1998). *Fourth Conference of the Parties, Convention on Biodiversity, Bratislava, Slovakia, 11 May 1998*: Oral Intervention, International Indian Treaty Council, Agenda item 10, Implementation of Article 8 (j) 'Working for the Rights and Recognition of Indigenous Peoples'.
Webber, E. (1976). *Peasants into Frenchmen: the modernization of rural France, 1870–1914*. Palo Alto, CA: Stanford University Press.
Wikipedia (n.d.). *Ernest Renan*. URL: http://en.wikipedia.org/wiki/Ernest_Renan. Accessed 21 November 2006.

Chapter 11
Political ecology perspectives on ecotourism to parks and protected areas

Lisa M. Campbell, Noella J. Gray, and Zoë A. Meletis

In many countries, parks and protected areas have 'become the cornerstone of tourism and recreation' (Task Force on Economic Benefits of Protected Areas of the World Commission on Protected Areas (WCPA) of IUCN, 1998: ix), and are a key attraction for ecotourists (Ceballos-Lasurain, 1996; Weaver, 1998; Honey, 1999). While the IUCN argues that 'the link between protected areas and tourism is as old as the history of protected areas' (Eagles *et al.*, 2002: xv), the importance of this relationship has undoubtedly grown with continued growth in tourism, and, more specifically, in ecotourism. Tourism is often described as the world's largest industry and, while a small component of this overall industry, ecotourism is believed to be one of the fastest growing sub-sectors (Weaver, 1999; The International Ecotourism Society (TIES), 2005).[1]

Definitions of ecotourism are many, and have proliferated since the term was popularized in the 1980s. In an often cited IUCN publication, ecotourism is defined as:

> environmentally responsible travel and visitation to relatively undisturbed areas, in order to enjoy and appreciate nature (and any accompanying cultural features – both past and present) that promotes conservation, has low visitor impact, and provides for beneficially active socio-economic involvement of local populations.
> (Ceballos-Lasurain, 1996: 20)

The International Ecotourism Society (TIES) defines ecotourism as 'responsible travel to natural areas that conserves the environment and improves the well-being of local people' (TIES, n.d.: ¶ 1). These two definitions reflect two components of ecotourism, with the former emphasizing the purpose of ecotourism and the latter emphasizing its impacts. Ecotourists are portrayed as seeking more than just leisure experiences, while the impacts of ecotourism are portrayed as beneficial to both local people and the environment. Both types of definitions reflect attempts to distinguish ecotourism and ecotourists from traditional forms of tourism and tourists. Ecotourism is part of 'The New Moral Tourism' that arose from 'angst-ridden discussion(s)' (Butcher, 2003: 6) about tourism and its negative impacts on host communities and environments, and the accompanying 'denigration of mass tourism' (Butcher, 2003: 7). While we recognize that tourism to parks and protected areas was taking place long before the term ecotourism was coined, and that ecotourism does not necessarily require the existence of protected areas,[2] in this chapter we are concerned with tourism to parks and protected areas that is generally conceived of as, or considered to be, ecotourism.

The literature on ecotourism to parks and protected areas is dominated by impact studies of particular cases which, in general, have shown the results of ecotourism in practice to be disappointing, with negative consequences resulting for the environment and local people (Ziffer, 1989; Cater, 1994; Bookbinder et al., 1998; Honey, 1999; Farrell and Marion, 2001). Thus, the focus of much work is on 'getting ecotourism right' and case studies are often assessed against existing best practice frameworks (e.g. Ross and Wall, 1999; Scheyvens, 1999; McDonald and Wearing, 2003). While these studies have undoubtedly contributed to our understanding of ecotourism, many lack wider theoretical frameworks that might help position ecotourism as a phenomenon both reflecting and reinforcing human–environment relations and tied to larger economic, political, and social processes. In this chapter, we address this gap in the literature by examining ecotourism to parks and protected areas through the lens of political ecology. Like West et al. (2003), we believe that without an improved theoretical understanding of ecotourism, case study research will keep rediscovering the disappointments of ecotourism in practice.[3]

In the first part of this chapter, we provide a brief overview of political ecology, focusing on the two dominant threads of research: a structural (neo-Marxist) concern with material practice and a poststructural concern with discourse. Both threads are relevant to the study of ecotourism to parks and protected areas. We also review some of the relevant research by political ecologists on parks and protected areas, ecotourism, and tourism. In the second part of the chapter, we consider three themes of interest to political ecologists – the social construction of nature, conservation and development narratives, and alternative consumption – and what researchers concerned with these contribute to our understanding of ecotourism to parks and protected areas. In the concluding section, we outline a political ecology *of* ecotourism to parks and protected areas, and suggest ways in which this can enhance studies of this growing phenomenon.

Political ecology

While there is no single definition of political ecology, there is general agreement that it is an approach to understanding environmental issues, conflicts, and problems. According to Watts (2000: 257), political ecology 'seeks to understand the complex relations between nature and society through a careful analysis of what one might call the forms of access and control over resources and their implications for environmental health and sustainable livelihoods'. Wilshusen (2003: 41) defines it as an 'overarching frame of inquiry for exploring the politics of natural resource access and use at multiple levels over time'. For Bryant (1998: 79), 'political ecology examines the political dynamics surrounding material and discursive struggles over the environment in the third world'. This last definition, by referencing both material and discursive struggles, highlights two branches of political ecology: the structural, which focuses on the interactions of 'ecology and a broadly defined political economy' (Blaikie and Brookfield, 1987: 17), and the poststructural, which focuses on the control over discourse, knowledge, and ideas (Blaikie, 1999; Watts, 2000).

The structural approach to political ecology emerged in the late 1970s and combined 'the concerns of ecology and a broadly defined political economy. Together this encompasses the constantly shifting dialectic between society and land-based resources, and also within classes and groups within society itself' (Blaikie and Brookfield, 1987: 17). Political ecology critiqued its precursor, cultural ecology, as insufficiently attentive to the broader political and economic forces impacting on local human–environment interactions. In doing so, it was influenced by the growth of neo-Marxism in the social sciences, which pointed to the role of the global capitalist system, and its attendant class relations and modes of production, in shaping local environmental conflicts. Thus, political ecology saw local communities as characterized by 'the presence of markets, deep social inequalities, enduring conflict, and forms of cultural disintegration associated with their integration into a modern world system'

(Watts, 2000: 261). While political ecology differentiated itself from cultural ecology by bringing 'into the analysis social relations that are not necessarily proximal to the ecological symptoms' (Paulson et al., 2003: 206), it retained the focus on in-depth, local environmental histories (Walker, 2005). Thus, structural political ecology situated environmental change and resource conflicts in political and economic contexts with multi-scalar dimensions, ranging from the local to the global, and emphasized the historical processes influencing environmental change.

The second phase of political ecology began in the 1980s, when authors began to question the structural determinism of neo-Marxist analyses (in which local people were largely cast as victims), the vague specification of politics, and the assumptions made about ecological reality. Using diverse theoretical constructs, researchers have focused on the various actors involved in environmental conflicts and the power relations between them (Bryant and Bailey, 1997). The most prominent theoretical influence in this phase has come in the form of poststructuralism and discourse theory (Escobar, 1994, 1996; Peet and Watts, 1996). Adger et al. (2001: 683) 'broadly define discourse as a shared meaning of a phenomenon', while Peet and Watts (1993: 228) suggest that discourse 'is a particular area of language use related to a certain set of institutions and expressing a particular standpoint'. Poststructuralists argue for the importance of discourse as more than a theoretical concern; Escobar (1996), for example, argues that understandings of sustainable development, transmitted through discourse, shape power relations by legitimizing some approaches to economy and environment and not others. Discourses are, therefore, reflections of power relations; those with power assert their discourses, thereby determining what will count as truth and knowledge for all of society.

Thus, poststructural political ecology has been concerned with plurality in knowledge, including ecological knowledge (Watts, 2000). In contrast to the 'taken for granted' ecology of early political ecology, poststructural political ecology requires a phenomenology of nature and recognizes that this is open to debate. Nature, itself, is identified as a social construction, embedded in discourse, and what is silenced in such discourse is as important as what dominates. Scientific experts, sometimes aligned in epistemic communities, are deeply implicated in the production and dissemination of dominant discourses (Fischer, 2000). The task of the discursive political ecologist is thus to map the ways in which knowledge and power disperse through complex networks to produce political-ecological outcomes. One important critique of the poststructural emphasis on discourse is the tendency to see discourses as monolithic, independently reproducing themselves. Both Moore (2000) and Leach and Fairhead (2000) call for greater attention to the agency of individuals as conscious participants in the uptake, transformation, and dissemination of environment and development discourses. The faulty determinism of structural political ecology should not be replaced with equally faulty discursive determinism, a point that we return to in the final section of this chapter.

A few caveats regarding the scope of this chapter: first, we focus on ecotourism to parks and protected areas in 'Third World' or 'developing' countries. One feature of political ecology has been an overriding emphasis on the Third World and marginalized groups, and a related concern for social justice. While there are convincing arguments for extending political ecology to the analysis of First World problems (McCarthy, 2002; Robbins, 2002), existing research on parks and protected areas and ecotourism has been undertaken largely in a Third World context, and our chapter reflects this focus. Second, we engage with both structural and poststructural approaches to political ecology when examining ecotourism to parks and protected areas, as discursive and material practice are coupled, often tightly. Finally, as Blaikie (1999) observes, the relative newness of political ecology means that reviewing past research often involves an *ex-post* re-labelling of work that initially did not self-identify as political ecology. While the majority of authors referenced in this chapter do identify their work as political ecology, we engage in some *ex-post* re-labelling where necessary. Regardless, some of the reviewed literature would not be labelled as political ecology (especially that cited in our section on alternative consumption). However, political ecology has typically drawn on various theories, fields, and disciplines (Neumann, 2005) and our chapter is in keeping with this tradition.

Political ecology perspectives on protected areas and ecotourism

Political ecologists have devoted some energy to the study of protected areas, which is unsurprising given political ecology's overall interest in forms of access to, and control over, resources; as spatially defined conservation units, parks and protected areas regulate resource use through controlling (and eliminating certain forms of) access. 'Political ecologists reveal how these spaces of conservation become arenas of conflict that result in distinctive patterns of resource management' (Zimmerer and Bassett, 2003: 5). Many studies focus on how parks limit activities of local people, and the resulting conflicts that ensue (e.g. Neumann, 1998; Sundberg, 2003; Nygren, 2004). However, few political ecologists have engaged in critiques of tourism,[4] ecotourism in general, or ecotourism to parks and protected areas more specifically.

Some political ecologists studying protected areas mention tourism or ecotourism as an activity when describing their case studies, but do not include it in their detailed analyses. For example, Brown (1998) examines biodiversity conservation in Royal Bardia National Park, Nepal. She looks at the meaning and use of biodiversity to and by different groups, and at the spatial and physical relations of users to park resources. While she refers to the conflict between local forms of resource use and tourism, her focus is on park management regimes and their impacts on local people. The tourism industry is implicated in some negative impacts, but indirectly through providing incentives for park protection and due to profit leakage. Similarly, a study by Daniels and Bassett (2002) examining conflict over resources in Lake Nakuru National Park, Kenya, 'one of the most visited parks in the country', focuses on conflicts between local people, NGOs, and the State. Few (2002) examines community participation in protected areas planning in Belize, and situates his work in political ecology's interest in power and actors. Ecotourism is identified as a motive for establishing protected areas, and local actors are differentiated according to their interests in ecotourism, but ecotourism itself is not questioned.

Perhaps the most explicit applications of political ecology to ecotourism are by Young (1999) and Belsky (1999, 2000) in their case studies in Baja California, Mexico, and in Gales Point, Belize, respectively. Young (1999) contrasts fishing and ecotourism in Baja California, with regards to the benefits of either activity to local resource users and conflicts over access to resources. She argues that the same local and state structures inhibiting effective fisheries management apply in the case of ecotourism; because local people are at a competitive disadvantage with outside investors and due to intra-community conflicts, local interests in long-term environmental sustainability are curtailed. Young (1999: 610) suggests that political ecology's multi-scalar and contextual approach, and its attention to how 'markets, policies, and political processes shape nature-society relations' makes it a useful framework to apply to her case study. Belsky (1999, 2000) critiques community-based ecotourism in Gales Point, arguing that, rather than empowering local people in their development and encouraging their support for conservation, ecotourism development has been subject to existing politics of class, gender, and patronage that result in the inequitable distribution of the costs and benefits of ecotourism. Central to her analysis is the concept of community, the tendency of outsiders to over-simplify this, and how such simplifications shape 'the design, practice, negotiations, and outcomes of community conservation projects' (Belsky, 2000: 645). For Belsky, political ecology's interest in power relations and in representations of nature and people is critical. While taking different approaches, both Young and Belsky effectively challenge the idea that ecotourism is inherently different from other forms of externally driven development. In contrast to the authors cited in the preceding paragraph, Young and Belsky focus their analysis on ecotourism rather than on the related protected areas to which ecotourists are drawn.[5]

Some researchers have used political ecology to examine more traditional forms of tourism, and their work highlights how parks and protected areas may be used strategically, even when ecotourism is not the focus of the industry. For example, Stonich (1998, 2003) uses political ecology to examine tourism to the Bay Islands, Honduras. Stonich (1998) focuses on the unequal distribution of the economic and environmental costs and benefits of tourism, concluding that the poorest residents bear the majority of the costs and receive few of the benefits. In addition, their livelihood activities are often blamed for environmental degradation, while the greater impacts by the tourism industry are ignored. In a second publication, Stonich (2003) specifically focuses on the creation of marine protected areas (MPAs). While existing tourism to the Bay Islands is not primarily ecotourism, MPA creation is seen as a way to attract this market segment, protect the beach and ocean resources that traditional tourists enjoy, and extend control by elites over tourism development. Again, Stonich's analysis shows how local people lose access to resources in MPA designation while the tourism industry's negative impacts on the marine environment remain mostly unaddressed. In a second example, Gössling (2003a) applies political ecology to what he calls 'high value conservation tourism' in the Seychelles. The Seychelles protects the highest percentage of its land in protected areas globally, and while tourism to the Seychelles is not necessarily ecotourism, the government of the Seychelles sees a clean, healthy, protected environment as a key competitive advantage when trying to attract high-end tourists. Parks and protected areas represent environmental conservation, and are something tourists can identify with to justify 'their stay in such exclusive environments' (Gössling, 2003a: 215). Furthermore, as a visible and measurable indicator of environmental progress, protected areas are one of the few means by which the Seychelles, with its high standards of living, can attract international financial assistance. The irony, of course, is that travel to the Seychelles and the maintenance of high-end tourism facilities lead to over-proportional energy and resource use by tourists and the industry (for an ecological footprint analysis of tourism to the Seychelles, see Gössling *et al.*, 2002).

The studies described above illustrate some of the ways that political ecology can be applied to case studies of parks and protected areas, ecotourism, and tourism more generally, and yet none of them focuses explicitly on the issue of ecotourism *to* parks and protected areas. We find a lack of critical attention to this subject by political ecologists surprising, for several reasons. First, given the increasing popularity of ecotourism and its reliance on parks and protected areas, it is insufficient to treat ecotourism as a mere by-product of park creation. Rather, ecotourism is often implicated in park creation. Thus, political ecology's concern with parks and protected areas should extend to ecotourism. Second, parks and protected areas and ecotourism are linked to capitalism; the parks movement in the US has been characterized as a romantic reaction against the frontier mentality associated with capitalist expansion in the mid- to late nineteenth century (McCormick, 1989; Cronon, 1995). Ecotourism, a more recent phenomenon, has been tied to late-stage (post-Fordist) capitalism and the increased interest in niche opportunities by sophisticated and demanding consumers (Mowforth and Munt, 1998). Additionally, the rise of ecotourism reflects green development thinking, where environmental conservation (and more specifically parks and protected areas) is expected to pay for itself (Adams, 1990; McAfee, 1999; and see the Task Force on Economic Benefits of Protected Areas of the World Commission on Protected Areas (WCPA) of IUCN, 1998). Thus, the traditional concerns of political ecologists with capitalism (its impacts on environment and people) are relevant. Finally, while protected areas themselves have been critiqued as reflecting dominant (Western) human–environment relations that separate humans from nature, ecotourism is part of this same process, with ecotourists seeking unspoiled pristine nature for their leisurely consumption (Urry, 1995; Mowforth and Munt,

1998; Ryan *et al.*, 2000; West and Carrier, 2004). Thus, both parks and protected areas and ecotourism are expressions of predominantly Western values that can be analysed with a more discursive approach to political ecology.

In order to move beyond case study analysis, in this chapter we examine the *idea* of ecotourism to parks and protected areas, using a political ecology approach and drawing on both the material and discursive traditions. We specifically focus on three areas of political ecology research: (1) the social construction of nature; (2) conservation and development narratives; and (3) alternative consumption.

The (socially constructed) nature of ecotourism

As illustrated in the definitions cited in the introduction, 'natural' destinations and attractions, pristine areas that show no sign of human activity, are critical to ecotourism. Parks and protected areas are the dominant way of establishing these natural areas as discrete and separate from human activity. Without objecting to nature conservation per se, political ecology helps to question the assumptions underlying particular ideas of 'nature' as they are produced and reinforced by ecotourism to protected areas. Too often, the nature being protected by parks or visited by ecotourists is taken as given. Following a brief review of the 'social construction of nature' arguments typical of the political ecology literature, this section will explore how the application of such arguments might help to develop a more theoretically informed understanding of ecotourism to parks and protected areas.

The 'social construction of nature' is a phrase 'commonly employed to stress the role of representation, discourse and imagery in defining and framing our knowledge of nature and the natural' (Neumann, 2005: 47). This idea has been vehemently challenged by conservation biologists among others (e.g. Soulé and Lease, 1995; Gandy, 1996), who are concerned that if nature is merely a social construction (i.e. a product of culture and language), rather than an independent entity with its own agency, then the ability to advocate for environmental protection is undermined (Eden, 2001). However, these challenges often misconstrue and simplify what is a complex, nuanced argument (Neumann, 2005). Social constructionism encompasses a range of philosophical positions with differing ontological and epistemological commitments regarding what constitutes nature and the means by which we can know it.[6] Moreover, as Bryant (1998) and Forsyth (2003) both emphasize, the aim of social constructivist arguments in political ecology is rarely to deny the existence of nature (or environmental problems, or biophysical reality), but rather to demonstrate that how nature is identified and depicted is a highly politicized process.

Engagements with social constructionism within the political ecology literature cover a range of positions, which Robbins (2004) organizes into two groups – the 'hard' or 'radical' constructivists and the 'soft' constructivists. Some political ecologists tend toward an ontologically idealist (i.e. radical or hard) position that sees 'language not as a reflection of "reality" but as constitutive of it' (Escobar, 1996: 46). Associated with the linguistic turn in the social sciences, this version of constructionism traces back to Foucault and post-structuralism more generally (Demeritt, 2002). Willems-Braun (1997), for example, takes this position in his study of how nature has been produced and enacted in both colonial and postcolonial British Columbia, through the discourses of colonial surveyors, contemporary forestry companies, and environmentalists. He argues that nature is never *mis*represented, as it can only ever be present *through* representation. Willems-Braun (1997: 5) is thus not interested as much in ideas *about* nature, so much as he seeks to document 'the *emergence* of "nature" as a discrete and separate object of aesthetic reflection, scientific inquiry, and economic and political calculation at particular sites and specific historical moments'.

While hard constructionists exist, most political ecologists invoke a softer form of constructionism (Robbins, 2004: 114) in which language is not constitutive of nature, but the 'subjective conceptual system' through which our knowledge of the objective world of nature is filtered. Much of the political ecology literature adopts a critical realist position (Forsyth, 2003; Neumann, 2005), rejecting more extreme constructionist approaches while still 'sharing post-structuralist concerns of the importance of discourse, representation, and imagery in structuring knowledge of the world' (Neumann, 2005: 47). Constructions are not just discourse, as they have consequences for political practices with associated material outcomes (such as the establishment of a national park). 'The imagined forest [or "nature"] becomes the real one, and vice versa, through the enforcement of such constructs by powerful people over time. In this way, the line between objects and ideas is blurred' (Robbins, 2004: 110). Softer constructionists thus call for greater attention to the agency of individuals as conscious participators in the uptake, transformation, and dissemination of ideas and discourses about nature (e.g. Leach and Fairhead, 2000).

Cronon's (1995) treatise on 'the trouble with wilderness', which is perhaps the best-known argument regarding the social construction of nature, has important implications for both historical and contemporary understandings of parks (and tourism to them). Drawing on historical evidence, he argues that wilderness is not a natural state, but a time/place/culture-specific idea; it is a product of the late nineteenth-century United States. Cronon traces the growth of the wilderness concept in American culture, from a wasted and hostile wilderness awaiting the productivity of human (European) civilization, to a threatened wilderness in need of saving from civilization gone too far. The early national parks were introduced to protect this new idea of a 'threatened wilderness', serving several critical social functions in the process. First, they were expressions of social class. 'Ever since the nineteenth century, celebrating wilderness has been an activity mainly for well-to-do city folk' (Cronon, 1995: 79). Second, they were spaces of leisure and consumption (versus production); wilderness was to be protected, not used (Cronon, 1995). Neumann (1998) similarly argues that wilderness represents a largely visual notion of nature, an Anglo-American aesthetic reinforced through 'centuries of painting, poetry, literature, and landscape design [and more recently, tourist brochures]' (Neumann, 1998: 10). The split between nature/culture, wilderness/civilization, and consumption/production dictates that the only acceptable role for humans in wilderness is as observer (Neumann, 1998). The upper class, aesthetic consumption of nature through ecotourism to parks is further discussed in the section on alternative consumption. The third social function performed by the first national parks was the exclusion of a whole group of people, the Native Americans, in order to create the people-free wilderness that parks were supposed to contain. The persecution and displacement of native peoples that occurred in the early history of the United States, generally as well as specifically in relation to the establishment of national parks as wilderness areas, is well documented (Spence, 1999; Burnham, 2000). As Spence (1999: 4) notes, 'uninhabited wilderness had to be created before it could be preserved'. Since Yellowstone was officially established as the first American national park in 1872 (Cronon, 1995), US parks, as symbols of human-free wilderness, have served as an international model for protected areas that displace and exclude local people (Guha, 1989; Neumann, 1998; Spence, 1999).[7] The power of this socially constructed idea of people-free wilderness is related to its incorporation into a dominant conservation narrative, as discussed in the next section. Nature and wilderness are not just ideas, they are policy prescriptions for protected areas that dictate 'the exclusion of people as residents, the prevention of consumptive use and minimization of other forms of human impact' (Adams and Hulme, 2001: 10).

The numerous, widespread detrimental effects for local people caused by the imposition of parks are well documented (e.g. West and Brechin, 1991; Ghimire and Pimbert, 1997). Ecotourism, in contrast, is often presented as an inclusive alternative that engages local people, providing them with benefits rather than restricting their livelihoods (e.g. Honey, 1999). However, even those forms of ecotourism that champion the rights and well-being of local people often seek to engage local people in the production and defence of a specific, Western view of nature, a view that has previously been used to justify their exclusion from traditionally inhabited land and that runs counter to their own worldview (Akama, 1996). Ecotourism to parks might provide some support for nearby residents, but it also continues to support a particular version of nature that tends to be divorced from local environmental concerns in developing countries (Guha, 1989). Ecotourism might be no more than eco-imperialism, demanding that host destinations supply and comply with a Western construction of people-free nature (Mowforth and Munt, 1998).

In cases where ecotourism does include local people in nature, it often invokes the image of the noble savage, tribal people living traditional lifestyles in harmony with their environment (Mowforth and Munt, 1998). Just as a socially constructed idea of nature underwrites an exclusionary model of protected areas, so too does a socially constructed idea of 'traditional' or 'indigenous' people delimit the manner in which local people might be readmitted to the nature promoted by ecotourism.[8] Urry (1995) discusses the idea of social pollution to refer to the presence of social groups that interfere with tourists' expectations of place; expectations that are constantly shifting. In Australia, for example, tourists have been 'increasingly finding that Aboriginal culture and practices are no longer "polluting" but are part (or even the most important part) of the exotic attractions of Australia' (Urry, 1995: 189). Mowforth and Munt (1998: 274) refer to this as zooification, a process that 'involves turning tribal peoples into one of the "sights" of a rainforest expedition or a trek'. While in some cases indigenous groups may cooperate with such constructions by 'staging authenticity' (Mowforth and Munt, 1998), in other cases worlds (or natures) may collide, as they do when ecotourists to the Arctic witness a local whale hunt (Hinch, 1998). In their argument for a renewed theoretical critique of ecotourism, West and Carrier (2004) suggest that the interaction of socially constructed ideas of nature and neoliberalism produce a set of common pressures, which they find in ecotourism to parks in Jamaica and Papua New Guinea. They argue that this interaction has a 'tendency to lead not to the preservation of valued ecosystems but to the creation of landscapes that conform to important Western idealizations of nature through a market-oriented nature politics' (West and Carrier, 2004: 485; see also Vivanco, 2001). By calling for analyses of ecotourism that account for both discourse and political economy, West and Carrier (2004) are inadvertently advocating a political ecology of ecotourism.

While some analysts (and even some ecotourists) might be aware of this critique of the nature underlying ecotourism, this 'awareness comes in spite of, rather than because of, the common image and presentation of ecotourism' (Carrier and Macleod, 2005: 329). The construction of nature as a pristine, people-free landscape (except for a few tourists), continues to be reproduced by travel brochures and advertisements, fuelling the geographical imagination of ecotourists (Norton, 1996; Mowforth and Munt, 1998). As Gössling (2003b) points out, ecotourism is both a result of, and reinforces, dominant Western visions of human–environment relations. However, it is important to note that despite the dominance of a particular Western construction of nature in international conservation (including discourses of protected areas and ecotourism), this is not the only model of nature in circulation, nor are its effects ever pre-determined (Olwig, 2004; West and Carrier, 2004). There are some examples of ecotourism ventures in Latin America where local values play an important role (Wesche, 1996; Stronza, 2001). As Vivanco (2001) asserts, based on research in Costa Rica, ecotourism should be analysed as an arena for the contestation of different views and values regarding nature, rather than the enforcement of one dominant view. Moreover, dominant social constructs of nature or indigenous can sometimes prove strategically useful to less powerful groups, who might consciously deploy them to strengthen their claims to resources (Brosius, 1997; Sundberg, 2003; cf. Li, 1996, on 'community'). A political ecology of ecotourism must document the evolution of different social constructions of nature, indigenous, and related concepts, as well as their circulation, contestation, strategic deployment by both more and less powerful groups, and material consequences for the people and landscapes associated with ecotourism to protected areas.

Talking about ecotourism: conservation and development narratives

One of the dominant themes in poststructural political ecology has been the concept of narratives. A narrative can be defined as a story with a 'beginning, middle, and end (or premises and conclusions, when cast in the form of an argument) and revolves around a sequence of events or positions in which something happens or from which something follows' (Roe, 1991: 288). Narratives justify and inform action to avert disaster or achieve gains. In his original argument that focused on development narratives, Roe (1991) suggested that narratives are often necessary, as they allow for decision-making in the face of uncertainty. The problem arises when narratives prove incorrect; embedded in institutions and with explanatory and descriptive power, they are difficult to displace. Specific examples of narratives failing to play out on the ground, even when numerous, are insufficient to displace a dominant narrative (Roe, 1991). This can only happen when a counter-narrative that tells a 'better story' develops, and Adams and Hulme (2001: 10) argue that counter-narratives must be as 'parsimonious, plausible and comprehensive' as the original.

Roe's (1991) concept of narrative has been applied by political ecologists (and others) to environmental policy and its impacts on local people. One of the most influential studies is by Fairhead and Leach (1995, 1996), who challenge narratives of desertification in West Africa that link deforestation with increasing human populations. Using aerial photos, historical archives, and ethnographic interviews with local inhabitants, the authors demonstrate how the agro-ecological practices of local people have actually generated forest islands around their settlements, in a landscape otherwise dominated by savanna. And yet the State and NGOs, engaged as they are in the narrative of deforestation, intervene to change these same agro-ecological practices (Fairhead and Leach, 1995, 1996). Thus, narratives are not just stories; they have material consequences.

Forsyth (2003) reviews several general environmental narratives (using the term orthodoxy), and works of political ecologists that challenge them. These include narratives of desertification, tropical deforestation, shifting cultivation, rangeland degradation, agricultural intensification, watershed degradation and water resources, and Himalayan environmental degradation. Forsyth's list reflects a traditional concern of political ecologists with marginalized people and their use of natural resources in pursuing their livelihoods. However, political ecologists have also turned their attention to narratives related to conservation of natural resources through parks and protected areas (e.g. Neumann, 1996, 1998). For example, Campbell (2002b) has described a traditional wildlife conservation narrative as follows. The 'problem' or 'crisis' is identified as local people who harvest wildlife and/or threaten it indirectly through competition for wildlife habitat needed to support increasing human populations. Unless human activity is checked, wildlife extinctions are inevitable. The 'solution' is to remove people from spaces in order to provide wildlife with a place where it is not subject to exploitation or competition. Protection is enforced by the State, and if local people do not respect the conditions of their removal, and return to hunt or harvest, they become 'poachers' and 'encroachers'. In doing so, they reconfirm original beliefs about the crisis and, as they are breaking the law, the solution becomes more and better enforcement (Campbell, 2002b: 30). Adams and Hulme (2001) describe how this traditional conservation narrative (labelled the fines and fences approach, coercive conservation, or fortress conservation) developed in Africa, linking it to a variety of forces, including the early imaginings of expatriate colonial men about what 'wild' Africa should be. Early parks and protected areas on the continent allowed for hunting by expatriates and foreign visitors, for whom the activity was characterized as noble and character building, while hunting by local people was characterized as barbaric (MacKenzie, 1988; McCormick, 1989; Neumann, 1996). Thus, the social construction of nature and of indigenousness, as discussed in the previous section, is implicated in the formation of the traditional narrative.

Parks and protected areas have been criticized over the years on a number of practical and philosophical fronts, including their biological utility, the costs of protection, and justice concerns associated with exclusionary protection, and their failure to combine conservation and development (the latter critique reflecting a general interest in the concept of sustainable development) (Western and Wright, 1994; Ghimire and Pimbert, 1997). As a result, a conservation counter-narrative has arisen to challenge the dominant narrative. Adams and Hulme (2001: 13) identify this as community-based conservation, with two elements: (1) the imperative to allow local people to participate in management of protected resources, and (2) the linkage of conservation objectives with local development needs. Alternatively, Campbell (2002b) argues that community-based conservation is one component of the counter-narrative, concerned primarily with local participation, while sustainable use is a second component, concerned primarily with providing wildlife and/or biodiversity with economic value so that there are incentives to conserve it. This separation of the components of the counter-narrative allows for sustainable use projects that do not include the participation of local people, and community-based conservation that takes place with very low levels of use.

Campbell (2002b) further divides the sustainable use component of the counter-narrative into consumptive and non-consumptive use. With consumptive use defined as the deliberate removal or killing of an organism (Freese, 1998), ecotourism is categorized by default as non-consumptive.[9] As reflected in the TIES definition cited in the introduction, ecotourism is often associated with local development in a way that mainstream tourism is not, with local people empowered and maintaining control over development and its associated economic benefits (e.g. Whelan, 1991). Due to its status as non-consumptive use and its emphasis on local benefits and involvement, i.e. its ability to mesh with the conservation counter-narrative, ecotourism has become a favoured solution of wildlife conservation experts, and Campbell (2002a) illustrates this with case studies from Costa Rica. She shows how marine turtle conservation experts at three sites in Costa Rica have strategically adopted the counter-narrative of community-based conservation and sustainable use in their promotion of ecotourism. By using the language of the conservation counter-narrative, experts appear to be concerned with local livelihoods as well as conservation. Yet, by promoting ecotourism, they are able to continue to support restrictive parks and protected areas, the tools of the traditional narrative, because parks are key ecotourist attractions. Likewise, experts can support prohibition on more consumptive forms of resource use, as these conflict with use by ecotourists (Campbell, 2002a).

While Roe (1991) suggests that narratives become embedded in institutions, Jeanrenaud (2002) argues against a monolithic conservation movement that promotes a single vision of conservation. In the case of World Wide Fund for Nature (WWF), she identified four groups operating within the organization in the 1990s: (1) cosmocentrics, focused on ecosystem and biodiversity conservation; (2) anthropocentric neoliberals, who emphasize economic and political processes, especially the role of the market; (3) radical anthropocentrics, focused on livelihood needs and rights of marginalized groups; and (4) anthropocentric elites, who promote a traditional conservation agenda based on anthropocentric and theocentric values (e.g. nation building, especially among elites from developing countries). In a similar way, Nygren (1998) identifies four dominant streams of environmentalism in Costa Rica: (1) environmentalism for profit; (2) environmentalism for nature; (3) environmentalism for people; and (4) alternative environmentalism. While recognizing this diversity, Gray (2003) and Campbell (2002b) argue that part of ecotourism's appeal lies in its ability to serve the needs of such diverse interests. For example, ecotourism is conceived as a way to make conservation pay for itself (environmentalism for profit), provide income for local people to meet development needs (environmentalism for people), and justify the creation of protected areas to serve the tourist industry (environmentalism for nature) (Gray, 2003). More specifically, Campbell (2002b) considers ecotourism to Costa Rica and the way it has influenced traditional political groups and alliances between them. The capital accumulation nexus, social reform nexus, and ecodevelopment nexus (first identified by Carriere, 1991) all find something to identify with in ecotourism (Campbell, 2002b). In establishing ecotourism as a component of a narrative, and the broad appeal it has to diverse interest groups, political ecology helps us to understand why ecotourism continues to be promoted in spite of failures to live up to expectations in practice.

Ecotourism as alternative consumption

Recently, political ecologists have turned their attention to the issue of 'alternative consumption'. Alternative consumption is 'the "new" activism', making consumption an important site for moral expression (Bryant and Goodman, 2004: 344). With the expansion of civil society into consumption (Butcher, 2003), and the continued expansion of the neoliberal agenda globally, consumers are depicted as powerful agents of change (West and Carrier, 2004). They use this power to demand fairer trade and more responsible producers and governments; thus, alternative consumption is a site for political voice and mobilization (Miller, 1995). With emphasis on the individual, consumers become 'the frontline' (Bryant and Goodman 2004: 344) and consumption the locus for resisting the exploitive elements of capitalism.

Focused as it is on consumers, alternative consumption is one means of bringing tourists into the analysis of ecotourism to parks and protected areas. As noted earlier, ecotourism and ecotourists have largely been overlooked by political ecologists. Furthermore, Bryant and Goodman (2004) argue that an analysis of consumption allows us to break free from the North/South dichotomies that have plagued political ecology. While political ecologists have traditionally been concerned with marginalized communities in the South, 'there is surprising little effort . . . devoted to assessing how social processes integral to the North may affect Southern political ecologies through a variety of geographical pathways' (Bryant and Goodman, 2004: 347).

Bryant and Goodman (2004) identify two commodity cultures within alternative consumption. The first is a conservation-seeking culture concerned for the environment and with an interest in preserving it. This translates into buying products such as organic food and green cleaning products. The second is a solidarity-seeking culture focused on social justice through fairer trade and labour practices. Concern for peoples (especially workers in the Global South) translates into buying products such as fair trade cocoa or coffee. While Bryant and Goodman (2004) contrast products that appeal to the conservation-seeking versus the solidarity-seeking cultures, ecotourism purportedly combines the two; it is about helping others use nature in a less destructive (and more profitable) way (West and Carrier, 2004), and doing so by expressing consumer preference. In this section, ecotourism, a pre-eminent form of alternative consumption, is analysed from three perspectives: (1) the moralization of consumption; (2) the consumption of aesthetic nature (and community); and (3) consumption and neoliberal capitalism.

As outlined in the introduction, ecotourism is an important component of 'The New Moral Tourism' and appeals to consumers searching for something better than traditional mass tourism (Mowforth and Munt, 1998; Butcher, 2003). Part of the appeal lies in consumer anxieties about environmental damage in First World nations (i.e. the homes of most ecotourists), or 'the projection of guilt from self onto others' (Heyman, 2005: 114). The ecotourism industry extracts profit from this guilt through the use of moral suasion (Heyman, 2005). For example, ecotourists are pitted against mass tourists; the former are altruistic and contributory while the latter are self-interested and damage-causing (Bryant and Goodman, 2004). Ecotourism is better because it is portrayed as non-consumptive (often replacing consumptive uses of wildlife and other resources), having minimal environmental impacts, and supporting local culture and/or being community-friendly (Boo, 1990; Wilson and Tisdell, 2001). Not only is ecotourism better than mass tourism, however, it is a desirable or positive activity that ecotourists can feel good about. Ecotourism can even be viewed as offering salvation, a way for tourists to help preserve people and places, notably people and places 'over there' (Mowforth and Munt, 1998; Bryant and Goodman, 2004). Putting ecotourism on this pedestal moves it beyond a form of consumption that benefits the consumer (Heyman, 2005) and gives it political and moral power.

While ecotourism has both environmental and socio-economic goals, and alternative consumption has both conservation-seeking and solidarity-seeking components, environmental features arguably dominate in ecotourism[10] and the aesthetics of ecotourism are critical. Given the traditional importance of aesthetics and the tourist 'gaze' to tourism in general (Urry, 1995, 2002; Ryan et al., 2000), this emphasis is hardly surprising. However, given ecotourism's claims about promoting local development, the ecotourism aesthetic warrants further scrutiny. Ecotourism to parks and protected areas is based on, and reinforces, an aesthetic of wilderness (Cronon, 1995) or nature (West and Carrier, 2004), and promotes Edenic myths (Bryant and Goodman, 2004; Nelson, 2005) to draw the ecotourist, as discussed in the section on the social construction of nature. Ecotourism destinations must exemplify 'nature', 'exotic', and/or 'simple' (West and Carrier, 2004: 491). For example, Costa Rica, a country highly successful in cultivating an ecotourism image, uses the marketing slogan 'All Natural Ingredients' (www.visitcostarica.com). Communities can be part of the ecotourism aesthetic, provided they remain natural or simple and not overdeveloped. Thus, the imposition of an ecotourism aesthetic on a community can work against the solidarity-seeking aspect of alternative consumption, by promoting a limited and/or static vision of development, one that favours traditional (often subsistence) livelihoods, architecture/infrastructure, and culture in general. Any ecotourism-related infrastructure that is developed (such as canopy walks, hiking trails, or souvenir shops) might have little value to local host communities, and yet might be prioritized over infrastructure improvements that detract from the aesthetic, such as paved roads or concrete buildings. Such restrictions might be at odds with local wants, needs, and aspirations, i.e. local notions of the right to develop (Thrupp, 1990; Urry, 1995; Mowforth and Munt, 1998; Scheyvens, 1999; West and Carrier, 2004). Through the establishment of parks and protected areas and the design of nearby facilities, the ecotourist aesthetic can act much like zoning or other regulatory planning tools do in terms of constraining local development.

The ecotourist aesthetic might also work against the conservation-seeking component of alternative consumption, with environments managed for key species of interest to tourists rather than for overall ecosystem function and health.[11] A case study by West and Carrier (2004) of Montego Bay, Jamaica, illustrates how both the conservation and solidarity-seeking components of alternative consumption may fail to materialize with ecotourism, at a number of levels. First, under a neoliberal agenda where conservation pays for itself, Montego Bay was selected as the site for Jamaica's first marine park, not because it was the most pristine or environmentally valuable location, but because the area had the tourist infrastructure in place to attract, house, and entertain would-be ecotourists, who were seen as critical to the park's success. Second, ecotourist beliefs about what a pristine marine environment should look like influenced park management. While the overall health of the bay depended on less visible elements (bacteria, sea urchins, sediment, and coral growth), managers had to spend time on more marketable features, so that tourists would pay to snorkel and dive there (i.e. so the park could pay for itself). Finally, in spite of the fact that fishing in the park by local people was legal, park managers were under increasing pressure to 'overwrite coastal waters with a new set of ecotourist meanings identifying certain sorts of people, fee-paying ecotourists, as properly in those waters – indeed as necessary to their survival – and Jamaicans in small boats as belonging elsewhere' (West and Carrier, 2004: 488). This example illustrates the way that ecotourism might be different from other alternative consumption products; because ecotourists go directly to the ecotourism product (rather than purchasing it from a shelf in a supermarket), the conflicts between conservation-seeking and solidarity-seeking components of alternative consumption come to the fore.

Alternative consumption has been critiqued for its acceptance of neoliberal economics and its focus on the individual; individual consumers in the developed North make choices that benefit environments and local producers in the South. With this emphasis on consumers expressing preferences in the market, alternative consumption poses few challenges to the global capitalist system. Rather, it accepts this as given and, by creating new products within it, encourages its continuance and extension (Manokha, 2004; West and Carrier, 2004). Capitalism is made 'nicer' through fairer trade relations and/or environmental practices, rather than questioned or overthrown (Goodman, 2004). Thus, alternative consumption relies on the commodification of nature. In the case of ecotourism, the establishment of parks and protected areas for tourists to visit represents the commodification of these green spaces and particular species within them (Dorsey et al., 2004). McAfee (1999) and Escobar (1996) describe this form of 'nature undisturbed' as postmodern ecological capital, in contrast with modern ecological capital (e.g. forests as lumber). Parks and protected areas become not simply plots of land set aside for conservation purposes that happen to be visited by ecotourists, but places created for and *by* ecotourists (Urry, 1995; West and Carrier, 2004), and that should accommodate the ecotourist gaze (Urry, 1995, 2002; Ryan et al., 2000; West and Carrier, 2004). Sites that are not of interest to the ecotourist, i.e. that are not in demand, might be overlooked. Furthermore, in the ecotourist search for authenticity, ecotourism often brings 'backstage' regions that ecotourists seek out to the 'frontstage' (MacCannell, 1973; Butcher, 2003), making once little-known places a 'must see' for growing numbers of ecotourists. By contributing to the commoditization of places, ecotourism arguably works against itself; by putting previously unknown destinations on the tourism map, it replaces the 'authentic' places that it is trying to preserve with created places (Urry, 1995; Mowforth and Munt, 1998; West and Carrier, 2004).

Political ecology's approach to ecotourism as a form of alternative consumption offers deeper insight into the ecotourist as a consumer, a political actor, and a socio-political identity, rather than a mere bystander in the drama that unfolds when parks and protected areas clash with local peoples' livelihood aspirations in ecotourism destinations. This addresses one of the weaknesses in the existing political ecology literature, where ecotourists are rarely a focus of analysis. Furthermore, by focusing on the consumptive aspects of ecotourism, alternative consumption also connects with political ecology's traditional interests in capitalism. A political ecology approach to ecotourism as a form of alternative consumption should be concerned with 'the distribution of power in consumption' (Heyman, 2005: 128) in order to help (re)focus studies of ecotourism back outward to the important (neoliberal) political and economic context in which it occurs, and where consumer wants are key (West and Carrier, 2004).

Towards a political ecology of ecotourism to parks and protected areas

Our chapter began by identifying two characteristics of the existing literature on ecotourism to parks and protected areas. First, studies of ecotourism are largely case study based and often atheoretical. Many reveal the ways in which the benefits of ecotourism bypass local communities or that ecotourism development negatively impacts the environment, and some use best practice frameworks to suggest ways in which such problems might be overcome. While such studies are useful and provide rich context-specific data, we agree with West *et al.* (2003) that without a larger theory of ecotourism, researchers will keep rediscovering the shortcomings of ecotourism in practice. Second, political ecologists have studied parks and protected areas, tourism, and ecotourism, but there are few studies that focus on ecotourism *to* parks and protected areas. In many cases, ecotourism is part of the context or background for political ecology's concern with the impacts of (state) resource management interventions on local people. Analysis of ecotourists is almost entirely absent. This might be explained by political ecology's traditional interests in marginalized peoples and the structural constraints (primarily political and economic) on their livelihood options, and state management interventions and alliances with major resource extraction companies (e.g. timber, oil). As an industry often dominated at the local level by small- to medium-sized private businesses, ecotourism and ecotourists might have simply slipped under the radar.

In this chapter, we outline a political ecology of ecotourism to protected areas by drawing on three thematic interests of political ecologists: the social construction of nature, conservation and development narratives, and alternative consumption. These are certainly not the only themes of interest to political ecologists, or the only ones relevant to the issue of ecotourism to parks and protected areas.[12] However, they were selected because of the number of ways in which they work together and can be integrated to provide an enhanced theoretical understanding of ecotourism to parks and protected areas, and it is to their integration that we now turn.

While Campbell (2002a, b) identifies ecotourism as the 'received wisdom' of a conservation counter-narrative that promotes sustainable use and community-based conservation, the widespread and growing popularity of ecotourism suggests that it may be more appropriate to consider a separate narrative *of* ecotourism. The narrative begins with acceptance (or celebration) of a neoliberal economic reality in which nature pays its way in order to survive. Local people are seen as having legitimate developmental needs and therefore they must be given incentives to save nature rather than convert it for other productive activities. Ecotourism, due to its non-consumptive status, is conceived of as a more economically beneficial and sustainable use of nature than traditional activities. Furthermore, local people are able to capture economic benefits due to the small-scale nature of ecotourism and the concerns of ecotourists with supporting local cultures and economies. To attract ecotourists, host countries must develop ecotourism products, and the most easily identified are parks and protected areas that have the added benefit of being able to charge entrance fees. When ecotourists come to these parks, everyone wins: local people provide services to tourists and earn more income than they would via other uses of resources, nature is protected in parks and protection is supported through entrance fees, and ecotourists contribute to causes they believe in while experiencing unique environments and peoples. This ecotourism narrative is a powerful one. Unlike many environmental narratives that are of crisis, the ecotourism narrative is one of salvation; nature and local people are saved through the actions of ecotourists, and parks are the temples to which the morally aware consumers flock to do good. As such, ecotourism is often the starting point of conservation projects, rather than one of many options to consider.[13]

This narrative persists in spite of evidence that ecotourism often fails to meet expectations in practice, and the elements of political ecology (individually, but more powerfully in combination) reviewed in this chapter can help to explain why. First, narratives are always resilient in the face of evidence that they are wrong, and our review of conservation and development narratives suggests that a narrative of ecotourism might be particularly resilient because it meets the needs of a variety of interest groups, regardless of their views on the best way to pursue conservation and development. In this way, the ecotourism narrative might be considered a supra-narrative, under which a variety of conservation and development narratives can peacefully co-exist. Second, the ecotourism narrative is supported by a deeply embedded (Western) social construction of nature that most often depicts nature as something separate from humans and in need of protection from the ravages of capitalist development. Parks and protected areas originated from this view, and, by participating in ecotourism, ecotourists both reflect and reinforce it. Third, ecotourism appeals to conscientious consumers who are interested in alternative options that are both more labour and environmentally sensitive than traditional forms of tourism. Ecotourism provides people with the opportunity to assuage their guilt over their resource-consumptive lifestyles by contributing to nature conservation and local development 'over there'. Since ecotourists need consumable products, parks and protected areas are a key ecotourist commodity. However, the dominant social construction of nature remains, and though ecotourists purport to be both solidarity-seeking and conservation-seeking alternative consumers, their vision of nature allows for solidarity with very limited forms of local development. There is an equally strong social construction of indigenous or traditional peoples, and local development needs and wants might conflict with this.

There are several ecological outcomes of ecotourism to parks and protected areas. By far the most widely recognized impacts concern the protection (or removal) of nature from traditional productive activities when it is set aside in parks and protected areas, and this protection occurs with or without ecotourism. Much of the political ecology work on parks and protected areas has focused on the impacts of protection on the livelihoods of local people. However, we argue that there are additional ecological outcomes associated specifically with ecotourism and ecotourists. First, the ecotourist 'gaze' might demand that parks are created in places that have features ecotourists want to see or, as shown in the case of Montego Bay, Jamaica, that have the infrastructure to support the ecotourists required to ensure financial viability (West and Carrier, 2004). Second, the same gaze focuses on the 'frontstage' environment and related infrastructure, e.g. charismatic species and landscapes, and the hiking trails, viewing platforms, and information displays required to enjoy them. Meanwhile, the 'backstage' environment and infrastructure, e.g. water quality, waste treatment and disposal facilities, might be neglected. Third, the ecotourist focus on the parks and protected areas product masks the broader impacts of ecotourism. For example, Carrier and MacLeod (2005) introduce the idea of the 'ecotourist bubble' to describe the limited context within which ecotourism is often viewed and presented. They recount the story of a tourist who had travelled to Antarctica and was careful to note that she had avoided stepping on the fragile plant life, even though a much larger concern might be the carbon dioxide emissions resulting from her air travel. Overall, if the 'nature' being supported by ecotourism is based on the ecotourist gaze and a particular wilderness construction, then the focus is on plant and animal life rather than emissions, resource consumption (e.g. fresh water), or other sorts of impacts.[14] Conservation becomes marine mammal conservation, or sea turtle conservation, or rainforest conservation (i.e. aesthetic conservation (Green, 1990 cited in Urry, 1995)), and concerns for the wider environment are lost.

In developing a political ecology of ecotourism to parks and protected areas, and more specifically a narrative of ecotourism, we are in no way arguing for the end of ecotourism to parks and protected areas. Such a call would be naive and, if heeded, the impacts would be undesirable in many places. We recall at this point one critique of discursive political ecology, i.e. the tendency to see discourses as monolithic and intractable. As Moore (2000: 655) argues: 'Far too often, contemporary analyses eclipse the micro-politics through which global development discourses are refracted, reworked, and sometimes subverted in particular localities ... The specificity of these struggles belies any single totalizing development discourse.' As suggested in our treatment of the social construction of nature, the existence of a dominant social construction does not mean it is the only one, and the effects of a related narrative of ecotourism are not always predetermined. Though far fewer in number than their critical counterparts, there are case studies of ecotourism to parks and protected areas where the narrative (or individual elements of it) plays out and is realized to some extent in practice. Resources can be protected, local people can benefit from ecotourism and agree that it is a superior form of development, local values can be respected, and tourists can experience nature and culture in meaningful ways (e.g. Colvin, 1996; Wesche, 1996; Wunder, 2003; Stronza, 2004). Thus, in particular places and for particular peoples, the results of the narrative can be good. Equally, however, there are ample case studies in the literature describing when the narrative remains just that: a story with little relation to what happens in practice. The strength of the narrative means that these far more numerous examples are treated as individual exceptions, the problems of which are to be corrected.

Our hope is that in recognizing an ecotourism narrative, the powerful values underlying it, its broad appeal to a variety of interest groups, and its links to neoliberalism, we can, like West et al. (2003), find ways to subvert it. Roe (1991) and Adams and Hulme (2001) would suggest that an important step in subversion is the development of a counter-narrative, that is as 'parsimonious, plausible, and comprehensible as the original' (Adams and Hulme, 2001: 10). We suggest that re-writing the ecotourism narrative, for example, in a way that opens it up to alternative constructions of nature and that recognizes and challenges the ecotourism aesthetic, might be as important/effective as trying to manage for impacts on site at individual parks and protected areas. Recognizing the inherent (rather than site-specific) challenges associated with ecotourism might lead, for example, to a different conceptualization of the role of ecotourism in national and international conservation strategies. Rather than trying to achieve environmental and socio-economic objectives via ecotourism at each park, we might instead envision a system of parks where some are sacrificed to ecotourists, some to local people, and some to 'nature'. Ecotourism thus becomes one of many options to engage local support for parks and protected areas, rather than the only one.

As discussed in the section on conservation and development narratives, part of the strength of the ecotourism narrative lies in its broad appeal to a variety of interest groups. However, there are signs that the attraction of the ecotourism narrative is waning for some. For example, Kiss (2004) questions the economic and conservation gains of ecotourism to parks and protected areas and Weaver (2002) argues that 'hard' ecotourism, i.e. ecotourism that meets its goals of protecting the environment, cannot provide sufficient revenue to ensure local support. More generally, Wilshusen et al. (2002) outline a resurgent protectionist movement concerned by the poor conservation outcomes associated with efforts to integrate local economic development into parks and protected areas, including efforts made via ecotourism. This movement calls for the return to people-free, strictly protected parks and protected areas (Oates, 1999; Terborgh, 1999), and has shaken the types of alliances supported by ecotourism, detailed by Campbell (2002b) and Gray (2003). Brockington et al. (2006) describe a resulting 'unproductive' discomfort between those interested in parks and those interested in peoples, and Redford et al. (2006) suggest this discomfort is leading to a 'brittleness' in our conception of parks, one that threatens to undermine their utility for conservation and/or local development. They argue that the dialogue on parks and protected areas needs to be opened up, with a recognition that 'parks' fit into a variety of categories that 'incorporate people and their economic endeavours in different ways' (Redford et al., 2006: 2). Such a dialogue, should it transpire, would also provide the much needed opportunity for reconceptualizing the role of ecotourism to parks and protected areas.

Redford et al. (2006) lament that the two sides in the parks-versus-people debate are engaged in a 'dialogue of the deaf', so intent are they in making their points and defending their views. 'Social scientists have set out bold and effective criticisms of the social dimensions (and especially social effects) of park creation, and content with their hostile critique they have not often engaged with the issue of policy reform' (Redford et al., 2006: 1). We reiterate here that failure to get beyond case studies of ecotourism to parks and protected areas to a more theoretically informed understanding of the political, economic, and social context in which ecotourism takes place, will make any attempts to 'do ecotourism better' superficial. The reform called for by Redford et al. (2006) will only be meaningful if it recognizes, engages with, and hopefully challenges the ecotourism narrative, and political ecology is one approach that can help to accomplish this.

Notes

1. A recent article in *The New York Times*, for example, identified ecotourism as the 'buzzword of the year' (Higgins, 2006).
2. Wearing and Neil (1999) and Weaver (1998) expand ecotourism to include travel to degraded natural sites, such as participation in an oil spill clean-up.
3. West *et al.* (2003) argue that there is a political economy of ecotourism that can explain, for example, why local people receive few ecotourism benefits. They then look for ways that local people may escape this political economic reality and capitalize on ecotourism opportunities.
4. An edited volume by Gössling (2003c) uses political ecology to examine case studies of tourism to tropical islands, but the extent to which individual chapters integrate political ecology into their analysis varies substantially.
5. Only one of Young's (1999) field sites lies in a protected area, but this biosphere reserve was not 'fully functioning' at the time of her research, and few local people realized it existed. While the issue of 'paper parks' is a compelling one, in this chapter we focus on ecotourism to parks and protected areas that experience some level of administration and enforcement.
6. Several authors have attempted to dissect the social constructionist argument, offering typologies of constructionism generally (Hacking, 1999), and in relation to 'nature' more specifically (Proctor, 1998; Demeritt, 2002).
7. As Schelhas (2001) notes, Native Americans have not been excluded from *all* US parks, and some US park experiences may, indeed, offer valuable lessons in the international context. Likewise, many parks in the US did not uphold the supposed elements of the US park model. However, the focus here is on the power of a discourse that upholds a particular, exclusionary model of a park based on an idea of 'people-free nature'. It is the power of this image, rather than the variation in experience, that is of interest.
8. More generally, Brockington *et al.* (2006) are critical of the overall focus among conservationists on relationships between 'indigenous' people and protected areas, while relatively little attention is given to non-indigenous local people.
9. The term consumption is used differently in different fields. In economic terms, consumption contrasts to production, with the latter referring to the transformation of natural resources into goods. In wildlife conservation, the discussion of sustainable use distinguishes between consumptive and non-consumptive use, with consumption referring to the direct removal of a species or its parts for use and non-consumptive referring to more passive viewing by tourists (some would argue this is an erroneous distinction, see Tremblay, 2001; Meletis and Campbell, 2007). The concept of production is absent in the wildlife conservation literature on sustainable use.
10. While ecotourists are supposed to be interested in local culture, studies of tourist preference show that local cultures often rank considerably lower than environmental features (Jacobson and Robles, 1992; Hvenegaard and Dearden, 1998).
11. Green (1990, cited in Urry, 1995: 186) defines this outcome as aesthetic conservation, i.e. 'to conserve an environment in accordance with pre-given conceptions of beauty and the sublime, conceptions which often depend upon what is being contrasted with the environment in question'.
12. For example, many political ecologists study the network of actors that influence resource management, and that work at various geographic scales. We could have considered the various actors involved in promoting ecotourism and parks and protected areas, incentives provided by donor agencies to national governments to encourage them to pursue these options, the interests of various state agencies and national elites in promoting them, and how local communities may be differentiated according to their interests in, and involvement with, ecotourism (this resembles the type of political economy of ecotourism suggested by West *et al.* (2003).
13. For example, a recent WWF publication on sea turtle conservation advocates tourism as the solution for turtle conservation programmes globally (Troëng and Drews, 2004).
14. Concern with the impacts of ecotourism have led some tourism analysts to question the categorization of ecotourism as more environmentally and socially responsible than mass tourism. Weaver (1991), for example, suggests that conventional mass tourism concentrates impacts and can develop the infrastructure required to deal with 'backstage' environmental impacts. In contrast, ecotourism

disperses impacts, and disperses them into areas that are often environmentally fragile. This dispersal in combination with an ecotourism aesthetic might limit the capacity to deal with environmental impacts.

Literature cited

Adams, W.M. (1990). *Green Development: environment and sustainability in the Third World*. London: Routledge.
—— and Hulme, D. (2001). Conservation and community: changing narratives, policies and practices in African conservation. In D. Hulme and M. Murphree (eds), *African Wildlife and Livelihoods: the promise and performance of community conservation*, pp. 9–23. Oxford: James Curry.
Adger, W.N., Benjaminsen, T.A., Brown, K., and Svarstad, H. (2001). Advancing a political ecology of global environmental discourses. *Development and Change* 32, 681–715.
Akama, J. (1996). Western environmental values and nature-based tourism in Kenya. *Tourism Management* 17 (8), 567–574.
Belsky, J.M. (1999). Misrepresenting communities: the politics of community-based rural ecotourism in Gales Point Manatee, Belize. *Rural Sociology* 64 (4), 641–666.
—— (2000). The meaning of the manatee: an examination of community-based ecotourism discourse and practice in Gales Point, Belize. In C. Zerner (ed.), *People, Plants and Justice*, pp. 285–308. New York: Columbia University Press.
Blaikie, P. (1999). A review of political ecology: issues, epistemology, and analytical narratives. *Zeitschrift für Wirtschaftsgeographie* 43 (3–4), 131–147.
—— and Brookfield, H. (1987). *Land Degradation and Society*. London: Methuen.
Boo, E. (1990). *Ecotourism: the potentials and pitfalls* (Vols 1 and 2). Baltimore, MD: World Wildlife Fund.
Bookbinder, M.P., Dinerstein, E., Rijal, A., and Cauley, H. (1998). Ecotourism's support of biodiversity conservation. *Conservation Biology* 12 (6), 1399–1403.
Brockington, D., Igoe, J., and Schmidt-Soltau, K. (2006). Conservation, human rights, and poverty reduction. *Conservation Biology* 20 (1), 250–252.
Brosius, J.P. (1997). Endangered forest, endangered people: environmentalist representations of indigenous knowledge. *Human Ecology* 25 (1), 47–69.
Brown, K. (1998). The political ecology of biodiversity, conservation and development in Nepal's Terai: confused meanings, means and ends. *Ecological Economics* 24, 73–87.
Bryant, R.L. (1998). Power, knowledge and political ecology in the third world: a review. *Progress in Physical Geography* 22 (1), 79–94.
—— and Bailey, S. (1997). *Third World Political Ecology*. London: Routledge.
—— and Goodman, M.K. (2004). Consuming narratives: the political ecology of 'alternative' consumption. *Transactions of the Institute of British Geographers* 29, 344–366.
Burnham, P. (2000). *Indian Country, God's Country: Native Americans and the national parks*. Washington, DC: Island Press.
Butcher, J. (2003). *The Moralisation of Tourism*. London and New York: Routledge.
Campbell, L.M. (2002a). Conservation narratives and the 'received wisdom' of ecotourism: case studies from Costa Rica. *International Journal of Sustainable Development* 5 (3), 300–325.
—— (2002b). Conservation narratives in Costa Rica: conflict and co-existence. *Development and Change* 33 (1), 29–56.
Carrier, J.G. and Macleod, D.V.L. (2005). Bursting the bubble: the socio-cultural context of ecotourism. *Journal of the Royal Anthropological Institute* 11, 315–334.
Carriere, J. (1991). The crisis in Costa Rica: an ecological perspective. In D. Goodman and M. Redclift (eds), *Environment and Development in Latin America*, pp. 184–204. Manchester: Manchester University Press.
Cater, E. (1994). Ecotourism in the third world: problems and prospects for sustainability. In E. Cater and G. Lowman (eds), *Ecotourism: a sustainable option?*, pp. 69–86. Chichester: Wiley.
Ceballos-Lascurain, H. (1996). *Tourism, Ecotourism and Protected Areas*. Gland, Switzerland: IUCN.
Colvin, J.G. (1996). Indigenous ecotourism: the Capirona programme in Napo Province, Ecuador [Electronic Version]. *Unasylva* 187. Retrieved 29 January 2006 from www.fao.org/documents/show_cdr.asp?url_file=/docrep/w2149e/w2149e07.htm.

Cronon, W. (1995). The trouble with wilderness; or, getting back to the wrong nature. In W. Cronon (ed.), *Uncommon Ground*, pp. 69–90. New York: W.W. Norton & Company.

Daniels, R. and Bassett, T.J. (2002). The spaces of conservation and development around Lake Nakuru National Park, Kenya. *The Professional Geographer* 54 (4), 481–490.

Demeritt, D. (2002). What is the 'social construction of nature'? A typology and sympathetic critique. *Progress in Human Geography* 26 (6), 767–790.

Dorsey, E.R., Steeves, H.L., and Porras, L.E. (2004). Advertising ecotourism on the internet: commodifying environment and culture. *New Media & Society* 6 (6), 753–779.

Eagles, P.F.J., McCool, S.F., and Haynes, C.D.A. (2002). *Sustainable Tourism in Protected Areas: guidelines for planning and management*. Gland, Switzerland and Cambridge, UK: IUCN.

Eden, S. (2001). Environmental issues: nature versus the environment? *Progress in Human Geography* 25 (1), 79–85.

Escobar, A. (1994). *Encountering Development: the making and unmaking of the third world*. Princeton, NJ: Princeton University Press.

—— (1996). Constructing nature: elements for a poststructural political ecology. In R. Peet and M. Watts (eds), *Liberation Ecologies*, pp. 46–68. London: Routledge.

Fairhead, J. and Leach, M. (1995). False forest history, complicit social analysis: rethinking some West African environmental narratives. *World Development* 23 (6), 1023–1035.

—— and —— (1996). Rethinking the forest-savanna mosaic. In M. Leach and R. Mearns (eds), *The Lie of the Land: challenging received wisdom on the African environment*, pp. 105–121. Oxford: International African Institute.

Farrell, T.A. and Marion, J.L. (2001). Identifying and assessing ecotourism visitor impacts at eight protected areas in Costa Rica and Belize. *Environmental Conservation* 28 (3), 215–225.

Few, R. (2002). Researching actor power: analyzing mechanisms of interaction in negotiations over space. *Area* 34 (1), 29–38.

Fischer, F. (2000). *Citizens, Experts, and the Environment*. Durham, NC: Duke University Press.

Forsyth, T. (2003). *Critical Political Ecology: the politics of environmental science*. London: Routledge.

Freese, C.H. (1998). *Wild Species as Commodities: managing markets and ecosystems for sustainability*. Washington, DC: Island Press.

Gandy, M. (1996). Crumbling land: the postmodernity debate and the analysis of environmental problems. *Progress in Human Geography* 20 (1), 23–40.

Ghimire, K.B. and Pimbert, M.P. (eds) (1997). *Social Change and Conservation*. London: Earthscan.

Goodman, M.K. (2004). Reading fair trade: political ecology imaginary and the moral economy of fair trade foods. *Political Geography* 23, 891–915.

Gössling, S. (2003a). 'High-value conservation tourism': integrated tourism development in the Seychelles? In S. Gössling (ed.), *Tourism and Development in Tropical Islands: political ecology perspectives*, pp. 203–221. Cheltenham, UK and Northampton, MA, USA: Edward Elgar.

—— (2003b). Tourism and development in tropical islands: political ecology perspectives. In S. Gössling (ed.), *Tourism and Development in Tropical Islands: political ecology perspectives*, pp. 1–37. Cheltenham, UK and Northampton, MA, USA: Edward Elgar.

—— (ed.) (2003c). *Tourism and Development in Tropical Islands: political ecology perspectives*. Cheltenham, UK and Northampton, MA, USA: Edward Elgar.

——, Borgströ-Hansson, C., Hörstmeier, O., and Saggel, S. (2002). Ecological footprint analysis as a tool to assess sustainability. *Ecological Economics* 43, 199–211.

Gray, N. (2003). Unpacking the baggage of ecotourism: nature, science, and local participation. *Great Lakes Geographer* 9 (2), 113–123.

Guha, R. (1989). Radical American environmentalism and wilderness preservation: a third world critique. *Environmental Ethics* 11, 71–83.

Hacking, I. (1999). *The social construction of what?* Cambridge, MA: Harvard University Press.

Heyman, J.M. (2005). The political ecology of consumption. In S. Paulson and L.L. Gezon (eds), *Political Ecology across Spaces, Scales, and Social Groups*, pp. 113–132. New Brunswick, NJ and London: Rutgers University Press.

Hinch, T. (1998). Ecotourists and indigenous hosts: diverging views on their relationship with nature. *Current Issues in Tourism* 1 (1), 120–124.

Honey, M. (1999). *Ecotourism and Sustainable Development: who owns paradise?* Washington, DC: Island Press.

Hvenegaard, G.T. and Dearden, P. (1998). Ecotourism versus tourism in a Thai National Park. *Annals of Tourism Research* 25 (3), 700–720.

Jacobson, S.K. and Robles, R. (1992). Ecotourism, sustainable development, and conservation education: development of a tour guide training program in Tortuguero, Costa Rica. *Environmental Management* 16 (6), 701–713.

Jeanrenaud, S. (2002). Changing people/nature representations in international conservation discourses. *IDS Bulletin* 33 (1), 111–122.

Kiss, A. (2004). Is community-based ecotourism a good use of biodiversity conservation funds? *Trends in Ecology and Evolution* 19 (5), 232–237.

Leach, M. and Fairhead, J. (2000). Fashioned forest pasts, occluded histories? International environmental analysis in West African locales. *Development and Change* 31, 35–59.

Li, T.M. (1996). Images of community: discourse and strategy in property relations. *Development and Change* 27, 501–527.

MacCannell, D. (1973). Staged authenticity: arrangements of social space in tourist settings. *American Journal of Sociology* 79 (3), 589–603.

MacKenzie, J.M. (1988). *The Empire of Nature: hunting, conservation and British imperialism.* Manchester: University of Manchester Press.

Manokha, I. (2004). Modern slavery and fair trade products: buy one and set someone free. In C. van den Anker (ed.), *The Political Economy of New Slavery*, pp. 217–234. New York: Palgrave Macmillan.

McAfee, K. (1999). Selling nature to save it? Biodiversity and green developmentalism. *Environment and Planning D: Society and Space* 17, 133–154.

McCarthy, J. (2002). First World political ecology: lessons from the Wise Use movement. *Environment and Planning A* 34 (7), 1281–1302.

McCormick, J. (1989). *The Global Environmental Movement: reclaiming paradise.* London: Belhaven.

McDonald, M. and Wearing, S. (2003). Reconciling communities' expectations of ecotourism: initiating a planning and education strategy for the Avoca Beach rock platform. In B. Garrod and J.C. Wilson (eds), *Marine Ecotourism: issues and experiences*, Aspects of Tourism 7, pp. 155–170. Clevedon, Buffalo, Toronto, Sydney: Channel View Publications.

Meletis, Z.A. and Campbell, L.M. (2007) Call it consumption! Reconceptualizing ecotourism as consumption and consumptive. *Geography Compass* 1 (4), 850–870.

Miller, D. (ed.) (1995). *Acknowledging Consumption.* London and New York: Routledge.

Moore, D.S. (2000). The crucible of cultural politics: reworking 'development' in Zimbabwe's eastern highlands. *American Ethnologist* 26, 654–689.

Mowforth, M. and Munt, I. (1998). *Tourism and Sustainability: new tourism in the Third World.* London: Routledge.

Nelson, V. (2005). Representation and images of people, place and nature in Grenada's tourism. *Geografiska Annaler* 87 B (2), 131–143.

Neumann, R.P. (1996). Dukes, earls, and ersatz Edens: aristocratic nature preservationists in colonial Africa. *Environment and Planning D: Society and Space* 14, 79–98.

—— (1998). *Imposing Wilderness: struggles over livelihood and nature preservation in Africa.* Berkeley, CA: University of California Press.

—— (2005). *Making Political Ecology.* London: Hodder Arnold.

Norton, A. (1996). Experiencing nature: the reproduction of environmental discourse through safari tourism in East Africa. *Geoforum* 27 (3), 355–373.

Nygren, A. (1998). Environment as discourse: searching for sustainable development in Costa Rica. *Environmental Values* 7, 201–222.

—— (2004). Contested lands and incompatible images: the political ecology of struggles over resources in Nicaragua's Indio-Maís Reserve. *Society & Natural Resources* 17, 189–205.

Oates, J.F. (1999). *Myth and Reality in the Rain Forest: how conservation strategies are failing in West Africa.* Berkeley, CA: University of California Press.

Olwig, K.F. (2004). Comments on West and Carrier. *Current Anthropology* 45 (4), 491–492.

Paulson, S., Gezon, L.L., and Watts, M. (2003). Locating the political in political ecology: an introduction. *Human Organization* 62 (3), 205–217.

Peet, R. and Watts, M. (1993). Introduction: development theory and environment in an age of market triumphalism. *Economic Geography* 69 (3), 227–253.

—— and —— (eds) (1996). *Liberation Ecologies: environment, development, social movements*. London: Routledge.

Proctor, J.D. (1998). The social construction of nature: relativist accusations, pragmatist and critical realist responses. *Annals of the Association of American Geographers* 88 (3), 352–376.

Redford, K.H., Robinson, J.G., and Adams, W.M. (2006). Parks as Shibboleths. *Conservation Biology* 20 (1), 1–2.

Robbins, P. (2002). Obstacles to a First World political ecology? Looking near without looking up. *Environment and Planning A* 34 (8), 1509–1513.

—— (2004). *Political Ecology*. Malden, MA: Blackwell.

Roe, E. (1991). Development narratives, or making the best of blueprint development. *World Development* 19 (4), 287–300.

Ross, S. and Wall, G. (1999). Ecotourism: towards congruence between theory and practice. *Tourism Management* 20, 123–132.

Ryan, C., Hughes, K., and Chirgwin, S. (2000). The gaze, spectacle and ecotourism. *Annals of Tourism Research* 27 (1), 148–163.

Schelhas, J.W. (2001). The USA national parks in international perspective: have we learned the wrong lesson? *Environmental Conservation* 28 (4), 300–304.

Scheyvens, R. (1999). Ecotourism and the empowerment of local communities. *Tourism Management* 20, 245–249.

Soulé, M.E. and Lease, G. (1995). *Reinventing Nature? Responses to postmodern deconstruction*. Washington, DC: Island Press.

Spence, M.D. (1999). *Dispossessing the Wilderness: Indian removal and the making of the national parks*. New York: Oxford University Press.

Stonich, S.C. (1998). Political ecology of tourism. *Annals of Tourism Research* 25 (1), 25–54.

—— (2003). The political ecology of marine protected areas: the case of the Bay Islands. In S. Gössling (ed.), *Tourism and Development in Tropical Islands: political ecology perspectives*, pp. 121–147. Cheltenham, UK and Northampton, MA, USA: Edward Elgar.

Stronza, A. (2001). Anthropology of tourism: forging new ground for ecotourism and other alternatives. *Annual Review of Anthropology* 30, 261–283.

—— (2004). Comments on West and Carrier. *Current Anthropology* 45 (4), 492–493.

Sundberg, J. (2003). Strategies for authenticity and space in the Maya Biosphere Reserve, Petén, Guatemala. In K.S. Zimmerer and T.J. Bassett (eds), *Political Ecology: an integrative approach to geography and environment-development studies*, pp. 50–69. New York: The Guilford Press.

Task Force on Economic Benefits of Protected Areas of the World Commission on Protected Areas (WCPA) of IUCN (1998). *Economic Values of Protected Areas: guidelines for protected area managers*. Gland, Switzerland and Cambridge, UK: IUCN.

Terborgh, J.W. (1999). *Requiem for Nature*. Washington, DC: Island Press.

The International Ecotourism Society (TIES) (2005). *Ecotourism Fact Sheet*. Retrieved 27 October 2006 from http://206.161.82.194/WebModules/WebArticlesNet/articlefiles/15-NEW%20Ecotourism%20 Factsheet%20Sept%2005.pdf.

The International Ecotourism Society (TIES) (no date). Home page [Electronic Version]. Retrieved 27 October 2006 from www.ecotourism.org/webmodules/webarticlesnet/templates/eco_template. aspx?articleid=95&zoneid=2.

Thrupp, L.A. (1990). Environmental initiatives in Costa Rica: a political ecology perspective. *Society & Natural Resources* 3, 243–256.

Tremblay, P. (2001). Wildlife tourism consumption: consumptive or non-consumptive. *International Journal of Tourism Research* 3, 81–86.

Troëng, S. and Drews, C. (2004). *Money Talks: economic aspects of marine turtle use and conservation*. Gland, Switzerland: WWF-International.

Urry, J. (1995). *Consuming Places*. New York: Routledge.

—— (2002). *The Tourist Gaze* (2nd edn). London, Thousand Oaks, New Delhi: SAGE Publications.

Vivanco, L.A. (2001). Spectacular quetzals, ecotourism, and environmental futures in Monte Verde, Costa Rica. *Ethnology* 40 (2), 79–92.

Walker, P.A. (2005). Political ecology: where is the ecology? *Progress in Human Geography* 29 (1), 73–82.

Watts, M. (2000). Political ecology. In E. Sheppard and T. Barnes (eds), *A Companion to Economic Geography*, pp. 257–274. Malden, MA: Blackwell.

Wearing, S. and Neil, J. (1999). *Ecotourism: impacts, potentials, and possibilities*. Oxford: Butterworth Heinemann.

Weaver, D.B. (1998). *Ecotourism in the Less Developed World*. New York: CAB International.

—— (1999). Magnitude of ecotourism in Costa Rica and Kenya. *Annals of Tourism Research* 26 (4), 792–816.

—— (2002). The evolving concept of ecotourism and its potential impacts. *International Journal of Sustainable Development* 5 (3), 251–264.

Wesche, R. (1996). Developed country environmentalism and indigenous community controlled ecotourism in the Ecuadorian Amazon. *Geographische Zeitschrift* 84, 157–168.

West, P. and Carrier, J.G. (2004). Ecotourism and authenticity: getting away from it all? *Current Anthropology* 45 (4), 483–498.

—— and Brechin, S.R. (eds) (1991). *Resident People and National Parks: social dilemmas and strategies in international conservation*. Tucson, AZ: University of Arizona Press.

——, Fortwangler, C.L., Agbo, V., Simsik, M., and Sokpon, N. (2003). The political economy of ecotourism: Pendjari National Park and ecotourism concentration in Northern Benin. In S.R. Brechin, P.R. Wilshusen, C.L. Fortwangler, and P.C. West (eds), *Contested Nature*, pp. 103–115. Albany, NY: State University of New York Press.

Western, D. and Wright, M.A. (eds) (1994). *Natural Connections: perspectives in community-based conservation*. Washington, DC: Island Press.

Whelan, T. (1991). Ecotourism and its role in sustainable development. In T. Whelan (ed.), *Nature Tourism: managing for the environment*, pp. 3–22. Washington, DC: Island Press.

Willems-Braun, B. (1997). Buried epistemologies: the politics of nature in (post)colonial British Colombia. *Annals of the Association of American Geographers* 87 (1), 3–31.

Wilshusen, P.R. (2003). Exploring the political contours of conservation: a conceptual view of power in practice. In S.R. Brechin, P.R. Wilshusen, C.L. Fortwangler, and P.C. West (eds), *Contested Nature*, pp. 41–57. Albany, NY: State University of New York Press.

——, Brechin, S.R., Fortwangler, C.L., and West, P.C. (2002). Reinventing a square wheel: critique of a resurgent 'protection paradigm' in international biodiversity conservation. *Society & Natural Resources* 15 (1), 17–40.

Wilson, C. and Tisdell, C. (2001). Sea turtles as a non-consumptive tourism resource especially in Australia. *Tourism Management* 22, 279–288.

Wunder, S. (2003). Native tourism, natural forests and local incomes in Ilha Grande, Brazil. In S. Gössling (ed.), *Tourism and Development in Tropical Islands: political ecology perspectives*, pp. 148–177. Cheltenham, UK and Northampton, MA: Edward Elgar Publishing.

Young, E.H. (1999). Balancing conservation with development in small-scale fisheries: is ecotourism an empty promise? *Human Ecology* 27 (4), 581–620.

Ziffer, K. (1989). *Ecotourism. An uneasy alliance*. Washington, DC: Conservation International.

Zimmerer, K.S. and Bassett, T.J. (2003). Approaching political ecology: society, nature, and scale in human-environment studies. In K.S. Zimmerer and T.J. Bassett (eds), *Political Ecology: an integrative approach to geography and environment-development studies*, pp. 1–28. New York: The Guilford Press.

Chapter 12
Summary and synthesis:
observations and reflections on parks and protected areas in a changing world

Douglas A. Clark, Kevin S. Hanna, and D. Scott Slocombe

This volume provides a judicious and critical examination of the rapidly evolving field of protected areas management, based on an assessment of recent experiences worldwide. Such an assessment needs to provide ideas that enhance the resilience of the discourse on parks and protected areas – and, ultimately, the resilience of the social-ecological systems they aim to protect – and ensure that this dialogue continues to work towards serving legitimate human and ecological needs. How well we have achieved this goal is evident in the extent to which the chapters contribute new insights into contemporary challenges of park management. In the first chapter we argued that parks and protected areas face two significant challenges. One is the growing complexity of managing them. Protected area managers today are facing qualitatively different problems from those that such institutions were originally designed to address.

The second challenge is the increasingly polarized 'people versus nature' discourse about exactly what the appropriate priorities for protected areas should be. In a sense this discourse is expressed primarily through exchanges in the academic literature, but it reflects real-world debates too. Such literature also often underpins and indeed justifies the decisions made on the ground; so these ideas can have substantive consequences for ecosystems and peoples' livelihoods. While the polarized nature of this discourse could be broadly interpreted as a protected area management trend in its own right, in this book we have chosen to focus on aspects of the complexity facing parks and protected areas. By emphasizing substantive developments on the ground – rather than academic debates – the chapters in this volume are well placed to contribute different and constructive ways to move beyond it.

The preceding chapters are divided into two main groups. The first group examined recent experiences and trends in the evolving nature of governance for protected areas. The second group presented an array of critical perspectives and insights into current and conventional approaches to park and protected area management. The chapters in each part reveal much about these important topics, and amply describe the complexity and multidimensional character of protected area management and policy in the early years of the twenty-first century.

Part I: The challenges of governance

Relationships between societies and protected areas are changing in important ways, and societies are not fixed constants either. As a society changes, so will the functions and services demanded from protected areas. Governance models are proliferating in various forms, a theme detailed by Eagles (Chapter 3) but apparent in many other chapters as well. The state management model is increasingly less dominant, yet, as Pollock *et al.* point out in Chapter 6, questions of power relationships remain, even though the relationships of interest may be new ones. That chapter highlights an important area for ongoing attention; the growing role of private institutions in protected area management, and the types of tensions that can arise with their involvement in what are still often thought of as public policy processes.

A polycentric approach may be necessary to function within a dynamic network of participants (McGinnis, 1999); which moves us beyond simplistic, dichotomous characterizations of the park governance question as 'people versus biodiversity' or 'top-down versus bottom-up' (Chapters 7, 8, and 11). Instead, polycentrism leads toward a more complex, thoughtful, and responsive dialogue on the role of parks and protected areas in the sustainability of natural and human systems, and the place of such areas in the very real and messy context of global environmental and political change. For example, ICDPs as a strategy were not set up to allow research, so little formal learning and adaptation occurred (Chapters 5 and 8). Overcoming individual and organization attitudes and values is a challenge, and without explicit learning strategies and approaches there is little reason to expect 'rational' management approaches to inherently catalyse such change.

Accordingly, we might need to temper our optimism about prospects for the adoption of approaches such as adaptive management or resilience management; an outcome consistent with Walters' (1997) pessimistic assessment of the implementation of adaptive management. Following from this, current lines of thinking in ecosystem management such as resilience theory, adaptive governance, and adaptive co-management all suggest that deliberate policy interventions might be self-defeating, and that we need to allow (and support) social-ecological systems to self-organize to respond to changes. This approach represents a major paradigm shift for protected areas and will be hard for many protected area systems to implement. It is also not clear what implications polycentric governance in its various forms may have for the science-management relationship (Chapter 9), or for research activities in protected areas.

Environmental change often results from new forms of social-ecological interaction, many of which might not be well understood. Change also creates new complex settings and conditions (Chapters 2, 4, 6, and 8). But there might be ways of learning from others' experiences, and developing templates for recognizing issues and addressing them – templates with broad application to a range of biophysical, cultural, and political contexts. Biosphere reserves seem to be a productive 'laboratory' for the testing of such new templates (Chapter 6). Less optimistically, a potentially important caveat must be observed with respect to learning and adaptation. Much of the discussion of the 'new paradigm' (Beresford and Philips, 2000) and polycentric governance for protected areas (and other forms of conservation) remains dependent on the assumption of a stable civil society that is capable of engagement with governments. This assumption is tenuous in many parts of the world, especially places experiencing severe socio-economic crises and those with a history of coercive state-driven conservation. Francis (Chapter 2) observes that it could become even more tenuous over time in other regions as well. Building some measures of robustness into protected area systems is a topic that should be given greater attention.

The challenges to governance practices are many. Simply defining conservation territories and setting boundaries does not suffice any more – if indeed it ever did (Zimmerer, 2000). Protected areas need to be managed in a way that lets them cope with change and complexity. Protected area managers need to be able to rethink their goals on an ongoing basis and thus require capacity to engage in 'double-loop learning' (Argyris and Schön, 1978). Neufeld's example of the non-story of the Chilkoot Trail to Aboriginal people (Chapter 10), for example, shows the depth that such re-examination can require. Though important, this kind of social learning process can pose considerable risk to established institutional interests, and is rarely easy (Diduck et al., 2005). Campbell et al.'s questions about the dominant ecotourism narrative (Chapter 11) clearly identify reasons why some participants benefiting from it might be unwilling – or even unable – to question it.

Part II: Critical perspectives

Chapters in this section explore and extend the established criticisms of park and protected area management in several useful ways. First, they question comfortable myths about how parks come into being, and the ongoing social and cultural costs of those creations. Second, insights into the behaviour of protected area management institutions are presented that can be used strategically to understand and enhance decision-making processes; not just in the cases at hand, but, given institutional commonalities, likely elsewhere as well. Third, the chapters offer observations and concepts aimed towards reconciling the normative and ethical gaps between idealized goals and complex, unfolding realities.

The role of information is much more than just instrumental decision support (Chapters 2, 4, 5, and 9), something long recognized in the policy analysis world (e.g. Healy and Ascher, 1995), but which seems to be a perpetual surprise to the scientific community. The role of science in many dimensions of parks and protected areas is problematic but, nevertheless, important (Chapters 8 and 9). It can also be that political and social forces might be more important in protected areas designation and even day-to-day management; which might be a difficult bit of knowledge for conservation scientists to accept. There is a clear need to examine the narratives about the role of science (largely biological) in protected areas, and understand that science is value laden, rarely neutral, and often implicitly, or explicitly, linked to established economic and social forces. But far from discarding science, we need to also understand that there are disciplines quite different from those usually concerned with protected areas, such as anthropology, that offer unique approaches and tools for integration of information from different sources (e.g. Chapter 5), and in many settings these 'new' realms are gaining currency in the parks and protected areas discourse. Increased disciplinary breadth in protected areas research is not only productive and efficient, it is now requisite.

We must also expect conflict, but conflict is not always bad (Chapters 7, 10, and 11). New designations, new and responsive approaches to management and new macro policy settings can emerge for, and from, conflictual settings. What is important is that protected areas management learn and develop better methods for addressing conflict within planning and management processes. Narratives are important too, and reflect different understandings of a place or situation, but they can be both a barrier to, and an enabler of, learning and adaptation (Chapters 5, 6, 8, 9, and 11). Understanding narratives might be the key to advancing change, but there is also a strategic need to approach such tactical problems carefully, and with much humility. Attention must be paid to human dimensions of problems, a task requiring skills and techniques that are often new to parks and protected areas managers and the institutional environments they usually work within.

There is a clear need to expand the range of subjects of, and contributions to, parks and protected areas management. There are numerous topics that have received little attention, but deserve more: e.g. critical analysis of park policies, the role of science in management processes, the politics of park establishment, and the role of parks in development and local contexts. There is a range of newer approaches that have gained currency in broader resource and environmental management that could usefully be applied more to parks and protected areas, e.g. political ecology. And there is a range of variously considered actors, understandings, and approaches that might be better integrated, e.g. natural science, social science, and non-science; experts, managers, and local people; risk perception, resource management and conservation, and visitor management. Overall, a constructively critical eye, multiple perspectives, and new, complexity-derived understandings could facilitate much useful change and development in parks and protected areas management.

Climate change and protected areas: an elephant in the room?

Climate change is emerging as a problem of global scope, posing particular challenges for conservation in protected areas (e.g. Welch, 2005). Projected effects on ecosystems and societies could render superfluous most notions of conservation to date (Schellnhuber, 2006). While this volume does not specifically address climate change's impacts on protected areas (see e.g. Scott and Suffling, 2000; Scott and Lemieux, 2005), its importance and recent prominence demand some comments in the context of this discussion. Climate change could challenge the current integrity (however defined) of many cherished parks and protected areas, and, indeed, render them into something quite different from the landscapes and qualities that made them special enough to warrant protection. Such perceived crises might be irreversible in ecological terms (or practically so) yet demands for ameliorative action will be swift and loud. In such situations pragmatic, contextual adaptation would possess the 'virtue of necessity'; and the value of such an approach even before crises occur is beginning to gain currency (Brunner and Lynch, 2007). Interactions between climate change and other governance challenges are also a topic requiring attention.

Final reflections

What do the chapters tell us about the parks-versus-people debate? Campbell *et al.* (Chapter 11) address it directly, arguing the need for new narratives, and Francis puts emphasis on biodiversity protection in a longer-term perspective, suggesting that today's imperative might seem less urgent in as-yet unknown future circumstances (Chapter 2). One lesson to draw from those chapters is that protected areas are dynamic systems, not inexorably or uniformly moving towards 'perfection' – or desecration (Chapter 8). From such a perspective, parks can be viewed as ongoing experiments in society's relationship with nature. To some degree though, this experimentation is relatively blind, with no controls, few deliberate comparisons, and the results – for good or ill – becoming institutionalized and diffused.

How do these chapters and lessons enhance the resilience of that discourse? Primarily by bringing together different kinds of disciplinary knowledge to reveal new insights, also possibly by provoking some small disturbances by questioning the status quo, and also as those insights are picked up and implemented by protected area managers and co-managers. The approach here has been to move beyond management prescriptions, and, instead, to highlight fundamental challenges in parks and protected area conceptualization, policy design, and applied management. Critical, difficult, and thorny topics are curiously absent from much of the parks literature, where, we would venture to say, there has been a tendency to focus on the minutiae of operations, management, and administration. While these are helpful, the parks and protected areas discourse has tended to avoid being too critical of policy processes and decision-makers; as has resource and environmental management generally.

The cases and settings discussed here are timely and will find relevance to a range of locales. In the Canadian and American instances – large, developed nation-states, yet not particularly homogeneous – protected areas face a range of regional issues, multifaceted jurisdictional approaches to parks and protected areas management, and inherent evolving tensions within and without management agencies. In the US, the ongoing struggle to maintain and expand the parks system at all levels faces complex social, political, and economic pressures coupled to the struggle between the parks of the American imagination, and the quiet subterfuge of recent political ideological shifts that have profoundly affected all facets of American life. Within Canada there is a dichotomy between the rhetoric and image of the pristine Canadian environment and the reality of an enduring (and arguably ongoing) legacy of weak federal environmental actions. Added to this are inconsistent policies across Canadian provinces and American states, where some have expanded and strengthened their protected areas systems, while others see them as a burden and barrier to development. Globally, parks and protected areas face common challenges, conflict in its many forms (war, political instability, the clash of ideas about who parks should serve), poor resourcing, the impacts of peripheral activities, the demands of growing or struggling economies, and a rising clamour for different – indeed, more inclusive – management approaches and an end to fortress parks.

Transforming parks and protected areas is probably necessary, in at least some contexts. In some cases, that might mean redefining boundaries, functions, and activities and infrastructure within them. More often, though, it will probably mean new policies and more diverse, and often more complex, governance arrangements. The changing world we consider includes biophysical change, and socio-economic, political, and cultural change. We believe the chapters in this book have illustrated the nature of at least some of the change we have seen and can anticipate. Most of all, we hope this book can contribute to at least a few adaptive and anticipatory transformations, instead of merely reactive ones.

Literature cited

Argyris, C. and Schön, D.A. (1978). *Organizational Learning: a theory of action perspective*. Reading, MA: Addison-Wesley.

Beresford, M. and Philips, A. (2000). Protected landscapes: a conservation model for the 21st century. *The George Wright Forum* 17, 15–26.

Brunner, R.D. and Lynch, A.H. (2007). Context and climate change: an integrated assessment for Barrow, Alaska. *Climate Change* 83 (1/2): 93–111.

Diduck, A., Bankes, N., Clark, D., and Armitage, D. (2005). Unpacking social learning in social-ecological systems in the north. In F. Berkes, H. Fast, M. Manseau, and A. Diduck (eds) (2005). *Breaking Ice: renewable resource and ocean management in the Canadian north*, pp. 269–290. Calgary: University of Calgary Press/Arctic Institute of North America.

Healy, R.G. and Ascher, W. (1995). Knowledge in the policy process: incorporating new environmental information in natural resources policymaking. *Policy Sciences* 28, 1–19.

Jones, K.R. and Wills, J. (2005). *The Invention of the Park: from the Garden of Eden to Disney's Magic Kingdom*. Cambridge: Polity.

McGinnis, M.D. (ed.) (1999). *Polycentric Governance and Development*. Ann Arbor, MI: University of Michigan Press.

Schellnhuber, H.J. (ed.) (2006). *Avoiding Dangerous Climate Change*. Cambridge: Cambridge University Press.

Scott, D. and Suffling, R. (2000). *Climate Change and Canada's National Park System: a screening level assessment*. Waterloo: University of Waterloo.

—— and Lemieux, C. (2005). Climate change and protected areas planning in Canada. *The Forestry Chronicle*, Sept./Oct., 696–703.

Walters, C.J. (1997). Challenges in adaptive management of riparian and coastal ecosystems. *Conservation Ecology* 1 (2), 1. Available online at: www.consecol.org/vol1/iss2/art1.

Welch, D. (2005). What should protected areas managers do in the face of climate change? *The George Wright Forum* 22 (1), 75–93.

Zimmerer, K.S. (2000). The reworking of conservation geographies: nonequilibrium landscapes and nature-society hybrids. *Annals of the American Association of Geographers* 90 (2), 356–369.

Index

Note: page references in italics indicate illustrations.

abalone 159
Aboriginal cultural landscape 188, 189, 190–1
Aboriginal Ownership and Government Management Model 51–2, 51
Aboriginal peoples 5, 190, 198n8, 223; of Australia 51, 207; defined 51; ecological integrity policy and 157, 159; governments of 40, 51–2, 51, 124, 147; interests of 51, 140–1; land claims of 51, 139, 163; Native Americans 206, 216n7; place names 191; rights of 115; see also BaAka; Bilos; First Nations; Indigenous peoples; Indigenous peoples and protected heritage areas (PHAs) of Canada
Adams, W.M. 206, 208, 215
adaptive cycles 25–6
adaptive frameworks 86
adaptive management 76–7, 98–100, 155, 223
Adger, W.N. 202
advocacy groups 40, 140; see also environmental movement organizations (EMOs)
aesthetics of ecotourism 211, 216n11, 217n14
Africa: conservation narratives 208; Parastatal Model 43, 44, 45; South Africa 45, 46, 49, 56, 162, 164n5; Uganda 98, 155, 164n5; see also Dzanga-Sangha Dense Forest Reserve (RDS)
agriculture 92, 101, 116, 120, 126, 128
Alexander, Shelley M. 7, 62–84; Banff-Bow Valley Study (BBVS) 63–5; data issues 66–8; GIS and species-based protected area management 68–70; global protected areas information 70–3, 72, 73; monitoring and adaptive management 76–7; protected areas system assessment and design 73–6; Web-based GIS and the public 77–8
alternative consumption, ecotourism as 210–12, 213–14
American Association for the Advance of Science (AAAS) 171
animal/human dynamic 103–4
anthropology 7, 85–6; of protected area management 94–103; research at RDS 91–3
art and nature 183, 184, 206
Australia 51, 207

BaAka 88–9, 92–3, 96, 97, 98, 100–1, 103
Baja California, Mexico 203
Banff-Bow Valley Study (BBVS) 63–5
Banff National Park 160–1
Barber, C.V. 15
Barnaby, Joanne 196
Bassett, T.J. 203
Bayanga 89, 92, 96, 100
Bay Islands, Honduras 204
bears 146
Belize 56, 203
Belsky, J.M. 203–4
Bilos 88–9, 98, 100
BIMS (Biological Information Management System) 75
biodiversity 21–2; Indigenous and local communities and 22, 193–4; information technology and 68–70, 74; resilience and 26, 27
Biological Research Division (BRD) 178
BioRap 75

227

Index

Biosphere Reserve Integrated Monitoring Programme (BRIM) 76
biosphere reserves 8, 20, 21, 113, 114–16, 223; see also Canadian biosphere reserves
birds 123–4
Blaikie, P. 201
British Columbia (BC) 4, 5–6, 47, 48, 137, 138–9, 139, 150; Carmanah and Walbran valleys 139, 143–5, 147, 148; Clayoquot Sound 139, 141–3, 147, 148; Commission on Resources and Environment (CORE) 145, 147–8; 'Great Bear Rainforest' 139, 145–7; Haida Gwaii Archipelago 159, 160; Protected Areas Strategy (PAS) 148–9; Stein Valley Nlaka'pamux Heritage Park 139, 140–1
Brookfield, H. 201
Bruce Trail 118, 127
Bryant, R.L. 201, 210
Buffer Zone at RDS 92, 95, 97, 98
buffer zones, biosphere reserve 115
Bunaken National Park, Indonesia 2, 3–4, 6
Burton, T.L. 39, 40

Campbell, C. 112
Campbell, Lisa M. 8, 200–21; alternative consumption 210–12; conservation and development narratives 208–10; definitions of ecotourism 200; political ecology 201–3; political ecology and ecotourism 203–5, 212–15; social construction of nature 205–7
Canada 4, 5–6, 112, 225–6; information technology 63–5; management of parks and protected areas 43, 44–5, 47–9, 50, 51–2, 54, 55; see also British Columbia; Canadian biosphere reserves; ecological integrity policy in Canadian national parks; protected heritage areas (PHAs) in Canada
Canada National Parks Act (CNPA) 156–7, 158, 160, 161, 162, 163, 164n3
Canadian biosphere reserves 111, 113, 115; Niagara Escarpment 116–20, 117, 127, 128; Redberry Lake (Saskatchewan) 116, 123–6, 123, 127, 128; see also Oak Ridges Moraine (ORM) (Ontario)
capacity building 100–1, 112, 126, 128
capitalism 204, 210; global 19, 29, 30, 201, 212
Carcross-Tagish First Nation 186, 187
Cariboo-Chilcotin Land Use Plan (CCLUP) 5
Carmanah and Walbran valleys 139, 143–5, 147, 148
Carrier, J.G. 207, 211
Catton, Ted 183
Central African Republic (CAR) 7, 86; see also Dzanga-Sangha Dense Forest Reserve (RDS)

change, climate 16, 27–8, 225
change, global 15, 16, 26–8
changes, social and ecological 1, 161–2, 161, 223
Chape, S. 71
Chase-Dunn, C. 29
Chilkoot Trail National Historic Site 186–7
Chilkoot Trail Oral History Project 187
chimpanzees 90, 95, 98
China 49
Churn Creek Protected Area 4, 5–6
citizen groups see community groups
civil society 17, 223
Clark, Douglas A. 1–11, 154–68, 222–6; complexity of protected area management 2–4, 6; discourses of ecological integrity 154–5, 161–3, 161; Haida Gwaii Archipelago 159, 160; 'people versus nature' controversy 4, 6–7; Wapusk National Park 158–9, 159; Wood Buffalo National Park and Banff National Park 160–1
class 201, 206
Clayoquot Sound 139, 141–3, 147, 148
climate change 16, 27–8, 225
Coalition on the Niagara Escarpment (CONE) 119
colonialism 184, 185, 195, 205
Commission on Resources and Environment (CORE) 145, 147–8
committees, volunteer 124–6
communities, defined 110–11; see also local communities
community-based conservation 4, 209, 213
community groups 23, 118–19, 120, 126, 127, 128, 129; see also committees, volunteer
complexity of protected areas management 2–4, 6
complex systems 16, 23–5, 156; see also complex systems thinking; social-ecological systems (SESs)
complex systems thinking 24, 25, 28–32
conflict see land use conflicts
conservation, community-based 4, 209, 213
conservation and development narratives 208–10, 213, 224
conservation and development projects, integrated see Integrated Conservation and Development Projects (ICDPs)
conservation capacity 86
conservation movement 209
conservation-seeking culture 210, 211
constructionism, social see social construction of nature and ecotourism

consumption 216n9; see also alternative consumption, ecotourism as
Convention Concerning the Protection of the World Cultural and Natural Heritage 193
Convention for the Safeguarding of the Intangible Cultural Heritage 194
Convention on Biological Diversity (CBD) 20
coral reef biodiversity conservation 3
core areas, biosphere reserve 115
Core Dzanga Sector Park 91–2, 95, 97–8
corporations, for-profit see for-profit corporations
Costa Rica 46, 209, 211
Crater Lake National Park 173–4
Cronon, W. 206
cross-cultural and ecological keystones 86, 101–3, 102
cultural diversity 193, 194
cultural ecology 201, 202
cultural heritage, defined 198n19; see also protected heritage areas (PHAs) in Canada
cultural landscape, Aboriginal 188, 189, 190–1
cultural pluralism and governance 194–6
cultural reintegration 190
culture and state 184, 185, 192–4
cultures 182; ecotourism and local 216n10; Indigenous 186, 188

data issues in spatial information technology 66–8; see also information technology (IT)
decolonization 185
Denali National Park 183
developing countries, ecotourism in 202, 207; see also ecotourism
development and conservation narratives 208–10, 213, 224
diamonds 94
digital elevation model (DEM) 66
discourses of ecological integrity 154–5, 161–3, 161
discourse theory 202, 206, 214
discriminant function analysis (DFA) 70
donations 41, 43, 45–6, 45, 53, 54, 55, 125
Draper, D. 110
Ducks Unlimited Canada (DUC) 125, 126
duikers 90, 91, 98
Dzanga-Sangha Dense Forest Reserve (RDS) 86–104, 86, 88; adaptive management 98–100; capacity building 100–1; collaborative anthropological research 91–4, 92; cross-cultural keystones 101–3, 102; integrated conservation and development projects (ICDPs) 96–7; wildlife species 95–6; zoning 97–8

Eagles, Paul F.J. 7, 39–61; evaluations of management models for parks and protected areas 53–6; good governance criteria 56–8; management models for parks and protected areas 42–52
Earth System Science Partnership (ESSP) 27
Ecolodge Model 46–7, 46, 56
ecological integrity policy in Canadian national parks 8, 154–64, 158, 161; discourses of ecological integrity 154–5, 161–3, 161; Haida Gwaii Archipelago 159, 160; protected heritage areas and 189; Wapusk National Park 158–9, 159; Wood Buffalo National Park and Banff National Park 160–1
Ecological Integrity Statement for Kluane National Park and Reserve 189, 190–2
Ecosystem-Based Management (EBM) 147
ecosystem management 26, 137, 156, 190
ecosystem-pluralistic discourse 155, 161, 162, 163
ecotourism 8–9, 46, 56, 200–17; as alternative consumption 210–12; conservation and development narratives 208–10; defined 200; literature on 201; narrative 213, 214–15; political ecology and 203–5, 212–15; social construction of 205–7, 213; see also tourism
effectiveness 53, 57, 58
efficiency 53, 57, 58
elephants 164n5; Dzanga-Sangha Dense Forest Reserve (RDS) 87, 90, 91–2, 92, 93, 95, 98, 102
Ellsworth, J.P. 111, 112
emparkment 157, 162, 164
entrance fees 44, 47, 213; see also user fees
environmentalism 209
environmental movement organizations (EMOs) 116, 127, 128; British Columbia 140–1, 142, 144–5, 146, 148, 149–50; Niagara Escarpment 118, 119, 120; Oak Ridges Moraine 121–3; Redberry Lake 125–6
environmental movements 138
equity 53, 55, 57–8, 100
Escobar, A. 205
Euro-American cultural views of nature 183, 206–7
Europe, nation-states in 182–3
expenditures for protected areas 31

facade management 170
Fairhead, J. 208
Federal Wildlife Refuges 55
First Nations 146, 147, 182, 189, 196; Carcross-Tagish 186, 187; Lytton 140, 141; Mikisew

Index

Cree 162; Mt Currie 140; Nuu-chah-nulth 142; Southern Tutchone 190–2; Yukon 187; see also Indigenous peoples and protected heritage areas (PHAs) of Canada
fishing 3, 170, 211, 203
Fluker, Shaun 8, 154–68; discourses of ecological integrity 154–5, 161–3, 161; Haida Gwaii Archipelago 159, 160; Wapusk National Park 158–9, 159; Wood Buffalo National Park and Banff National Park 160–1
for-profit corporations 41; Ecolodge Model 46, 56; Parastatal Model 43, 44; Public and For-profit, Private Combination Model 47–50, 47; resource ownership 40, 53
fragmentation see institutional fragmentation
Francis, George 7, 15–38; biosphere reserve governance 115–16; complex open systems 23–5; complex systems thinking and globalization 28–31; institutional fragmentation 113, 127; Niagara Escarpment 117; overview perspective on governance 16–20; protected areas and governance challenge 20–3; regimes 114; social-ecological systems (SESs) 16, 25–8
Friends of Clayoquot Sound 142
F-TRAC (Florida Forever Tool for Efficient Resource Acquisition and Conservation) 75

Gales Point, Belize 203
gathering: Aboriginal people of Australia 51; Dzanga-Sangha Dense Forest Reserve (RDS) 87, 88, 91, 92, 95; effects of tourism and research on 100
geographic information systems (GIS) 7, 64, 65; data issues 66–8; monitoring programmes 76; NATURE-GIS 71, 73; species-based protected area management and 68–70; Web-based GIS and the public 77–8; see also information technology (IT)
georeferencing 67
global capitalism 19, 29, 30, 201
global change 15, 16, 26–8
globalization 28–31
Global Positioning Systems (GPS) 66, 77
global protected areas information 70–3, 72, 73
Global South 210
global warming 27; see also climate change
Glover, T.D. 39, 40, 57
Golden Era National Park Model 42–3, 43
gold rush 186, 187
Goodman, M.K. 210
gorillas 87, 89–90, 91–2, 92, 93, 95–6, 98, 102–3, 102
Gössling, S. 204

governance 7, 15, 20–3, 222–4; of biosphere reserves 115; criteria 56–8; cultural pluralism and 194–6; defined 17; landscape-scale 112–14, 127, 129; local communities and 110–11; overview of 16–20; protected heritage areas and 189; regimes 110–11, 114, 116, 117–18, 121, 126–8; steering 112, 118–19, 120, 121–2, 124–5, 126–7; see also Canadian biosphere reserves
governments 40, 41, 42, 111, 112; Aboriginal Ownership and Government Management Model 51–2, 51; British Columbia 138, 141, 142, 144–5, 146–7, 149; Canadian biosphere reserves 118–19, 124–6, 128; combination models 47–50, 47, 50; debates around role of 52; ecological integrity 156–7, 160; Golden Era National Park Model 42–3, 43; Oak Ridges Moraine 122; protected heritage areas (PHAs) 187, 188, 196; sources of income 54; see also Parastatal Model
Graham, J. 39–40, 42, 56–7
graph theory 76
Gray, John 184, 193, 194, 195, 196
Gray, Noella J. 8, 200–21; alternative consumption 210–12; conservation and development narratives 208–10; definitions of ecotourism 200; political ecology 201–3; political ecology and ecotourism 203–5, 212–15; social construction of nature 205–7
'Great Bear Rainforest' 139, 145–7
Greenpeace 142, 146
Gwaii Haanas National Park 51
Gwichya Gwich'in 188–9

Haida Gwaii Archipelago 159, 160
Hall, T.D. 29
Hanna, Kevin S. 1–11, 137–53, 222–6; Carmanah and Walbran valleys 139, 143–5; Churn Creek Protected Area 4, 5–6; Clayoquot Sound 139, 141–3; Commission on Resources and Environment (CORE) 145, 147–8; complexity of protected areas management 2–4; 'Great Bear Rainforest' 139, 145–7; 'people versus nature' controversy 4, 6–7; Protected Areas Strategy (PAS) 148–9; Stein Valley Nlaka'pamux Heritage Park 139, 140–1
Hannah, L. 58
Hardin, Rebecca 7, 85–109; adaptive management 98–100; capacity building 100–1; cross-cultural keystones 101–3, 102; collaborative anthropological research at RDS 91–4, 92; Dzanga-Sangha Dense

Forest Reserve (RDS) 87–90; integrated conservation and development projects 96–7; zoning 97–8
Heimstra, Mary 181
Heyman, J.M. 212
Historic Sites and Monuments Board of Canada (HSMBC) 185, 188, 195, 197n2
history of protected areas 31, 182
Honduras 204
Hulme, D. 206, 208, 215
human/wildlife interactions 103–4, 170, 176, 208; see also people-versus-nature debate
hunting 87–8, 89, 90, 91, 95–6, 98, 100; Aboriginal people of Australia 51; BaAka camps 92–3; effects of tourism and research on 100; safari 87, 93, 94, 99; snow geese 158–9, 159
hybrid organizations 17
Hydén, Göran 114

identity: of Aboriginal peoples 157; cross-cultural keystones and 101; nation-states and 192, 195, 196; protected heritage areas (PHAs) and 182, 187, 189
imperialism 184, 185, 195, 207
income for parks and protected areas, sources of 41, 54–6
Indians of Canada Pavilion 188, 197n5
indigenous knowledge 100; see also traditional knowledge
Indigenous peoples 4, 8, 193–4, 195–6; socially constructed idea of 207; see also Aboriginal Ownership and Government Management Model; Aboriginal peoples; BaAka; Bilos; First Nations; Indigenous peoples and protected heritage areas (PHAs) of Canada; Native Americans
Indigenous peoples and protected heritage areas (PHAs) of Canada 182, 183, 184–5, 195–6; contacts between Parks Canada and Indigenous peoples 185–92
individuals, rights of 192–3
Indonesia 2, 3–4
information management 62, 75, 78, 99, 174
information technology (IT) 7, 62–78; Banff-Bow Valley Study (BBVS) 63–5; data issues 66–8; GIS and species-based protected area management 68–70; global protected areas information 70–3, 72, 73; monitoring and adaptive management 76–7; protected areas system assessment and design 73–6; Web-based GIS and the public 77–8
institutional fragmentation 111, 113, 126, 127, 128, 129

Integrated Conservation and Development Projects (ICDPs) 4, 31, 86, 104n3, 223; Dzanga-Sangha Dense Forest Reserve (RDS) 94, 96–7, 99; socio-ecological integrity and 163
Interactive Map Center (IMC) 77
International Convention on Biodiversity 193–4
International Covenant on Economic, Social and Cultural Rights 193
International Ecotourism Society, The (TIES) 200
International Union for the Conservation of Nature and Natural Resources (IUCN) 4, 9n1, 15; ecotourism 200; governance 18, 20–1, 22–3; information technology 62
inventory of protected areas, global 20, 71, 72
Isle Royale National Park 176–7

Jamaica 211
Jamison, A. 129
Jessop, B. 17, 19, 20, 112, 114
Jojkic, Dushan 8, 137–53; Carmanah and Walbran valleys 139, 143–5; Clayoquot Sound 139, 141–3; Commission on Resources and Environment (CORE) 145, 147–8; 'Great Bear Rainforest' 139, 145–7; Protected Areas Strategy (PAS) 148–9; Stein Valley Nlaka'pamux Heritage Park 139, 140–1
Jones-Walters, L. 111, 112
judicial decisions 160–1, 162, 170

Kluane National Park and Reserve 51, 189, 190–2
Knight, B. 111
knowledge: anthropology of 100; naturalized 189, 198n8; political ecology and 202; traditional 188, 190–1, 195; see also science
Kongana 92, 93
Kreutzwiser, R. 111
Kruger National Park 46, 56, 162, 164n5
Kutas, Brian 8, 137–53; Carmanah and Walbran valleys 139, 143–5; Clayoquot Sound 139, 141–3; Commission on Resources and Environment (CORE) 145, 147–8; 'Great Bear Rainforest' 139, 145–7; Protected Areas Strategy (PAS) 148–9; Stein Valley Nlaka'pamux Heritage Park 139, 140–1

land and resource management planning (LRMP) 146, 147, 148, 149
landscape connectivity 76
landscape-scale governance 112–14, 127, 129; see also Canadian biosphere reserves

land use conflicts 8, 137–8, 224; Carmanah and Walbran valleys 139, 143–5; Clayoquot Sound 139, 141–3; Commission on Resources and Environment (CORE) 145, 147–8; four stages of 149–50; 'Great Bear Rainforest' 139, 145–7; Protected Areas Strategy (PAS) 148–9; Stein Valley Nlaka'pamux Heritage Park 139, 140–1
Laurentian thesis 184, 186, 195
Leach, M. 208
legislation 115; Canada National Parks Act (CNPA) 156–7, 158, 160, 161, 162, 163, 164n3; Niagara Escarpment 118, 119; Oak Ridges Moraine 121, 122; Redberry Lake 123, 124; US national parks 169–70, 174
Leopold, A. Starker 171
linear regression 69
local communities: biodiversity and 22, 193; biosphere reserves and 8, 114; capacity building and 100–1, 112; community-based conservation 4, 209; conflict in British Columbia 142; conflicts between conservationists and 100; ecological integrity and 161; governance and 110–11, 113; Integrated Conservation and Development Projects (ICDPs) and 31; tourism and 204; see also Aboriginal peoples; BaAka; Bilos; Canadian biosphere reserves; community groups; Indigenous peoples; local people and ecotourism
local people and ecotourism 200, 201, 206–7, 213, 214, 216nn3, 8; alternative consumption and 211; conservation and development narratives and 208, 209; Gales Point, Belize 203
Loë, R. de 111
logging: in British Columbia 138–9, 142–3, 144–5, 146; in Dzanga-Sangha Dense Forest Reserve (RDS) 87, 89, 90, 92, 94, 97, 98
logistic regression 69
Louis, C. 119
Lytton First Nations 140, 141

McCarthy, J. 30
McKinney, C. 114
Macleod, D.V.L. 207
McMillan Bloedel (MB) 142–3, 144
McNeely, Jeffrey A. 15
Madikwe Game Reserve 49
Mahalanobis distance 70
management, national parks see science/management interface in national parks

management models for parks and protected areas 7, 41; Aboriginal Ownership and Government Management Model 51–2, 51; debates around 52–3; Ecolodge Model 46–7, 46; evaluations of 53–6; Golden Era National Park Model 42–3, 43; Non-profit Organization (NPO) Model 45–6, 45; Parastatal Model 43–5, 43; Public and For-profit, Private Combination Model 47–50, 47; Public and Non-profit, Private Combination Model 50; see also governance
managers, park see park managers and scientists, differences between
Manuel-Navarette, David 154
marine conservation 2, 3–4, 159, 211
marine protected areas (MPAs) 204
Meletis, Zoë A. 8, 200–21; alternative consumption 210–12; conservation and development narratives 208–10; definitions of ecotourism 200; political ecology 201–3; political ecology and ecotourism 203–5, 212–15; social construction of nature 205–7
meta-governance 18, 114
Mexico 203
MICOSYS (Minimum Conservation System) 74–5
Mikisew Cree First Nation 162
Millennium Development Goals (MDGs) 19, 20, 71, 163, 164n1
models of governance see management models for parks and protected areas
monitoring activities 76–7, 96, 119, 122, 128
Montego Bay, Jamaica 211
Monteverde Cloud Forest Reserve 46
Moore, D.S. 214
moose 176
moralization of consumption 210–11
More, T. 39, 40
Mowforth, M. 207
Mt Currie First Nations 140
multiculturalism 189, 195, 196
multivariate linear regression 69
Munt, I. 207

Nagwichoonjik National Historic Site 188
narrative, ecotourism 213, 214–15
narratives, conservation and development 208–10, 213, 224
National Biological Survey (NBS) 177–8
national parks, North American 183–4; see also ecological integrity policy in Canadian national parks; protected heritage areas (PHAs) in Canada; science/management interface in national parks; United States National Park Service (USNPS)

Index

nation-states 182–5, 192, 193, 194–6
Native Americans 206, 216n7
natural ecological integrity 161; see also wilderness-normative discourse
natural heritage, defined 198n10; see also protected heritage areas (PHAs) in Canada
Natural Resources Management Project (NRM), USAID 3
nature: commodification of 212; Euro-American cultural views of 183, 206–7; social construction of 205–7, 213
Nature Conservancy of Canada (NCC) 125, 126, 129n4
NATURE-GIS 71, 73
negotiation 149–50
Negrave, Roderick W. 8, 137–53; Carmanah and Walbran valleys 139, 143–5; Clayoquot Sound 139, 141–3; Commission on Resources and Environment (CORE) 145, 147–8; 'Great Bear Rainforest' 139, 145–7; Protected Areas Strategy (PAS) 148–9; Stein Valley Nlaka'pamux Heritage Park 139, 140–1
neoliberalism 4, 18–20, 207, 212, 213
neo-Marxist approach to political ecology 201
networks, ecological 22
networks, governance 18
Neufeld, David 8, 181–99; constructing the nation-state in North America 182–5; contacts between Indigenous peoples and Parks Canada 185–92; governance and cultural pluralism 194–6; state and culture after the Second World War 192–4
Neumann, R.P. 205, 206
Newman, Candace 3–4
new paradigm for protected areas 1, 4, 9, 21, 22, 156, 223
Niagara Escarpment (Ontario) 116–20, 117, 127, 128
Niagara Parks Commission, The 44–5
non-governmental organizations (NGOs) 17
non-profit institutions 40, 41, 45–6, 45, 55, 50, 53
Non-profit Organization (NPO) Model 45–6, 45, 55
Nuu-chah-nulth First Nations 142
Nygren, A. 209

Oak Ridges Moraine (ORM) (Ontario) 116, 117, 120–3, 120, 127, 128
Ontario 47–9, 50, 54, 55; see also Niagara Escarpment (Ontario); Oak Ridges Moraine (Ontario)
Ontario Conservation Authority 45

open systems 111, 113, 126, 128, 129; complex 23–5
Organic Act 169–70
otters 159
ownership of resources 40–1, 40, 53–6

Palliser Regional Park 182
Pan-European Ecological Network 22
Parastatal Model 41, 43–5, 43
Park Friends' Groups 50
park managers and scientists, differences between 172–3, 174–5; see also science/management interface in national parks
parks, defined 9n1
Parks Canada Agency (PCA) 51, 156, 163, 164n2, 185–92, 195
parks versus people debate 215, 225; see also people versus nature debate
park wardens 2
Paulson, S. 202
peace 192
peace parks 22, 66
Peet, R. 202
people-versus-nature debate 4, 6–7, 112, 222; constructing nation-states and 183; ecological integrity and 154–6, 160, 161–2; Integrated Conservation and Development Projects (ICDPs) and 31; people versus parks 215, 225; US national parks 172; see also Integrated Conservation and Development Projects (ICDPs); political ecology
Phillips, A. 22
place names, aboriginal 191
Plaunt, M. 118
policy, protected area 137–8, 147–9, 150; see also ecological integrity policy in Canadian national parks
policy, wildlife management 171
policy monitoring 119, 122; see also monitoring activities
political ecology 8–9, 201–3; alternative consumption and 210–12; conservation and development narratives 208–10; defined 201; ecotourism and 203–5, 212–15; social constructionism and 205–7
politics of establishing protected areas 138, 224; see also land use conflicts
Pollock, Rebecca M. 8, 110–33; Niagara Escarpment (Ontario) 116–20, 117; Oak Ridges Moraine (Ontario) 120–3, 120; Redberry Lake (Saskatchewan) 123–6, 123
polycentric governance 223
poststructural political ecology 202, 206, 208

233

Index

power: of consumers 210, 212; culture and 185, 194–5; local communities and 112; of place 138, 150; political ecology and 202; science as 100; state 182
predator control 170
principle components analysis (PCA) 70
progress, belief in 184, 185, 186, 194
protected area designation 137, 138, 140, 149, 150
protected areas, defined 9n1
protected areas management, growing complexity of 2–4, 6
Protected Areas Strategy (PAS), British Columbia 148–9
protected areas system assessment and design 73–6
protected heritage areas (PHAs) in Canada 181–92, 195–6; constructing the nation-state in North America 182–5; contacts between Indigenous peoples and Parks Canada 185–92
protectionist paradigm 4, 215
protest movements 138, 140, 142, 144–5, 146, 148, 150
Prudham, S. 30
Public and For-profit, Private Combination Model 47–50, 47
Public and Non-profit, Private Combination Model 50
public goods 52
public-private partnerships 52–3
purpose of parks 9n1, 172
purpose of protected areas 9n1, 42
purpose of United States National Park Service (USNPS) 169–70

Quinn, Michael S. 7, 62–84; Banff-Bow Valley Study (BBVS) 63–5; data issues 66–8; GIS and species-based protected area management 68–70; global protected areas information 70–3, 72, 73; monitoring and adaptive management 76–7; protected areas system assessment and design 73–6; Web-based GIS and the public 77–8

Rapport, D.J. 113
Raz, J. 198n12
reason 183, 184
Redberry Lake (Saskatchewan) 116, 123–6, 123, 127, 128
Redford, K.H. 6, 215
Reed, Maureen G. 8, 110–33; Niagara Escarpment (Ontario) 116–20, 117; Oak Ridges Moraine (Ontario) 120–3, 120; Redberry Lake (Saskatchewan) 123–6, 123

regime formation 111, 116, 129
regimes, defined 114
regimes, governance 110–11, 114, 116, 117–18, 121, 124, 126–8
religion and nature 184
Remis, Melissa J. 7, 85–109; adaptive management 98–100; capacity building 100–1; collaborative anthropological research at RDS 91–4, 92; cross-cultural keystones 101–3, 102; Dzanga-Sangha Dense Forest Reserve (RDS) 87–90; integrated conservation and development projects 96–7; zoning 97–8
Remote Buffer Zone at RDS 93–4
Remote Park Sector at RDS 93
remote sensing (RS) 66, 67, 68–9
Renan, Ernest 182
research 91, 94, 97, 99–100, 126, 128, 129; US National Park Service and 171–9, 176
reserves, dynamic, ephemeral and mid-succession 27
resilience 6–7, 26–8, 161, 222, 223, 225
resource ownership 40–1, 40, 53–6
rights, aboriginal 115
rights, human 192–3
Risby, Lee 8, 154–68; discourses of ecological integrity 154–5, 161–3, 161; Haida Gwaii Archipelago 159, 160; Wapusk National Park 158–9, 159; Wood Buffalo National Park and Banff National Park 160–1
Robbins, P. 206
Robbins, W.J. 171
Roe, E. 208
Root, T.L 27
Royal Society for the Protection of Birds (RSPB) 46
Runte, Alfred 182

Sabi Sabi Private Game Reserve 46
safari hunting 87, 93, 94, 99; see also hunting
Said, Edward 184
Saskatchewan 116, 123–6, 123, 181, 182
Save the Oak Ridges Moraine (STORM) Coalition 122, 127
science 17, 100, 128, 163, 198n8, 202, 224; see also science/management interface in national parks
science/management interface in national parks 8, 169–79, 176; Crater Lake National Park 173–4; differences between park managers and scientists 172–3, 174–5; history of US National Park Service management 170–2; Isle Royale National Park 176–7
sea otters 159
Selman, Paul 113

Seychelles 204
Sierra Club 144, 148
SI/MAB Biodiversity Program 76
SITES 75
Slocombe, D. Scott 1–11, 222–6; complexity of protected area management 2–4; 'people versus nature' controversy 4, 6–7
Smithsonian Institution (SI) 76
snow geese 158–9, 159
social and ecological changes 1, 161–2, 161, 223
social construction of nature and ecotourism 205–7, 213
social-ecological interactions 7, 157
social-ecological systems (SESs) 16, 25–8, 32, 155, 161, 223
social justice 6, 20, 185, 202, 210
societal taxes 41, 42, 43, 47, 50, 51, 54
socio-ecological integrity 161–3, 161
solidarity-seeking culture 210, 211
South, Global 210
South Africa 46, 49, 56, 162, 164n5
South African National Parks (SANParks) 45
Southern Tutchone 190–2
spatial information technology 66–8; see also information technology (IT)
species-based protected area management 68–70
Spence, M.D. 206
state power 182; see also nation-states
statistical approaches in wildlife conservation and parks management 69–70
steering governance 112, 118–19, 120, 121–2, 124–5, 126–7; see also governance
Stein Valley Nlaka'pamux Heritage Park 139, 140–1
Stoltmann, Randy 144
Stonich, S.C. 204
structural political ecology 201–2
sustainable development 30
Sweden 42
Swift, J. 112
systemic-normative discourse 155
systems: complex open 23–5; complex systems thinking and globalization 28–31; open 111, 113, 126, 128, 129; social-ecological (SESs) 16, 25–8, 32; socio-ecological integrity and 161–2, 161

Tanzania 44
taxes 41, 42, 43, 47, 50, 51, 54
taxonomies 28
technology see information technology (IT)
Texas State Parks 55

The International Ecotourism Society (TIES) 200
Third World, ecotourism in 202; see also ecotourism
thought experiments 32
toleration 196
tourism 31, 100; Dzanga-Sangha 87, 91, 95, 96, 97, 103; ecotourism and 200, 204, 210, 216n14; management models of parks and protected areas 40, 42–3, 44–5, 46–50; political ecology and 204; resource ownership 53; sources of income 54–6; see also ecotourism
traditional knowledge 100, 188, 190–1, 195
transboundary protected areas ('peace parks') 22, 66
transition zones, biosphere reserve 115
transpersonal collaborative discourse 155, 164n1
Tully, James 194, 196
Turner, Frederick Jackson 183

Udall, Stewart 171
Uganda 98, 155, 164n5
UNESCO biosphere reserves 111, 114–15, 116; Niagara Escarpment (Ontario) 116–20, 117; Redberry Lake (Saskatchewan) 116, 123–6, 123, 127, 128
UNESCO/Man and the Biosphere (MAB) Program 20, 21, 76
United Kingdom 46, 50, 77
United Nations: International Convention on Biodiversity 193–4; Millennium Development Goals (MDGs) 19, 20, 71, 163, 164n1; Universal Declaration of Human Rights 192, 193
United States 50, 225; construction of nation-state 182–3; ecological integrity 158; information technology 63, 77; management of parks and protected areas 42, 50, 55; Native Americans 206, 216n7; wilderness concept 206; see also United States National Park Service (USNPS)
United States National Park Service (USNPS) 169–79, 176; Crater Lake National Park 173–4; history 169–72, 182, 183, 206; Isle Royale National Park 176–7
Universal Declaration on Cultural Diversity 194
Universal Declaration of Human Rights (UDHR) 192, 193
urchins 159
USAID Natural Resources Management Project (NRM) 3
user fees 30, 41, 54, 55; typical management models of parks and protected areas 43, 44, 46, 47, 48, 50, 51; see also entrance fees

235

Index

Vancouver Island see Carmanah and Walbran valleys; Clayoquot Sound
Vancouver Island Land Use Plan (VILUP) 145, 147–8
visitor management 77–8, 170
Vreugdenhil, D. 74

Walbran valley see Carmanah and Walbran valleys
Walker, B. 25–6
Wapusk National Park 158–9, 159
Waterton/Glacier International Peace Park 66
Watts, M. 201–2
Web-based GIS 77–8
West, P. 207, 211
Whitelaw, Graham S. 8, 110–33; Niagara Escarpment (Ontario) 116–20, 117; Oak Ridges Moraine (Ontario) 120–3, 120; Redberry Lake (Saskatchewan) 123–6, 123
wilderness concept 206
wilderness-normative discourse 154, 155, 156, 158, 159, 161, 163
wildlife/human interactions 103–4, 170, 176, 208; see also people versus nature debate
wildlife management policy 171
Willems-Braun, B. 205
Wilshusen, P.R. 201
Wolong Panda Reserve 49
wolves 65, 158, 170, 176–7

women 100, 101
Wood Buffalo National Park 160–1
World Commission on Protected Areas (WCPA) 200
World Conservation Union 4, 15; see also International Union for the Conservation of Nature and Natural Resources (IUCN)
World Database on Protected Areas (WDPA) 70–1, 72
world-systems 28–30
World Wildlife Fund for Nature (WWF) 75, 209, 216n13
Wright, George 170
Wright, R. Gerald 8, 169–80; Crater Lake National Park 173–4; differences between park managers and scientists 172–3, 174–5; history of US National Park Service management 170–2; Isle Royale National Park 176–7
Wurman, R.S. 62
WWF-Canada Assessment of Representation (AoR) Analyst 75

Young, E.H. 203–4
Yukon First Nation 187

Zimmerer, K.S. 203
zonation plans 3
zoning 97–8